CAMBRIDGE LIBRARY COLLECTION

Books of enduring scholarly value

Botany and Horticulture

Until the nineteenth century, the investigation of natural phenomena, plants and animals was considered either the preserve of elite scholars or a pastime for the leisured upper classes. As increasing academic rigour and systematisation was brought to the study of 'natural history', its subdisciplines were adopted into university curricula, and learned societies (such as the Royal Horticultural Society, founded in 1804) were established to support research in these areas. A related development was strong enthusiasm for exotic garden plants, which resulted in plant collecting expeditions to every corner of the globe, sometimes with tragic consequences. This series includes accounts of some of those expeditions, detailed reference works on the flora of different regions, and practical advice for amateur and professional gardeners.

A Botanical Arrangement of All the Vegetables Naturally Growing in Great Britain

This two-volume milestone work, published in 1776, was the first major publication of William Withering (1741–99), a physician who had also trained as an apothecary (his *Account of the Foxglove, and Some of its Medical Uses* is also reissued in this series). The first systematic botanical guide to British native plants, the present work uses and extends the Linnaean system of classification, but renders the genera and species 'familiar to those who are unacquainted with the Learned Languages'. Withering offers 'an easy introduction to the study of botany', explaining the markers by which the plants are classified in a particular genus, and giving advice on preserving specimens, but the bulk of the work consists of botanical descriptions (in English) of the appearance, qualities, varieties, common English names, and uses of hundreds of plants. The book continued to be revised and reissued for almost a century after Withering's death.

Cambridge University Press has long been a pioneer in the reissuing of out-of-print titles from its own backlist, producing digital reprints of books that are still sought after by scholars and students but could not be reprinted economically using traditional technology. The Cambridge Library Collection extends this activity to a wider range of books which are still of importance to researchers and professionals, either for the source material they contain, or as landmarks in the history of their academic discipline.

Drawing from the world-renowned collections in the Cambridge University Library and other partner libraries, and guided by the advice of experts in each subject area, Cambridge University Press is using state-of-the-art scanning machines in its own Printing House to capture the content of each book selected for inclusion. The files are processed to give a consistently clear, crisp image, and the books finished to the high quality standard for which the Press is recognised around the world. The latest print-on-demand technology ensures that the books will remain available indefinitely, and that orders for single or multiple copies can quickly be supplied.

The Cambridge Library Collection brings back to life books of enduring scholarly value (including out-of-copyright works originally issued by other publishers) across a wide range of disciplines in the humanities and social sciences and in science and technology.

A Botanical Arrangement of All the Vegetables Naturally Growing in Great Britain

With Descriptions of the Genera and Species,
According to the System of the Celebrated Linnaeus

VOLUME 1

WILLIAM WITHERING

CAMBRIDGE
UNIVERSITY PRESS

CAMBRIDGE
UNIVERSITY PRESS

University Printing House, Cambridge, CB2 8BS, United Kingdom

Cambridge University Press is part of the University of Cambridge.

It furthers the University's mission by disseminating knowledge in the pursuit of
education, learning and research at the highest international levels of excellence.

www.cambridge.org
Information on this title: www.cambridge.org/9781108075879

© in this compilation Cambridge University Press 2015

This edition first published 1776
This digitally printed version 2015

ISBN 978-1-108-07587-9 Paperback

Selected botanical reference works available in the
CAMBRIDGE LIBRARY COLLECTION

al-Shirazi, Noureddeen Mohammed Abdullah (compiler), translated by Francis Gladwin: *Ulfáz Udwiyeh, or the Materia Medica* (1793) [ISBN 9781108056090]

Arber, Agnes: *Herbals: Their Origin and Evolution* (1938) [ISBN 9781108016711]

Arber, Agnes: *Monocotyledons* (1925) [ISBN 9781108013208]

Arber, Agnes: *The Gramineae* (1934) [ISBN 9781108017312]

Arber, Agnes: *Water Plants* (1920) [ISBN 9781108017329]

Bower, F.O.: *The Ferns (Filicales)* (3 vols., 1923–8) [ISBN 9781108013192]

Candolle, Augustin Pyramus de, and Sprengel, Kurt: *Elements of the Philosophy of Plants* (1821) [ISBN 9781108037464]

Cheeseman, Thomas Frederick: *Manual of the New Zealand Flora* (2 vols., 1906) [ISBN 9781108037525]

Cockayne, Leonard: *The Vegetation of New Zealand* (1928) [ISBN 9781108032384]

Cunningham, Robert O.: *Notes on the Natural History of the Strait of Magellan and West Coast of Patagonia* (1871) [ISBN 9781108041850]

Gwynne-Vaughan, Helen: *Fungi* (1922) [ISBN 9781108013215]

Henslow, John Stevens: *A Catalogue of British Plants Arranged According to the Natural System* (1829) [ISBN 9781108061728]

Henslow, John Stevens: *A Dictionary of Botanical Terms* (1856) [ISBN 9781108001311]

Henslow, John Stevens: *Flora of Suffolk* (1860) [ISBN 9781108055673]

Henslow, John Stevens: *The Principles of Descriptive and Physiological Botany* (1835) [ISBN 9781108001861]

Hogg, Robert: *The British Pomology* (1851) [ISBN 9781108039444]

Hooker, Joseph Dalton, and Thomson, Thomas: *Flora Indica* (1855) [ISBN 9781108037495]

Hooker, Joseph Dalton: *Handbook of the New Zealand Flora* (2 vols., 1864–7) [ISBN 9781108030410]

Hooker, William Jackson: *Icones Plantarum* (10 vols., 1837–54) [ISBN 9781108039314]

Hooker, William Jackson: *Kew Gardens* (1858) [ISBN 9781108065450]

Jussieu, Adrien de, edited by J.H. Wilson: *The Elements of Botany* (1849) [ISBN 9781108037310]

Lindley, John: *Flora Medica* (1838) [ISBN 9781108038454]

Müller, Ferdinand von, edited by William Woolls: *Plants of New South Wales* (1885) [ISBN 9781108021050]

Oliver, Daniel: *First Book of Indian Botany* (1869) [ISBN 9781108055628]

Pearson, H.H.W., edited by A.C. Seward: *Gnetales* (1929) [ISBN 9781108013987]

Perring, Franklyn Hugh et al.: *A Flora of Cambridgeshire* (1964) [ISBN 9781108002400]

Sachs, Julius, edited and translated by Alfred Bennett, assisted by W.T. Thiselton Dyer: *A Text-Book of Botany* (1875) [ISBN 9781108038324]

Seward, A.C.: *Fossil Plants* (4 vols., 1898–1919) [ISBN 9781108015998]

Tansley, A.G.: *Types of British Vegetation* (1911) [ISBN 9781108045063]

Traill, Catherine Parr Strickland, illustrated by Agnes FitzGibbon Chamberlin: *Studies of Plant Life in Canada* (1885) [ISBN 9781108033756]

Tristram, Henry Baker: *The Fauna and Flora of Palestine* (1884) [ISBN 9781108042048]

Vogel, Theodore, edited by William Jackson Hooker: *Niger Flora* (1849) [ISBN 9781108030380]

West, G.S.: *Algae* (1916) [ISBN 9781108013222]

Woods, Joseph: *The Tourist's Flora* (1850) [ISBN 9781108062466]

For a complete list of titles in the Cambridge Library Collection please visit:
www.cambridge.org/features/CambridgeLibraryCollection/books.htm

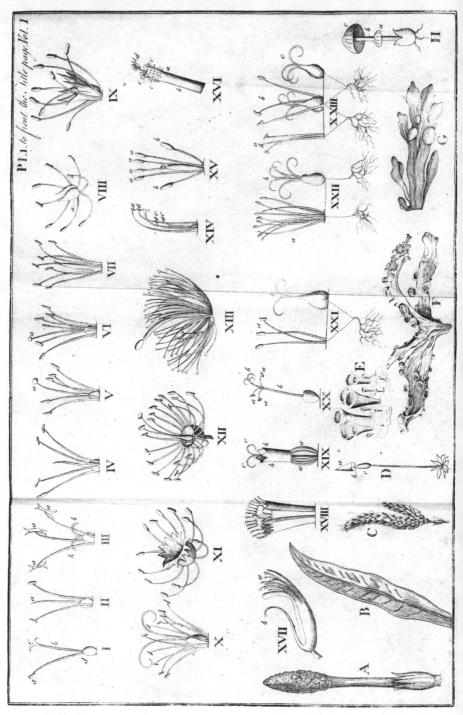

Pl.1 to front the Title page.Vol. 1

A

Botanical Arrangement

OF ALL THE

VEGETABLES

Naturally growing in GREAT-BRITAIN.

V O L I.

ESSE QUAM VIDERI

Robert Washington Oates

A

Botanical Arrangement

OF ALL THE

VEGETABLES

Naturally growing in GREAT BRITAIN.

WITH DESCRIPTIONS OF THE

GENERA and SPECIES,

According to the Syftem of the celebrated Linnæus.

Being an Attempt to render them familiar to thofe who
are unacquainted with the LEARNED LANGUAGES.

Under each SPECIES are added,

The moft remarkable VARIETIES, the Natural PLACES of
GROWTH, the DURATION, the TIME of FLOWERING, the
PECULIARITIES of STRUCTURE, the common *Englifh* NAMES ;
the NAMES of *Gerard, Parkinfon, Ray* and *Bauhine.*

The USES as MEDICINES, or as POISONS ;
as FOOD for Men, for Brutes, and for Infects.

With their Applications in OECONOMY and in the ARTS.

WITH AN EASY

INTRODUCTION TO THE STUDY OF BOTANY.

SHEWING

The Method of inveftigating PLANTS, and Directions how
to Dry and Preferve SPECIMENS.

The whole Illuftrated by COPPER PLATES and a copious GLOSSARY.

By WILLIAM WITHERING, M.D.

Ornari res ipfa negat, contenta doceri.

IN TWO VOLUMES.

BIRMINGHAM: Printed by M. SWINNEY,

For T. CADEL and P. ELMSLEY in the Strand, and
G. ROBINSON, in Pater-nofter-row, LONDON.

MDCCLXXVI.

T H E

D E S I G N.

NOTWITHSTANDING the very rapid pro-
grefs of Science fince the middle of the laft
century, it is only fince the beginning of this that
the ftudy of Natural Hiftory hath attracted the general attention of mankind.

Botanical enquiries in particular have been confined to a few individuals; partly from the difficulties which attended them, and partly from an opinion that they were only ufeful to the profeffors of
Medicine; but the eftablifhment of the LINNÆAN
SYSTEM hath called forth a number of votaries to
cultivate this amiable Science; and the pupils of
that great man have taught mankind, that the Medical properties of Plants, are far from being the
only circumftances worthy their attention.

Still however difficulties remain. An acquaintance with *one* of the Learned Languages is hardly sufficient to enable us to understand his works. The novelty of the subject compelled him to invent a variety of new terms; and without the advantages of an Academical Education, it is a very laborious task to attain precisely that idea which he wishes them to express.

In this situation of things, what can be done by those who from nature or from accident, find themselves possessed with a taste for Botanical knowledge? They study Botanical plates: but bad plates convey false or insufficient ideas; and good ones are too expensive for general use. If they apply to old voluminous Herbals; the want of method, the deficiencies, and the requisite long continued attention, fatigue the most industrious dispositions. If they seek for information in more modern productions; they find such a multiplicity of terms, formed from the Greek and Latin languages, with nothing English but the terminations; that they first begin to despair, and then relinquish in disgust.

A Desire to remove these difficulties, and to render the path in this walk of science, as easy as it is delightful, first gave rise to the present undertaking. The intention was laudable; but how far the execution corresponds with that intention, the public must decide.

From

From an apprehenfion that Botany in an Englifh
drefs would become a favourite amufement with
the Ladies, many of whom are very confiderable
proficients in the ftudy, in fpite of every difficulty;
it was thought proper to drop the fexual diftincti- Sexual Di-
ons in the titles to the Claffes and Orders, and to ftinctions.
adhere only to thofe of Number, Situation, &c.
Thofe who wifh to know the curious facts which
gave birth to this celebrated fyftem, may confult
the *Philofophia Botanica*; the *Sponfalia Plantarum*,
in the firft volume of the *Amœnitates Academicæ*;
Lee's Introduction, and *Milne's Botan. Dict.* where
they will meet with many other very interefting
Philofophical difquifitions. All controverfies about Of Syftem.
fyftem are ftudioufly avoided : Mankind are weary
of fuch unprofitable difputes : every Syftem yet invented, undoubtedly may glory in its peculiar
beauties, and with no lefs reafon, blufh for its particular defects.

It is fufficient for the prefent purpofe that the
fyftem of LINNÆUS is now very univerfally adopted;
and though confeffedly imperfect, it approaches fo
near to perfection, that we may perhaps never expect to fee any other improvements than fuch as
will be founded upon this plan.

With refpect to the language, fome apology is Language.
undoubtedly neceffary. It would be no difficult
matter to render the fentences more connected.
This would be lefs awkward, but at the fame time

a 2 lefs

leſs perſpicuous. Uſe will take off from the former defect, but never can remove the latter. The
beſt Syſtematic Naturaliſts are thoſe who communicate preciſe and well-defined ideas, and in ſuch a
manner, that the characteriſtic features of what
they are deſcribing ſhall obtrude themſelves upon
the eye *.

No part of this undertaking demanded more attention, or occaſioned more labour, than the ſe-
lection of the terms. The old Engliſh Botaniſts
afforded very conſiderable aſſiſtance : ſtill, however,
it muſt be confeſſed, that many words foreign to
the Engliſh language, are neceſſarily retained : but
wherever theſe occur, or wherever an Engliſh word
is taken in a more ſtrict acceptation than uſual, it
is to be found in the Gloſſary; and is there explained, either by deſcription, by the aſſiſtance of
engravings, or by reference to ſome well-known
example.

In quoting examples to explain the terms, it
was thought expedient to uſe the moſt common
and well-known names of Plants, and likewiſe to
take inſtances from Plants that are frequent in our
gardens, whether naturally growing in this country
or not.

Terms.

* LINNÆUS has great merit in the judicious diſpoſition of
his Matter ; ſo that a glance of the eye catches in a moment the
ſubject in queſtion, without the fatigue of reading page after
page. In this work, his mode of printing is adopted.

The

THE DESIGN. vii

The Latin terms ufed by L INNÆUS are thrown into alphabetical order; partly to demonftrate how little the Fnglifh language ftands in need of foreign terms, and partly with a view of affording affiftance when there is occafion to confult other writers upon Botanical fubjects.

The Synopfis of the Genera at the head of each Genera. Clafs, and the fubdivifions of the Orders, will be found greatly to fhorten the labour of inveftigation. But the fubdivifions are not without exceptions: Thefe, however, as well as the exceptions of particular fpecies, are noted in their proper places, in Italic characters. The Generic defcriptions are tranflated from the fixth edition of the *Genera Plantarum* of LINNÆUS; but they fometimes vary a little from the original, where truth and nature feemed to require it: And in fome inftances where the original was evidently more adapted to the foreign, than to the Britifh fpecies, it was neceffary to naturalize it to our own country, by additional obfervations.

The characters of the Species are tranflated from Species. the twelfth edition of the *Syftema Naturæ*: but as we have a right to expect that the Flora of a particular country, fhould be more full and perfect than one that is meant to contain nothing lefs than the productions of the whole furface of the earth, additional defcriptions are added to almoft every Species. Thefe additional defcriptions are printed in
a 3 Italics,

Italics, and are either taken from actual obfervation, or from authors of indubitable veracity.

Peculiarities. The ftructure of every leaf and every flower,
exhibits proofs of Divine Wifdom : but we are too
apt to pafs without attention, the phænomena that
are daily prefented to our eyes. In fome parts of
the vegetable kingdom, there are fuch peculiarities
of ftructure, and fuch amazing contrivances to anfwer the wife purpofes of the Creator, that a total
difregard to them would have been unpardonable:
Thefe therefore are mentioned, in a concife manner, yet fufficiently full to excite the attention of
the ingenious.

Sleep of Plants. Many Plants in the night-time, from a change
in the direction of the leaves and leaf-ftalks, affume
an appearance fo totally different from what is ufual
to them, that the moft expert Botanift would
hardly be able at firft fight to recognize his old acquaintance. This is called the SLEEP OF PLANTS;
and is noticed under fome of rhe Species; as is
Wakeing of likewife the WAKEING or opening of flowers.
Flowers. Moft flowers when once expanded, continue fo
night and day. Some clofe againft rainy weather,
but fome have their ftated times of opening and
clofing independent of any fenfible changes in the
atmofphere. Thus the white WATERLILY
opens at feven in the morning, and clofes again at
four in the afternoon. The yellow GOATSBEARD
has long been known to have this property, and is
therefore

therefore called by the country people, JOHN GO
TO BED AT NOON. If kept in a bottle of water,
it will be found to open at three or four o'clock in
the morning, and to clofe again before noon.

At a time when the names of Plants are already Names.
too numerous, nothing lefs than the plea of indi-
fpenfible neceffity can juftify the introduction of
new names for any of the Genera : But the intelli-
gent reader will immediately perceive the impoffi-
bility of forming an Englifh fyftem, without mak_
ing Englifh Generic names. Innovations of this
kind are never admitted without an abfolute neceffity; and wherever that neceffity appeared, recourfe
was had to the old Englifh Botanifts; and if they
afforded no affiftance, fuch a name was invented as
might be expreffive of fome ftriking characteriftic
of the Plant. To prevent confufion, the common
Englifh names are added to the Species, and where
the common name of a Plant cannot be retained for
the Genus, it is generally adopted as the trivial
name of the Species.

The Latin Synonyms of *Gerard, Parkinfon* and Synonyms.
Ray, are fubjoined to each Species; and likewife
the Synonyms of *Cafpar Bauhine*; the latter principally with a view to medical ftudents, becaufe
writers upon the Materia Medica generally make
ufe of *Bauhine*'s names. In the 24th Clafs the Synonyms of *Dillenius* are given, and references to
his plates.

x   THE DESIGN.

Soil and Situation.

In mentioning the places of growth, it was thought more eligible to give the Soil and Situation in general terms, than to specify the particular places where such and such Plants have been found; as the latter method would probably tend to retard the progress of the Science by damping the expectations of its votaries.

Oeconomical Uses.

The Oeconomical Uses of Vegetables, have been hitherto but little attended to by men eminent for Botanical Knowledge. The Theory of the Science, and the Practical Uses have been too much disjoined. At length, however, the generality of mankind are tired with the disputes about Systems, and the vegetable productions of Europe are pretty well arranged: It is time therefore to think of turning our acquisitions to some useful purpose. Our own countryman, the ingenious and indefatigable Mr. *Ray*, in his journeys through the different parts of this Island, collected some facts; but it is to the industry of LINNÆUS and his Disciples that we are indebted for the greater part of our knowledge upon this subject. Mr. *Stillingfleet*, with a zeal truly commendable, attempted to render the Botanical Study of the Grasses subservient to the great purposes of agriculture *; but whilst the farmers wanted the means of distinguishing the different Genera and Species, it was in vain to look for an adoption of his plan.

* See his Miscellaneous Tracts.

In

In years of fcarcity we are often told of mankind fuffering, and fometimes perifhing by famine; but furely this would never be the cafe, if men were apprized of the very great number of efculent Vegetables that grow wild and unnoticed in the fields. Some of them might be ufed with advantage, even amidft the luxury of plenty; and others may eafily be gathered in quantities fufficient to fupport life, when better or more agreeable food cannot be had. With this view, the part of the plant to be preferred, and the ufual mode of dreffing it, are mentioned under thofe Species that have been found the fitteft for foodful purpofes.

It is certainly a matter of the greateft confe- quence to determine what Species of Plants are preferred by particular Animals; for what is noxious to one Animal is often nutritious to another. Thus the WATER COWBANE is a certain poifon to cows; whereas the goat browfes upon it greedily. MONKSHOOD kills goats, but will not hurt horfes. Bitter Almonds are poifonous to dogs, but not to men. PARSLEY is fatal to fmall birds, while fwine eat it fafely; and Pepper is mortal to fwine but wholefome to poultry. Many experiments made with this view by LINNÆUS and his pupils at *Upfal*, are given in the fecond vol. of the *Amœnitates Academicæ*. The refult of thofe that relate to the Britifh Plants, will be found under the particular Species. Many of them were repeated ten and even twenty different times, and with

with the precautions mentioned below, which are here added for the fake of thofe who have lerfure and opportunity to purfue thefe interefting enquiries further. When horfes, cows, &c. are faid to eat the plant, it means that they eat the *Leaves* of the plant. Thus horfes and cows eat a great variety of Graffes; but they only eat the leaves; for when left to their own choice they never touch the flowering *Stems*. Some Plants are eaten early in the fpring, whilft young and tender, but are rejected in the fummer. The Animals chofen for thefe experiments muft not be over-hungry, for an empty ftomach will compel them to feed upon plants which nature never defigned for them : Nor fhould they be taken immediately out of the houfe; becaufe after living a confiderable time upon dry food they devour greedily every green vegetable that comes in their way. The Plants offered them fhould be handled as little as poffible, for fome Animals are very nice, and will refufe the moft agreeable food when defiled by fweaty hands. Throw the Plant in queftion upon the ground; if the Animals refufe to eat it, mix it with others that they are known to like, and if they ftill refufe it the point is clear, efpecially if the experiment is repeated with different individuals.

Ufes for Infects.

The Catalogue of Infects that feed upon the different Species of Plants, is confeffedly very imperfect; but it will ferve as a foundation for future obfervations of the fame kind. Perfection in thefe

THE DESIGN.

thefe matters is not to be attained at once. In its prefent ftate it may not be without its ufe. It will be feen that Infects live chiefly upon the products of the taller Trees which grow out of the reach of Quadrupeds, fuch as the ELM, the PEAR, the LIME and the OAK; or upon the thorny and ftinging Plants which repel the attacks of other Animals, as the ROSE, the THISTLE and the NETTLE; or upon fuch plants as exift every where in great abundance as the GRASSES; or laftly, upon fuch Plants as other Animals will not eat, as the DOCK, the WATERLILY, the FIGWORT and the HENBANE.

Derham fuppofes, and with a degree of probability on his fide; that the Virtues of Plants may be difcovered by obferving what Infects feed upon them. Thus the Green Tortoife Beetle feeds upon LYCOPUS and upon MENTHA, which are plants of the fame *Natural Order*, and poffeffed of the fame Virtues. The Swallow-tail Butterfly feeds upon feveral of the Rundle-bearing plants of the fifth Clafs, which we know are endowed with fimilar qualities; and the Orange-tip Butterfly devours the Common LADY-SMOCK and the MITHRIDATE, which are Plants of the fame *Natural Clafs* and have nearly the fame properties.

Many Plants change the appearance of fome of their parts to accommodate certain Infects with convenient

Changes.

convenient lodgings : Some of the moſt remarkable of theſe are noticed.

Uſes in the Arts.

It is beyond a doubt, that the Inhabitants of different countries, and artiſts of different kinds, know how to apply a number of Plants to anſwer a variety of purpoſes ; many more than we are at preſent aware of Such of theſe as could be ſelected from good authorities, or obtained from private information, are ſubjoined in their reſpective places,

Medical Virtues.

Many people will be ſurprized to find ſo little ſaid upon the Medical Virtues of Plants ; but thoſe who are beſt enabled to judge of this matter, will perhaps think that the greater part of that little might have been omitted. The ſuperſtition of former ages, operating upon the ignorance of mankind, gave riſe to miracles of every denomination ; and the faſhion of combining a great variety of ingredients with a deſign to anſwer any particular purpoſe, rendered the real efficacy of any of them extremely doubtful. The dreadful apprehenſions that men formerly entertained of poiſons, made them fearful of employing ſubſtances that were capable of doing miſchief, and therefore they rejected thoſe that were moſt likely to do good. A number of Vegetables fit only for food, were ſuppoſed capable of producing the greateſt alterations in the human body ; and at length every common Plant was eſteemed a cure for almoſt every diſeaſe. In this ſituation of things, little advantage can be reaped

reaped from the experience of former times : we
fhall fooner attain the end propofed, if we take up
the fubject as altogether new, and rejecting the
fables of the antient Herbalifts, build only upon
the bafis of accurate and well conceived experiments. To facilitate the work as much as poffible,
the following obfervations are added.

Certain Plants, capable of producing very fudden
and remarkable effects upon the Human Body, are
called Poifons. But Poifons in fmall dofes are the
beft medicines, and the beft medicines in too large
dofes are poifonous. Even the moft innocent ali-
ments in certain quantities are noxious.

We muft not difdain to learn the Medicinal ufes
of Plants from the meaneft of Mankind ; efpecially where they ufe their remedies in an uncompound-
ed form ; for what are thofe celebrated medicines
we import from the *Indies* at a confiderable annual
expence, as the Ipecacuanha, the Contrayerva and
the Sarfaparilla, but remedies, by long experience,
approved amongft the common people in the countries from which we purchafe them?

Plants of the fame Genus have fomething of the
fame Virtues: Thofe of the fame *Natural Order*
have ftill a nearer refemblance ; and thofe of the
fame *Natural Clafs* the neareft of all. Thefe properties are remarked in the Introductions to the
Claffes.

Plants

Plants having honey-cups detached from the petals are generally poifonous, as the HELLEBORE, the COLUMBINE and the DAFFODIL.

Plants with a milky juice are frequently poifonous ; as fome of the MUSHROOMS, the SPURGES, and CELANDINE.   Others are fo but in a lefs degree as HAWK-WEED, GOATSBEARD, SUCCORY and LETTUCE.

Plants with a fimple jointed ftem, fword-fhaped leaves and flefhy roots are generally acrid : thus the juice of the yellow FLAG excites vomiting and fneezing.

The pleafant-tafted and fweet-fmelling Plants are generally wholefome.

Plants that have an ungrateful fmell produce difagreeable effects, as the ELDER, and many of the FUNGUSSES. Thofe that fmell naufeoufly, as the HENBANE are generally poifonous.

Plants that are bitter to the tafte are ftomachic, and deftroy acidity ; as GENTIAN and CENTAURY.

Acid Plants abate heat and thirft, and refift putrefaction ; as the fruit of the CURRANT or the leaves of the WOODSORREL.

Aftringent

Aftringent Vegetables are difcovered by the tafte.

A red colour indicates acidity; a yellow, bitter-nefs, or acrimony.

Plants in dry feafons and dry fituations, have moft tafte; in moift ones they are more infipid. Thus the aromatic Plants, as THYME, SAGE and MARJORAM are by far the beft in dry foils; and every one knows the infipidity of Fruits in wet foils and wet feafons.

Plants that grow in watery fituations are often corrofive; as CROWFOOT and WATERLILY. But the corrofive aquatic Plants lofe much of their acrimony when cultivated in a dry foil: Of this the Garden CELERY is a fufficient example.

ROOTS are in greateft perfection when the leaves firft begin to put forth.
LEAVES are beft gathered when the bloffoms are beginning to open.
FLOWERS are in greateft perfection when mode-rately expanded.
SEEDS muft be gathered when they have attain-ed their full fize and are nearly ripe.
BARKS are beft in the winter before the fap rifes.

Different parts of the fame Vegetable often ma-nifeft very different properties: Thus the leaves of
WORM-

WORMWOOD are bitter, whilft the roots are aromatic. The Seed-veffels, or Heads of POPPIES are narcotic, but the feeds have no fuch quality.

It will readily be allowed that thefe rules are by no means univerfal: The exceptions to moft of them are numerous. We muft be content to confider them as rude and imperfect out-lines, which the induftry of future ages muft correct and compleat.

A N

E A S Y

INTRODUCTION

To the STUDY of

B O T A N Y.

WE muſt take it for granted that the Botanical Student will be at no loſs to diſtinguiſh a Vegetable, at firſt ſight, from an Animal or a Foſſil. It muſt likewiſe be believed that all Vegetables are capable of producing Flowers, and Seeds.

As the Linnæan Syſtem of Botany is chiefly founded upon the number, ſhape, and ſituation of the parts compoſing a Flower, we ſhall immediately enter upon a deſcription of theſe parts, which are the elements of the

b   ſcience;

science; and therefore a knowledge of them is indispensibly neceſſary.

Flowers conſiſt of the
{
EMPALEMENT.
BLOSSOM.
CHIVES.
POINTALS.
SEED-VESSELS.
SEEDS.
}

To theſe may be added, the HONEY-CUP and the RECEPTACLE.

Some Flowers poſſeſs all theſe different parts, whilſt others are deficient in ſome of them; but the Chives, or the Pointals, or both, are to be found in all.

The EMPALEMENT is formed of one or more green, or yellowiſh green leaves, placed at a ſmall diſtance from, or cloſe to the bloſſom.

The different kinds of Empalement are (1) Cups. (2) Fences. (3) Catkins. (4) Sheaths. (5) Huſks. (6) Veils. (7) Caps; but the moſt common is the CUP. For an explanation of theſe ſee the gloſſary; or look at a roſe, and the green covering that incloſes and ſupports the bloſſom is called the CUP. Pl. 3. fig. 1. (a. a. a. a. a.) The Cup of a Polyanthus is repreſented in pl. 3. fig. 10.

Linnæus ſays the Empalement is formed by the outer bark of the plant.

The BLOSSOM is that beautifully coloured part of a flower, which commands the attention of everybody. If it is entire and undivided, as in the Polyanthus, or Auricula; it is ſaid to be a bloſſom of *one Petal*; but if it is compoſed of ſeveral parts, it is accordingly ſaid to be

a bloſa bloffom of *one*, *two*, *three*, &c. or many parts or *Petals*. Thus the Bloffom of the Tulip is formed of fix Petals; and the Garden Rofes bear Bloffoms compofed of many Petals. The Bloffom is fuppofed to be an expanfior of the inner bark of the plant.

The CHIVES are flender thread-like fubftances, generally placed within the Bloffom, and furrounding the Pointals. A Chive is compofed of two parts, the *Thread* and the *Tip*, but the Tip is the effential part. Chives are formed of the woody fubftance of the plant.

The POINTALS are to be found in the center of the flower. Some flowers have only one Pointal; others have two, three, four, &c. and fome have more than can eafily be counted. Linnæus fays the Pointals are formed of the pith of the plant. A Pointal is compofed of three parts, the *Seedbud*, the *Shaft* and the *Summit*; but the Shaft is often wanting.

The SEED-VESSEL. In the newly opened flower, this part was called the *Seed-bud*; but when it enlarges and approaches to maturity, it is called the Seed-veffel. Some flowers have no Seed-veffels: in which cafe the Empalement inclofes and retains the Seeds until they are ripe.

SEEDS, are fufficiently well known; the fubftance to which they are fixed within the feed-veffel is called the *Receptacle of the Seeds*.

Honey-cups are thofe parts of a Flower which are found to contain honey. The tube of the Bloffom ferves the purpofe of a Honey-cup in many Flowers, as in the honey-fuckle: but in other flowers there is a peculiar organization deftined to this purpofe. See pl. 5. fig. 1, 2, 3, 4.

The Receptacle is that part to which the above-mentioned parts of a Flower are fixed. Thus if you take a Flower and pull off the Empalement, the Bloffom, the Chives, the Pointals and the Seeds or Seed-veffels, the remaining part at the top of the Stalk is the Receptacle. In many Flowers the Receptacle is not a very ftriking part, but in others it is very large and remarkable: thus in the Artichoke, after we have taken away the leaves of the Empalement, the Bloffoms, and the briftly fubftances; the part remaining, and fo much efteemed as food, is the Receptacle.

Having thus briefly mentioned the different parts that enter into the compofition of Flowers, let us for the fake of illuftration examine fome well known inftance. Suppofe it to be a flower of the Crown Imperial.

CROWN IMPERIAL.

ENPALEMENT. None.

BLOSSOM. - - Six Petals. (Pl. 3. fig. 2. *a. a. a. a. a. a.*)

CHIVES - - - Six. (Pl. 3. fig. 2. *b c. b c. b c. b c. b c. b c.*)
> *Threads* fix ; fhaped like an awl. (Pl. 3.
> fig. 2. *b. b. b. b. b. b.*)
> *Tips* oblong; four-cornered. (Pl. 3. fig. 2.
> *c. c. c. c. c. c.*)

POINTAL - - Single.
> *Seed-bud* oblong ; three-cornered. (Pl. 3.
> fig. 2. *d.*)
> *Shaft* longer than the Chives. (Pl. 3.
> fig. 2. *e.*)
> *Summit* with three divifions. (Pl. 3.
> fig. 2. *f.*)

SEED-VESSEL An oblong capfule, with three cells and
three valves. (Pl. 3. fig. 4, reprefents
the Seed-veffel cut a-crofs to fhew the
three cells in which the Seeds are contained.

SEEDS - - - - Numerous ; flat.

By confidering this defcription with fome attention,
and comparing it with the Flower itfelf, and likewife
with the engraved figures, we fhall foon attain a pretty
good idea of the different parts of a Flower. If a Crown
Imperial is not at hand, a Tulip or a Lily will correfpond pretty well with the above defcription. But if
we examine the Crown Imperial we fhall find at the
bafe of each Petal a hole, which is the Honey-cup. In
pl. 3. fig. 3. is a reprefentation of one of the Petals feparated from the reft, to fhew the Honey-cup at (*k,*) and
one of the Chives (*h. i.*)

It

It is natural to afk the ufes of thefe different parts—A full reply to fuch a queftion would lead us to a long difquifition, curious in itfelf, but quite improper in this place. Let it therefore fuffice to obferve that the production of perfect Seed is the obvious ufe of the flower; that for this purpofe the Seed-bud, the Summit and the Tips are all that are effentially neceffary; and perhaps the Summit might be difpenfed with. The fine duft, or meal that is contained in the Tips, is thrown upon the Summit of the Pointal: This Summit is moift and the moifture acting upon the particles of the duft, occafions them to explode, and difcharge a very fubtile vapour. This vapour paffing through the minute tubes of the Pointal, arrives at the Embryo Seeds in the Seed-bud, and fertilizes them. The Seeds of many plants have been obferved to become, to all appearance, perfect without this communication, but thefe Seeds are inca-pable of vegetation. In pl. 3. fig. 5 at *f*, one of the Tips is reprefented difcharging its duft; and at fig. 8 you fee a particle of duft greatly magnified and throw-ing out its vapour. The Empalement and the Petals feem primarily defigned as covers to protect the more effential parts; and perhaps it is not too vain an imagi-nation to believe, that a difplay of beauty was in fome meafure the defign of the Creator.

Independant however of thefe ufes defigned by Na-ture, the Botanift takes advantage of the different num-ber, figure, fize and fituation of thefe parts and affumes them as the foundation of a fyftematic arrangement. He divides all the vegetable productions upon the furface of the globe, into Claffes, Orders, Genera, Species and Varieties. The Claffes are compofed of Orders; the Orders are compofed of Genera, the Genera of Species, and the Species of Varieties.

Men

INTRODUCTION.

Men ufually confider the productions of Nature as forming three diftinct parts, called the Animal, the Vegetable and the Foffil or Mineral Kingdom.

Therefore taking the matter up in this familiar language, let us endeavour to attain an idea of Claffes, Orders, &c. by continuing the allufion. Let us compare

The
{
VEGETABLE KINGDOM to the KINGDOM of ENGLAND;
CLASSES - - - - - - - - to the COUNTIES;
ORDERS - - - - - - to the HUNDREDS;
GENERA - - - - - - to the PARISHES;
SPECIES - - - - - - to the VILLAGES;
VARIETIES - - - - to the HOUSES.
}

Some authors have aptly enough compared

A CLASS - - to an Army;
An ORDER - - to a REGIMENT;
A GENUS - - to a COMPANY;
And a SPECIES to a SOLDIER;

But no comparifon can be more in point, than that which confiders the vegetables upon the face of the globe, as analagous to the inhabitants; thus—

VEGETABLES refemble INHABITANTS;
CLASSES - - refemble NATIONS;
ORDERS - - refemble TRIBES;
GENERA - - refemble FAMILIES;
SPECIES - - refemble INDIVIDUALS;
And VARIETIES are the same *Individuals* in different dreffes,

All the vegetables in Great-Britain are divifible, according to the Syftem of Linnæus, into twenty-four Claffes.

The

The characters of the Claſſes are taken either from the *number*, the *length*, the *connexion*, or the *ſituation* of the Chives.

The characters of the Orders, are moſt frequently taken from the number of the Pointals; but ſometimes from ſome other circumſtances either of the Chives or Pointals, which will be noticed hereafter.

The characters or marks of the Genera, are taken from ſome particulars in the flower, before unnoticed; but the generic deſcriptions are deſigned to contain an account of *all* the moſt obvious appearances in the flower.

The ſpecies are moſtly characterized from peculiarities in the ſtem or leaves; ſometimes from parts of the flower; rarely from the roots.

Varieties. Both leaves and flowers are ſubject to variations; ſome of them evidently dependant upon ſoil and ſituation: but others owing to cauſes which are hitherto un-aſcertained. Thus the leaves of the Water Crowfoot that grow beneath the ſurface of the water, are much more divided than thoſe that grow above the ſurface: ſo that a perſon unacquainted with this circumſtance, would hardly believe they belong to the ſame plant. Again, the leaves of the perennial Snake-weed in wet ſituations are ſmooth; but in dry and warm ſituations hairy. Some authors therefore have reckoned them as diſtinct ſpecies; but let them change ſituations, and the appearances will be changed likewiſe. But why the leaves of Mint are ſometimes curled, thoſe of Holly or Mezereon variegated with white, &c. is a more difficult matter to determine; ſeeing that ſlips from theſe plants, though growing in different ſoils, do not loſe their peculiarities: but young ones raiſed from ſeeds

return

return to their original form. It is evident therefore that thefe, however different in appearance, are not to be confidered as diftinct fpecies, but as varieties.

No variations are more common than thofe of colour; but defirable as thefe changes are to the Florift, they have little weight with the Botanift, who confiders them as variable accidental circumftances, and therefore by no means admiffible into the character of a fpecies

Many flowers, under the influence of garden culture, become double; but double flowers are monfters, and therefore can only rank in a Syftem of Botany, as varieties. When we confider that every plant is compofed of an outer bark, an inner bark, a wood, and a heart or pith; and that flowers are formed by an expanfion of thefe parts; when we recollect too that the Chives are formed of the woody fubftance, and told that this woody fubftance was originally formed by many coats of the inner bark condenfed; we fhall not be at a lofs to account for the production of double flowers. The woody fubftance inftead of being formed into Chives is expanded into Petals. This feems to be effected by too much fucculent nourifhment preventing the wood being properly confolidated. Hence it is that the flowers with many Chives are more apt to become double, and to a greater degree, than thofe that have few; as appears in the Anemone, the Ranunculus, the Poppy and the Rofe. Where the Petals are fo much multiplied as to exclude all the Chives, the flowers neceffarily become barren.

CHARACTERS of the CLASSES.

CLASS.			PL. 1. (fig.)	EXAMPLES.
I.	One Chive.		1 (a)	Mares-tail.
II.	Two Chives.		2	Privet. Lilac.
III.	Three Chives.		3 (a. a. a.)	Saffron. Grasses.
IV.	Four Chives.	(All of the same length.)	4	Plantain. Teasel.
V.	Five Chives.	(Tips not united.)	5	Honeysuckle. Primrose.
VI.	Six Chives.	(All of the same length.)	6	Sparagus. Tulip.
VII.	Seven Chives.		7	Horse-Chesnut.
VIII.	Eight Chives.		8	Mezereon. Heath.
IX.	Nine Chives.		9	Gladiole. Bay-tree.
X.	Ten Chives.	(Threads not united.)	10	Campion. Fraxinella.
XI.	Twelve Chives.	(Chives fixed to the Receptacle.)	11	Houseleek. Purslane.
XII.	Twenty Chives.	(Chives fixed to the Cup or to the Petals.)	12	Hawthorn. Apple. Rose.
XIII.	Many Chives.	(More than 12; fixed to the Receptacle.)	13	Poppy. Larkspur.
XIV.	Two Chives longer.	(And two shorter.)	14 (a. a.) (c. c. c.)	Fox-glove. Toad-flax.
XV.	Four Chives longer.	(And two shorter.)	15 (a.a.a.a.) (b.b.)	Wall-flower. Cabbage.
XVI.	Threads united.	(At the bottom, but separate at the top.)	16 (c. c. c.)	Geranium. Mallow.
XVII.	Threads in two sets.	(Sometimes in one. Bluff. Butterfly-shaped.)	17 (a. b.)	Pea. Gorze. Broom.
XVIII.	Threads in many sets.	(Three or more sets.)	18	Orange. St. John's-wort.
XIX.	Tips united.	(Five Chives.)	19 (a. a.) (b. b.)	Violet. Dandelion. Sunflower
XX.	Chives on the Pointal.		20 (a. a. a.)	Orchis. Cuckow-pint.
XXI.	Chives and Pointals sep.	(In separate Flowers.)	21 (a.)	Oak. Nettle. Cucumber.
XXII.	Chives and Pointals dist.	(Upon distinct plants.)	22 (a.)	Hop. Misletoe. Yew.
XXIII.	Various situations.	(Chives alone, Pointals alone or Chives and Pointals together.)	23 (a. b. c.)	Cross-wort. Ash. Orache.
XXIV.	Flowers inconspicuous.		A. B. C. D. E. F. G. H.	Ferns. Mosses. Liverw. Mushr.

INTRODUCTION. xxvii

By looking over this table, by referring to the plate, and fometimes by having recourfe to the plants mentioned as examples, the learner will foon commit the character of the Claffes to memory, fo that upon the firft fight of a flower it will be no difficult matter to refer it to its proper Clafs; and a knowledge of the Orders will very readily be attained, by obferving, that

In the 14th Clafs the Orders depend upon the Seeds having a Seed-veffel or not.
- - - - 15th upon the fhape of the Seed-veffel.
- - - - 19th upon the ftructure of the Florets. (*See the introduction to the 19th Clafs.*)
- - - - 20th upon the number of Chives.
- - - - 21ft. and 22d. upon the number and fituation of the Chives, or the union of the Tips.
- - - - 23d. upon the fituation of the Chives and Pointals.
- - - - 24th upon the natural affemblages of plants refembling one another.

And in all the other Claffes not particularly fpecified, the Orders depend upon the *Number* of the Pointals.

Before we can underftand the Characters of Genera, we muft again confider the different parts that enter into the ftructure of the flowers, and learn how thefe different parts may be modified. As for inftance,

The Empalement is either a
{ Cup - - *as in* Auricula or Polyanthus.
Fence - - *in* Hemlock or Carrot.
Catkin - *in* Willow or Hazle.
Sheath - *in* Narciffus or Snowdrop.
Husk - - *in* Oats; Wheat; or Graffes.
Veil - - *in* Moffes.
Cap - - - *in* Mufhrooms.

For

For a more full explanation of thefe terms and references to the plates, examine the Gloffary.

The
Blossom
is either
{ One Petal *as the* Lily or Polyanthus.
Many Petals *as the* Rofe or Tulip.
Altogether wanting *as in* Narrow-leaved
 Dittander. }

For a more full explanation of the modifications of
Petals and Bloffoms fee the Gloffary, and likewife plate 4.

The CHIVES and POINTALS have been
fufficiently explained before.

A
Seed-vessel
is either a
{ Capsule, *as in* Poppy and Convolvulus.
Pod, - - - *in* Wallflower and Honefty.
Shell, - - *in* Pea and Broom.
Berry, - - *in* Elder; Goofeberry.
Fleshy - - *in* Apple or Pear.
Pulpy - - - *in* Cherry or Peach.
Cone, - - *in* Fir or Pine. }

Thefe terms will be found more fully explained in the
Gloffary, and illuftrated in plate 5.

A RECEPTACLE is either peculiar to one flower as
in the Rofe, Lily, and Polyanthus ; or common to many
flowers, as in Dandelion, Hawkweed and Artichoke.
(*See the Gloffary.*)

Flowers may be collected into
{ Spikes.
Panicles,
Broad-topped Spikes,
Bunches,
Rundles,
Tufts,
Whorls,
Catkins. }

Each

Each of thefe terms may be found in the Gloffary, where they are explained by familiar examples and references to the plates.

For a proper underftanding of Compound Flowers the reader is likewife referred to the Gloffary and to the explanation of the 4th plate.

Having now attained tolerably precife ideas of the conftitution of Claffes and Orders, and likewife of the parts upon which the Generic Charaƈters are founded ; let us feleƈt a few inftances of well known plants, and after inveftigating them fyftematically, we fhall be at no lofs to inveftigate others that are entirely unknown.

RULES for INVESTIGATION.

Firft, When a plant offers itfelf to our infpeƈtion, the firft thing to be determined is the Clafs to which it belongs. This is to be done by examining the Chives, and referring to the Table of the Claffes. Having fixed upon that which we believe to be right, let us turn to that Clafs in the book; and if the Introduƈtion to that Clafs gives us no reafon to alter our opinion, we are pretty certain of being fo far right. It is beft not to truft to the examination of one flower only ; for we fhall fometimes find that the number of Chives is different in different flowers upon the fame plant. In that cafe the claffic charaƈter muft be taken from the terminating flower.

Second. We muft next look how many Orders the Clafs confifts of; and after obferving the circumftances by which the Orders are determined, we muft compare thefe with the fubjeƈt in hand. If the Order we refer it to hath any fubdivifions, we fhall foon perceive under
which

which of the fubdivifions we muft expect to find the Genus.

Third. After comparing the Flowers with the different Genera, contained in the Order, or in the particular fubdivifion of the Order, we fhall readily perceive with which of them it correfponds; and looking forward to the defcription of that Genus, if the defcription agrees pretty exactly with our fpecimen, we conclude that we are now certain of the Genus. Doubtful matters will fometimes arife; but thefe are for the moft part made clear by obfervations fubjoined to the generic defcriptions.

Fourth. If none of the Generic Characters at the beginning of the Clafs agree with the Flower; we muft then look at the bottom of the Order or fubdivifion of the Order, and fee what plants are noticed under it in Italic Characters. Some of thefe muft be the plant; therefore looking for thefe in the index and comparing the generic defcriptions with the fpecimen in hand, we fhall not only difcover the Genus, but the circumftance likewife that perplexed us.

Fifth. Having now got the Genus, you will obferve that when the Species under that Genus are numerous, they are fubdivided. Confider then which of thefe fubdivifions it beft agrees with, and having determined that, compare it with the feveral Species; your plant muft agree with fome one of thefe: and if it is a Variety, you will probably find that variety mentioned. In determining upon the Species it is not abfolutely neceffary to read the whole of the fpecific character. The former part, printed in Roman Letters and ending with this mark — is fufficient; but the additional defcriptions in Italics are not without their ufe.

Sixth. Make it an invariable rule, not to pafs over a term you do not thoroughly underfland, without confulting the Gloffary. By this means you will very foon learn to do without confulting it at all.

Seventh. When you gather plants for examination colle&t a confiderable number of the Flowers, and if poffible, fome juft opening and others with the Seedveffels almoft ripe.

It was thought neceffary to give a variety of examples for inveftigation, 1. Becaufe only fome of them are to be found at any one feafon : 2. Becaufe plants common in one County are not equally common in all. 3. Becaufe the ftudent is not fuppofed previoufly to be acquainted with many plants, and thofe he does know are only a few of the moft common kind. 4. He is not defired to examine and compare *all* the examples : perhaps it will be better he fhould fometimes try his ftrength at unknown Flowers that he chances to pick up in his walks.

EXAMPLE I. PRIVET.

The Privet is a fhrub common enough in the hedges in many parts of England. It generally bloffoms in June, and its bloffoms are white. Let us fuppofe a branch of it in bloffom before us : that we are ignorant what plant it is ; and are required to inveftigate it. We look into feveral of the Bloffoms, and find two Chives in each Bloffom. This circumftance informs us it belongs to the fecond Clafs. Turning to the beginning of the fecond Clafs at page 5, we find it contains two Orders, and that the Orders depend upon the number of Pointals : therefore looking again at the Flowers we find one Pointal in each ; fo that our plant belongs to the firft Order of the

fecond

fecond Clafs. This Order we find fubdivided into four parts, and that thefe fubdivifions depend upon the regular or irregular form of the Petal, and upon the Bloſſom being fixed above or below the Seed-bud. In our ſpecimens the Bloſſom is one *regular* Petal fixed *below* the Seed-bud. Thefe circumftances correſpond only with the firft fubdivifion, and that fubdivifion contains only one Genus; fo that there can be no doubt but the plant is Privet. We find too that the Bloſſom is cloven into four parts, and that it is fucceeded by a Berry containing four Seeds. Looking forward thereſore to Privet, No. 4. page 6. we compare it with the generic defcription, and have the fatisfaction to find it agree pretty exactly. As this Genus contains only one Species, we foon determine that it is the Common Privet, or the Liguftrum vulgare of Linnæus.

EXAMPLE II. REED.

Upon the of banks rivers, in wet ditches, and upon the borders of pools, the Reed is fufficiently common. It is a fort of large Grafs, five or fix feet high and flowers in June. Having got a ſpecimen of this, we proceed to examine it fyftematically. At firft fight we obferve that the Flowers grow in panicles, and that each Flower contains three Chives. We therefore turn to the beginning of the third Clafs (page 17.) and find *that* Clafs divided into three Orders which depend upon the number of Pointals. Each of our Flowers contains two Pointals, which brings us to the fecond Order. This Order is divided into four parts. The firft fubdivifion contains the plants with Flowers fcattered, or irregularly difpofed, one only in each Empalement. Our plant agrees with the firft circumftance, but not with the laft, for we find five Flowers in each Empalement. The fecond fubdivifion viſion contains only two flowers in each Empalement, therefore we paſs that over, and come to the third, with ſcattered flowers, and ſeveral in each Empalement. Before we proceed further, we juſt look at the laſt ſub-diviſion, but finding the flowers without fruit-ſtalks fixed to a long toothed ſeat or Receptacle we immediately recur to the third ſub-diviſion at the top of the 19th page. This ſub diviſion contains ſix Genera, and we compare the Characters with the plant in hand. The want of an Awn, and the wool-lineſs at the Baſe of the Bloſſoms determines us to call it a Reed. Turning therefore to Reed (page 39) we com-pare it accurately with the Generic Deſcription, and find it correſpond with that. But as the parts conſtituting the Flowers of Graſſes are frequently very minute, we make uſe of the Botanical Microſcope and the Diſſecting Needles to diſplay them more clearly to the eye ; * and likewiſe take the advantage of comparing them with the figures in the plate fronting page 33. Having determined it to be a Reed, the only difficulty remaining is to aſcer-tain the Species. We ſee that only four ſpecies of Reed are natives of Great Britain; and the circumſtances of the five Florets in each Empalement added to the flexibility of the Panicle, which we had obſerved whilſt growing to be waved about with every wind, leave us no longer room to doubt that it is the Common Reed, or the Arundo Phragmitis of Linnæus.

EXAMPLE III. PLANTAIN.

The Plantain Flowers in June and July. It is very common in mowing Graſs, and on the ſides of roads. It is frequently ſtuck in the cages of Linnets and

c

Canary

---

* N.B. The Botanical Microſcope and Diſſecting Inſtruments may be had of the Publiſhers, or of the Country Bookſellers, price Seven Shillings.

Canary Birds, who are fond of the seeds. Upon examining a specimen of this, we find each Flower contains four Chives, nearly of the same length, and therefore we refer it to the fourth Clafs. At page 70 we find this Clafs includes three Orders, dependant upon the number of Pointals. Each of our Flowers contains only one Pointal, and therefore belong to the firft Order. This Order admits of seven fub-divifions. The fpecimen we have, contains Bloffoms of one Petal; this Petal is fixed below the Seed-bud, and there is but one Seed-veffel in each Flower. From thefe circumftances we look for it in the fecond fubdivifion, and finding by cutting acrofs the Seed-veffel, that it is divided into two cells, we call it a Plantain. At No. 45. page 76 we compare it with the Generic Defcription, and finding *that* agree in moft particulars, we try to determine the fpecies. There are fix fpecies of Plantain natives of Great Britain. Thefe Species are not fubdivided, therefore we begin with the firft; the Great Plantain, but the Leaves are not egg-fhaped; nor are the fialks cylindrical. The Hoary Plantain which is the fecond, agrees pretty well; but the Leaves are not downy, nor is the fpike of the Flowers cylindrical. With the third Species it agrees in every particular; therefore we call it the Ribwort Plantain, or the Plantago lanceolata of Linnæus; the Plantago anguftifolia major of Bauhine; the Plantago quinquenervia of Gerard: the Plantago quinquenervia major of Parkinfon; and in thofe authors we fhall find it defcribed under the refpective names.

EXAMPLE IV. HONEYSUCKLE.

This Plant is very common in our hedge-rows, and is very univerfally known; but let us fuppofe a Foreigner who never faw it before, ftruck with the beauty and

the

the fragrance of its Bloſſoms, carrying a piece of it home to examine it. Finding five Chives in each Flower, and the Tips not united, he refers it to the fifth Claſs. The Orders in that Claſs being determined by the number of Pointals, he knows it belongs to the firſt Order, for he obſerves only one Pointal in each Flower. This Order is ſubdivided into ſeven parts. The four naked Seeds, and the rough Leaves, immediately determine him to reject the firſt ſub-diviſion. The bloſſom being fixed beneath the Seed-bud, not correſponding with his Flower, he rejects the ſecond and paſſes on to the third ſubdiviſion, where he finds

* * * *Flowers of one Petal* ; *ſuperior.*

His Flower conſiſts of one Petal, and this Petal *is* fixed above the Seed-bud. This ſubdiviſion containing four Genera, he obſerves the three firſt have Capſules ; but in the laſt the Seed-veſſel is a Berry with two Cells ; this circumſtance added to a ſmall inequality of the Bloſſom, and the Knob at the top of the Pointal, induces him to believe it to be a Honeyſuckle. He looks for No. 87 (p. 131) and comparing the Flower with the Generic Deſcription, is confirmed in his opinion. Under this Genus he finds two Species. At the firſt peruſal of the ſpecific character, he diſcerns his plant to be the Woodbine Honey-ſuckle, or the Lonicera periclymenum of Linnæus.

EXAMPLE V. CARROT.

We ſelect this as an example of the *Umbelliferous* or Rundle-bearing plants, (*See the introduction to the 5th Claſs.*)

The five Chives with Tips not united, and the two Pointals evident in each Floret, determine us to look

c 2   for

for it in the second Order of the 5th Class. This Order
admits of four subdivisions, 1. *Flowers of one Petal : beneath.* But our plant hath five Petals ; therefore we go
to the 2d, *Flowers of five Petals* ; *beneath.* The Florets in
hand have five Petals, but the Petals are not placed beneath the Seed-bud. The third subdivision contains imperfect Flowers, or Flowers without Blossoms ; but our
Florets have Blossoms composed of five Petals ; therefore we proceed to the * * * * *Flowers of five Petals ; superior,
and two Seeds: In* Rundles. All these circumstances
agreeing with the plant before us, we must look for it
here ; but observing that this subdivision of the Order is
farther divided into plants that have the Fence *both general
and partial* ; into plants with the Fence *only partial* ; and
into Plants *without any* Fence ; we examine the specimen, and find a Fence to each Rundle and likewise a
Fence to each Rundlet. The unequal size of the Petals ;
the winged Fence, and the prickly Seeds, agreeing with
No. 109 Carrot, we turn to that Genus. Finding our
plant agree with the Generic Description, and only
one Species under that Genus, we know it to be the wild
Carrot, or the Daucus Carota of Linnæus.

EXAMPLE VI. CUCKOW-FLOWER.

White or Red Campion; Batchelors Buttons : Lychnis.
It grows wild in woods and Ditch-banks, flowering
in June and July. After examining several of the Flowers and finding ten Chives in each, and the Threads not
united ; observing too no Vestige of any Pointals, we
begin to suspect that it belongs either to the twenty-first
Class, where the Chives and Pointals are contained in
separate Flowers ; or to the twenty-second Class where the
Chives and Pointals are found not only in separate Flowers, but these Flowers grow upon distinct Plants. In this
dilemma

dilemma we go to the place where the Plant was gathered, and after examining feveral, at length find that the Flowers containing Chives, and the Flowers containing Pointals grow upon diſtinct Plants. We therefore turn to the twenty-ſecond Claſs, and finding the Orders of that Claſs founded upon the number of Chives, we look for it in the ninth Order, as the Flowers of that Order are ſaid to contain ten Chives. But this Order contains only the names of two plants in Italic Characters. By the Direction of the fourth rule for inveſtigation, we ſearch for theſe two plants, and find at page 270 the Generic Deſcription of the Cuckow-flower agreeing exactly with our ſpecimen, except in the circumſtance of the Chives and Pointals being in diſtinct Flowers. This exception however is noticed in the obſervations ſubjoined to the Generic Deſcription, and we find our plant to be the third Species, called the Campion Cuckow-flower.

EXAMPLE VII CRAB-TREE.

Finding about twenty Chives in each Bloſſom, we turn to the twelfth Claſs, which is the Claſs of twenty Chives. The Introduction to this Claſs informs us that the number of Chives alone will not diſtinguiſh it from the preceding and enſuing Claſs ; we therefore attend to the directions there delivered, and finding in our plant that the Cup is formed of a ſingle concave Leaf; that the Petals are fixed to the ſides of the Cup, and that the Chives do not ſtand upon the Receptacle, we conclude that we have claſſed it right ; and ſeeing each Flower furniſhed with five Pointals, we look for the Genus under the fourth Order. This Order contains three Genera. In the lowermoſt Genus the Cup is fixed below the Seed-bud, but in our Plant it is above the Seed-bud.

So

So far it correfponds with the two firſt Genera. The Cup being cloven into five parts, and the Bloſſom being compoſed of five Petals, are circumſtances common to both. But the Fruit of the firſt is a Berry, containing five Seeds, and the fruit of the ſecond is an Apple with five Cells and many Seeds. Hence it appears that our plant is undoubtedly the Apple, No. 203; and turning to the generic deſcription at page 295, we are confirmed in this opinion. We next compare it with the only two Britiſh Species, and ſoon perceive that we have got the Crab apple, or the Pyrus Malus of Linnæus.

EXAMPLE VIII. CROWFOOT.

The beautiful ſhining yellow Bloſſoms of Crowfoot, and the frequency of it in paſtures in the months of June and July will probably attract our notice; eſpecially as cattle leave it untouched, even when the paſture is bare. We therefore collect ſome of it; and finding a great number of Chives in each Bloſſom, we refer it to the Claſs of many Chives. The Introduction to this Claſs tells us, that the Chives ſtand upon the Receptacle and not upon the Cup. As this appears to be the caſe, we next examine the pointals, and finding them more than can readily be counted, we refer to the 7th Order of many Pointals. This Order includes eight Genera. The ſix firſt that occur have no Cup; but our Flower has a Cup of five Leaves. It is clear then that it muſt be either No. 228, or No. 229. Upon an accurate examination we obſerve the Honey-cup at the claw of each Petal, and therefore turn to the generic deſcription of the Crowfoot, page 330. Quite ſatiſfied about the Genus, we obſerve the Species are numerous, and arranged according as the Leaves are *divided*, or *not divided*. In our ſpecimen the Leaves are divided. We then compare it with each of the Species, and find it to be the upright Crowfoot, or the Ranunculus Acris of Linnæus.

EXAM-

EXAMPLE IX. ARCHANGEL.

Or white or red Deadnettle. It grows every where upon ditch-banks, amongſt rubbiſh, and in orchards.

Upon opening the Bloſſom we find four Chives, and as two of the Chives are conſiderably longer than the other two, we expect to find it in the 14th Claſs. After reading the Introduction to that Claſs, we have no doubt of having claſſed it right. We then obſerve that the two Orders in this Claſs are characteriſed from the Seeds being naked, or covered. In our ſpecimen we find four naked Seeds at the bottom of the Cup; ſo that it belongs to the firſt Order. This Order admits of two ſubdiviſions, founded upon the clefts of the Cup; our plant arranging under Cups with five Clefts, we carefully compare it with each of the generic characters; and after ſome difficulty; guided by the briſtle-ſhaped tooth on each ſide the lower lip of the Bloſſom; we pronounce it to be Archangel, No. 237. We ſhall get but little further information from the Generic Deſcription; for this Claſs being a natural aſſemblage of plants greatly reſembling each other, the Generic Differences are not very obvious. Upon reading the characters of the three Britiſh Species, we are ſoon determined by the taper-pointed, heart-ſhaped Leaves, to call it the White Archangel, or the Lamium Album of Linnæus.

EXAMPLE X. WALLFLOWER.

This plant is very generally known. It grows wild upon old walls, and is frequently cultivated in gardens.

Carefully remove the Empalement and the Petals, and you will find ſix Chives; two of which are ſhorter

than

than the other four. It belongs therefore to the fifteenth
Clafs. The Orders of this Clafs depend upon the form
of the Seed-veſſel ; and after examining the ſpecimen, you
refer it to the firſt ſubdiviſion of the ſecond Order ; for
the Seed veſſel is long, and the Leaves of the Cup ſtand
upright and cloſe to the Bloſſom. It is poſſible you muſt
diſſect ſeveral Flowers before you can aſcertain the Genus ; for this Clafs is compoſed of a natural aſſemblage
of plants, whoſe Flowers bear a ſtrong reſemblance to
each other, and the differences when this is the caſe, are
not very obvious. At length however, the ſmall glandular ſubſtance on each ſide the baſe of the Seed bud
determines you to refer it to No. 272, Wall-flower. Upon a compariſon with the Generic Deſcription at page
401, you find it accurately deſcribed ; and the ſhape of
the Leaves puts it beyond a doubt that it is the yellow
Wall-flower, or the Cheiranthus Cheiri of Linnæus.

EXAMPLE XI. MARSHMALLOW.

It naturally grows in ſalt marſhes, but upon account
of its medical uſes it is cultivated in moſt gardens, and
is pretty generally known.

Upon examining the Flower, we find the Chives numerous, and the Threads all united at the Baſe. Recol-
lecting that this circumſtance characteriſes the Flowers
of the ſixteenth Clafs, we find the Orders in that Clafs
depend upon the number of Chives ; and obſerving
that the Flowers before us contain more than ten Chives,
we muſt expect to find the plant in the ORDER of *many*
Chives. The three Genera contained in that Order
nearly reſemble each other; but the outer Cup being
cloven into nine parts, we *muſt* ſuppoſe it No. 287. Under
that Genus at page 430, we find only one Species, and

as

as our plant agrees both in the generic defcription and
the fpecific character, we pronounce it to be the common
Marſhmallow, or the Althæa Officinalis of Linnæus.

EXAMPLE XII. BROOM.

From the appearance of the Chives, which are all
united by the Threads, we ſhould be at a loſs whether to
expect this plant in the fixteenth or in the feventeenth
Claſs; but the butterfly-ſhape of the Bloſſom, determines
us to the latter. After reading the Introduction to that
Claſs *(at page* 431) we obſerve that the Orders depend
upon the number of Chives. The Flowers of our plant
contain ten Chives; and as the Threads are all united,
we are at no loſs to fee, that it belongs to the firſt ſub-
diviſion of the fourth Order. We now compare it with
the Generic Characters; but as the Genera of this Claſs
are a natural aſſemblage, and from their fimilarity admit
of one general Natural Character, the differences betwixt
each Genus muſt depend upon minute circumſtances,
and therefore demand a good deal of attention. At
length we perceive than it can be no other than the
Broom, No. 290. Comparing it therefore with the Generic Defcription (page 440) and ſtill further confirmed
by the character of the fpecies, we pronounce it to be
the commom Broom, or the Spartium Scoparium of
Linnæus.

EXAMPLE XIII. DANDELION.

This plant is in bloſſom during great part of the
fpring and ſummer; it grows in paſtures, road-fides, and
the uncultivated parts of gardens. At the firſt view we perceive its ſtructure to be very different from any we have
examined before; we hardly know what to call Chives

or what Pointals. The fact is this; this is a true compound flower. or a flower formed of a number of flowers sitting upon one common Receptacle, and inclosed by one common Empalement. Turning to Compound Flowers and Florets in the Gloffary, and reading the explanation of Compound Flowers in the fourth plate, we foon attain a true idea of the matter; and therefore feparating one of the Florets and examining it carefully, we find five Chives with the Tips united, and the Pointal paffing through the cylinder formed by the union of the Tips. We therefore refer it to the nineteenth Clafs. By reading the Introduction to that Clafs, we underftand ftill more clearly the nature of Compound Flowers and the Florets that compofe them. We learn too how the Orders are conftituted ; and upon examining the Flower before us, and finding that all the Florets are furnifhed with Chives and Pointals, we perceive that it belongs to the firft Order. From the fhape of the Bloffoms of the Florets, which are all long and narrow, we know that we muft look in the firft fubdivifion of that Order. Perceiving, that the Receptacle is an important affair in the character of thefe fort of Flowers, we pull off all the Florets in one of the Flowers, and expofe the Receptacle to view. We find it naked; *that is* not befet with chaffy or briftly fubftances. We find too a downy Feather adhering to the Seeds : and obferve the Scales of the Empalement flexible. Thefe characters correfpond with no other genus but Dandelion No. 312. Therefore we look for the generic defcription (p. 482) and comparing it with the Flower, experience a pleafing fatisfaction from their accurate agreement. There are four fpecies of this genus natives of Great-Britain ; we read the fpecific characters, and find our plant to be the common Dandelion, or the Leontodon Taraxacum of Linnæus.

It

INTRODUCTION.

It will be very proper for the learner thus to examine several more Genera of this Clafs; as the Burdock, the Thiftle, the Tanfy, the Daifie and the Groundfel; for by this means he will foon overcome every difficulty.

We are at firft furprized to find the Violet, and other fimple Flowers, in the fame Clafs with the compound Flowers; but we foon recollect that the claffic character is not taken from the appearance of the Bloffoms, but from the circumftance of the TIPS being UNITED.

By paying a proper attention to the nature of compound Flowers, we foon learn to diftinguifh them from double Flowers: and when by accident or cultivation any of the true compound Flowers become double, we fhall always find it depends upon the multiplication of fome of the parts, and the exclufion of others.

EXAMPLE XIV. CUCKOWPINT.

Not unfrequent in ftiff foils. It generally grows in fhady places; in roughs and at hedge bottoms. It is fometimes called Wake robin, or Lords and Ladies. It flowers in May.

There is fomething fo very peculiar and unufual in the ftructure of this Flower, that we find ourfelves altogether at a lofs how to fet about the inveftigation of it. What fhall we call this purplifh long fubftance that ftands upright within the Empalement? We remove the fheathing Empalement to examine the lower part of it, and there we find it furrounded by the Seed-buds. It muft therefore be a fort of Fruit-ftalk, or an elongated Receptacle. Upon a clofer infpection we obferve a number of hair-like Fibres or Threads, but without Tips. We now difcover fomething like Tips without Threads, and

from

from their fituation we begin to think the plant belongs perhaps to the twentieth Clafs. As the Tips are numerous and the Orders depend upon the number of Chives, we look for it in the Order of many Chives, where we find two Genera; and from their characters we think it moft likely to be the Cuckowpoint No. 355; but we are by no means certain. We turn therefore to that Genus (p. 552) and here we feel every doubt difpelled. The generic defcription, the fpecific character, and the obfervations, all confpire to identify the plant before us; and we congratulate ourfelves that we were not puzzled by any common circumftance.

The firft Order in this Clafs requires a particular examination, and if we know the Orchis, or any other Genus contained therein, it will be well to compare it with the defcriptions. The Orchis is frequent in moift clayey foils: it bears large Spikes of purplifh Flowers, and bloffoms in April, May, June and July.

EXAMPLE XV. BIRCH.

The Birch is a tree common enough in Great-Britain, and very generally known. The Flowers are difpofed in Catkins; fome of the Catkins are compofed of Florets which contain only Chives, but we find others upon the fame tree. whofe Florets contain only Pointals. This circumftance makes us refer it to the twenty-firft Clafs. As each of the barren Florets contains four Chives, we look for it in the fourth Order. The Catkins, and the number of Flowers or Florets in each fcale correfponding with our fpecimen, we fuppofe it the Birch, No. 365. We turn forward to that number, and the generic defcription at page 576, confirms our fuppofition. The Species are two, but the fhape of the leaves directs us to call our plant the White Birch, or the Betula alba of Linnæus.

EXAM-

E X A M P L E.   XVI.   H O P.

The Hop flowers in June. It grows common in the hedges in many parts of England, though cultivated in few.

We examine many of the Flowers and find five Chives in each, but no appearance of Pointals. We believe therefore that the Chives and Pointals grow upon diftinct plants, and accordingly turn to the twenty-fecond Clafs. As we have five Chives in each Flower and the Tips not united, we find it in the fifth Order. That Order containing only one Genus, there can be no room for doubt. We therefore call it Hop, No. 383. The generic defcription at page 611 correfponds ; and if we afterwards meet with a plant bearing only fertile Flowers, we find that defcribed likewife. There is only one Species, viz. the Brewer's Hop, or the Humulus lupulus of Linnæus.

The plants of the twenty-third Clafs are invefligated nearly in the fame manner; fo that particular illuftrations of that Clafs would be unneceffary. The common Afh, and the yellow Crofs-wort are frequent enough, and eafily attainable, fo that the ftudent may take thefe as examples upon which to exercife his fkill.

It ftill remains to fay fomething of the twenty-fourth Clafs. The plants in this Clafs are not arranged like the other parts of the fyftem, and therefore cannot be invefligated in the fame manner. They are divided into four natural affemblages, Ferns, Mosses, Thongs, and Fungusses. Thefe partake of nothing in common, that can with propriety bring them into the fame Clafs, unlefs it be the difficulty of difcerning their minute and in-

confpi-

INTRODUCTION.

conspicuous Flowers. The generic and specific characters being taken from their general external appearance, we can only recommend a careful perusal of the Introduction to the Class, and an intimate acquaintance with the terms. This being done, there will be no great difficulty in ascertaining the Species.

AFTER conducting my Pupils in this familiar manner through the different parts of the System, they no longer stand in need of my assistance. They find themselves equal to the investigation of every unknown British plant that comes before them. But this is not all: They will find that the Study of Nature is ever attended with pleasing reflections; that the Study of Botany in particular, independant of its immediate use, is as healthful as it is innocent. That it beguiles the tediousness of the road, and furnishes amusement at every footstep of the solitary walk.

I cannot bid them a more cordial farewel, than by wishing they may experience in this pursuit, a pleasure equal to that felt by

BIRMINGHAM,
May 22, 1776.   The AUTHOR.

☞ The Author will thankfully receive any communications that can tend to render this book more perfect ; and in case the public ever calls for a second edition, such new facts or observations as occur, shall be inserted, with suitable acknowledgements.

DIREC·

DIRECTIONS

FOR

DRYING and PRESERVING

SPECIMENS of PLANTS.

MANY methods have been devifed for the prefervation of plants ; we fhall relate only thofe that have been found moft fuccefsful.

Firft prepare a prefs, which a workman will make by the following directions.

Take two planks, of a wood not liable to warp. The planks muft be two inches thick, eighteeen inches long, and twelve inches broad. Get four male, and four female fcrews ; fuch as are commonly ufed for fecuring fafh windows. Let the four female fcrews be let into the four corners of one of the planks. and correfponding holes made through the four corners of the other plank for the male fcrews to pafs through, fo as to allow the two planks to be fcrewed tightly together. It will not be amif to face the bearing of the male fcrews upon the wood, with iron plates ; and if the iron plates went acrofs from

corner

corner to corner of the wood, it would be a good security against the warping.

Secondly, get half a dozen quires of large soft spongy paper; such as the stationers call Blossom blotting paper, is the best; and a few sheets of strong paste-board.

The plants you wish to preserve should be gathered in a dry day, after the sun hath exhaled the dew; taking particular care to collect them in that state wherein their generic and specific characters are most conspicuous. Carry them home in a tin box nine inches long, four inches and a half wide; and one inch and a half deep. Get the box made of the thinnest tinned iron that can be procured; and let the lid open upon hinges. If any thing happens to prevent the immediate use of the specimens you have collected, they will be kept fresh two or three days in this box, much better than by putting them in water. When you are going to preserve them suffer them to lie upon a table until they become limber, and then they should be laid upon a paste-board, as much as possible in their natural form; but at the same time with a particular view to their Generic and Specific characters. For this purpose it will be adviseable to separate one of the flowers and to display the generic character. If the specific character depends upon the flower or upon the root, a particular display of that will be likewise necessary. When the plant is thus disposed upon the paste-board, cover it with eight or ten layers of spongy paper, and put it into the press. Exert only a small degree of pressure for the first two or three days; then examine it, unfold any unnatural plaits, rectify any mistakes, and after putting fresh paper over it, screw the press harder. In about three days more, separate the plant from the paste-board, if it is sufficiently firm to allow of a change

of

of place; put it upon a fresh pasteboard, and covering it with fresh blossom-paper, let it remain in the press a few days longer. The press should stand in the sun-shine, or within the influence of a fire.

When it is perfectly dry, the usual method is to fasten it down with paste or gum water, on the right hand inner page of a sheet of large strong writing paper. It requires some dexterity to glue the plant neatly down, so that none of the gum or paste may appear to defile the paper. Press it gently again for a day or two, with a half sheet of blossom-paper betwixt the folds of the writing paper. When it is quite dry, write upon the left hand inner page of the paper, the name of the plant; the specific character; the place where, and the time when it was found; and any other remarks you think proper. Upon the back of the same page, near the fold of the paper, write the name of the plant, and then place it in your cabinet. A small quantity of finely powdered arsenic, or corrosive sublimate, is usually mixed with the paste or gum water, to prevent the devastations of insects; but the seeds of Staves-acre finely powdered, will answer the same purpose, without being liable to corrode or to change the colour of the more delicate plants. Some people put the dried plants into the sheets of writing-paper, without fastening them down at all; and others only fasten them by means of small slips of paper, pasted across the stem or branches. In the twenty-fourth Class some of the Genera contain a great number of species; where this is the case, and the specimens are small, several of them may be put into one sheet of paper.

Another more expeditious method is to take the plants out of the press, after the first or second day; let them remain upon the pasteboard; cover them with five or six

leaves

leaves of bloſſom paper, and iron them with a hot
ſmoothing iron, until they are perfectly dry. If the iron
is too hot it will change the colours; but ſome people,
taught by long practice, will ſucceed very happily. This
is quite the beſt method to treat the Orchis and other
ſlimy mucilaginous plants.

Another method is to take the plants, when freſh gathered, and inſtead of putting them into the preſs, immediately to faſten them down to the paper with ſtrong
gum-water: then dip a camel-hair pencil into ſpirit-var-
niſh, and varniſh the whole ſurface of the plant two or
three times over. This method ſucceeds very well with
plants that are readily laid flat, and it preſerves their colours better than any other. The ſpirit varniſh is made
thus. To a quart of highly rectified ſpirit of wine, put
five ounces of gum ſandarach; two ounces of maſtich in
drops; one ounce of pale gum elemy and one ounce of
oil of ſpike lavender Let it ſtand in a warm place, and
ſhake it frequently to expedite the ſolution of the gums.

Where no better convenience can be had, the ſpecimens may be diſpoſed ſyſtematically in a large folio book;
but a vegetable cabinet is upon all accounts more eligible.
In plate the XII, you have a ſection of a cabinet, in the
true proportions it ought to be made, for containing a
compleat collection of Britiſh plants. By the aſſiſtance
of this drawing, and the adjoining ſcale, a workman
will readily make one. The drawers muſt have backs
and ſides, but no other front than a ſmall ledge. Each
drawer will be fourteen inches wide, and ten inches from
the back to the front, after allowing half an inch for the
thickneſs of the two ſides, and a quarter of an inch
for the thickneſs of the back. The ſides of the drawers,
in the part next the front, muſt be ſloped off in a ſerpen-
tine

tine line, fomething like what the workmen call an ogee. The bottoms of the drawers muft be made to flide in grooves cut in the uprights, fo that no fpace may be loft betwixt drawer and drawer. After allowing a quarter of an inch for the thicknefs of the bottom of each drawer, the clear perpendicular fpace in each muft be as in the following table.

I. Two tenths of an inch.
II. One inch and two tenths.
III. Four inches and fix tenths.
IV. Two inches and three tenths.
V. Seven inches and eight tenths.
VI. Two inches and two tenths.
VII. Two tenths of an inch.
VIII. One inch and four tenths.
IX. Two tenths of an inch.
X. Two inches and eight tenths.
XI. One inch and two tenths.
XII. Three inches and five tenths.
XIII. Two inches and four tenths.
XIV. Three inches and eight tenths.
XV. Three inches and four tenths.
XVI. One inch and three tenths
XVII. Two inches and eight tenths.
XVIII. Six tenths of an inch.
XIX. Ten inches.
XX. One inch and nine tenths.
XXI. Four inches and four tenths.
XXII. Two inches and fix tenths.
XXIII. One inch and two tenths.
XXIV. Seventeen inches.

This cabinet fhuts up with two doors in front, and the whole may ftand upon a bafe, containing a few drawers, for the reception of duplicates and papers.

ABBREVIATIONS

In the GENERIC Characters.

B. Flowers without Pointals.
F, Flowers without Chives.
H. Flowers with Chives and Pointals.

In the SPECIFIC Characters.

A. Annual:
B. Biennial.
P. Perennial.
S. A Shrub or a Tree.

I N D E X.

☞ *The* English *Genera are expressed in Capitals, the* Latin *Generic names in Italic, and the Synonyms in Roman Characters.*

e

Row-

f. ELECAM.

h 2

i

Sen-

Star

Thlaspi

C L A S S   I.

O N E   C H I V E.

O R D E R   I.

O N E   P O I N T A L.

1. Marestail. *Empalement* none. *Bloſſom* none.

2. Glasswort. *Empalement* of one leaf. *Bloſſom* none.

† *Parſley-piert.*

O R D E R   II.

T W O   P O I N T A L S.

3. Stargrass. *Empal.* none. *Bloſs.* two petals. *Capſule* two cells.

ONE CHIVE.

1.MARESTAIL. 11 Hippuris. *Linnæi Syst. Natur e.*

EMPALEMENT, None.
BLOSSOM, None.
CHIVE, *Thread* single, growing upon the receptacle of the flower. *Tip* slightly cloven.
POINTAL, *Seedbud* oblong; superior. *Shaft* single; awl-shaped; upright; longer than the chive; situated betwixt the chive and the stem. *Summit* sharp; simple.
SEED VESSEL, None.
SEED, Single, roundish, naked.

Common
Vulgaris

MARESTAIL. As there is only one species known, Linnæus gives no description of it.—*Leaves narrow; growing in whorls round the joints, twelve or more at each joint.* Flowers *equal in number to the leaves.* Stem *straight; jointed.*
Limnopeuce. *Ray's Syn.* 136.
Equisetum palustre brevioribus foliis, polyspermum. *Bauh. pin.* 15.
Cauda equina fæmina. *Gerard.* 1114.
Equisetum palustre alterum, brevioribus setis. *Park* 1200.
In muddy ponds and ditches. P. May.
The Flower of this plant is found at the base of each leaf, and is as simple as can be conceived; there being neither empalement nor blossom; and only one chive, one pointal, and one seed.—It is a very weak astringent.—Goats will eat it; but Cows, Sheep, Horses and Swine refuse it.

2 GLASSWORT. 10 Salicornia.

EMPAL. Four-edged; lopped; distended; permanent.
BLOSS. None.
CHIVE. *Thread* single, simple; longer than the cup. *Tip* double; oblong; upright.
POINT. *Seedbud* oblong, egg-shaped. *Shaft* simple, standing under the chive. *Summit* cloven.
S.VESS None, the cup distended and permanent, includes the seed.
SEED Single.

OBS. *The number of chives in this genus is not very certain; some having found two in each flower.*

GLASSWORT

ONE POINTAL. 3

GLASSWORT with spreading, jointed, herbaceous stems Jointed
Each joint compressed and cloven at the top.—*Flowers three to-* Herbacea
gether, placed on each side the stem, in the clefts of the joints.
 Salicornia geniculata annua. *Rays Syn.* 136.
 Kali geniculatum annuum. *Bauh. pin.* 289. *Gerard.* 535.
  Park. 280.
 The varieties are,
1. Branches somewhat woody and trailing. *Ray's Syn.* 136.
2. With very long trailing shoots. *Ray's Syn.* 137.
3. Much branched and trailing. Leaves short, purplish. *Ray's Syn.* 137.
4. Upright with short leaves resembling those of Cypress. *Ray's Syn.* 137.
 Marsh Samphire. Saltwort.
 Common on the sea shore. A. August. Sept.
From the ashes of this plant a fossile alkaly is obtained, which
is in great request for making soap and glass. It is chiefly
made on the coast of the Mediterranean sea, and is called Soda.
The green plant steeped in salted vinegar makes an excellent
pickle, very little inferior to Samphire. The whole plant hath
a saltish taste, and is greedily devoured by cattle.

Order II. Two Pointals.

3 STARGRASS. 13 Callitriche.

EMPAL. None.
BLOSS. Two Petals bowed inwards, channelled; taper-
 ing to a point, standing opposite to each other
CHIVE. *Thread* single, long, crooked. *Tip* simple.
POINT. *Seedbud,* nearly round. *Shafts,* two, hair-like,
 crooked. *Summit* sharp.
S.VESS. *Capsule* roundish, a little compressed; with four
 angles and two cells.
SEEDS. Oblong: two in each cell.
OBS. *In the first species the chive and pointal are in separate flowers.*

STARGRASS with the upper leaves oval; the chive and Vernal
pointals in separate flowers—*Stems feeble, numerous.* Blossoms *small,* Verna
*white. Upper leaves growing near together in form of a star Lower
leaves in pairs.*
 Stellaria. *Ray's Syn.* 289.
 Stellaria aquatica. *Bauh. pin.* 141. *Park* 1282.
 Alsine palustris serpyllifolia. *Gerard.* 614.
 Star-headed Water Chickweed.
1. Small and creeping. *Ray's Syn.* 289.
 Very common in ditches and stagnant waters. A. April. Aug.

B   STARGRASS

Autumnal   STARGRASS with all the leaves ſtrap-ſhaped and cloven at
Autumnalis   the end. The chive and pointals in the ſame flower.—Bloſſoms
yellowiſh white.

     Stellaria aquatica longifolia. *Ray's Syn.* 290.
     In ditches and ſtagnant waters. A. Sept.
     It ſometimes grows ſo thickly matted together as to allow
one to walk upon it without ſinking.

CLASS

CLASS II.
TWO CHIVES.

ORDER I. ONE POINTAL.

‡ *Narrow-leaved Dittander. Shining Willow. Common Ash.*

* *Blossoms of one regular petal; beneath.*

4. PRIVET. *Blossom* with four clefts. *Berry* with four feeds.

** *Blossoms of one irregular petal; beneath. Seeds in a capsule.*

5. SPEEDWELL. *Blossom* with four divisions in the border; the lower segment narrower than the others.

6. BUTTERWORT. *Blofs.* gaping; furnished with a spur. *Empalement* cloven into five parts.

7. BLADDERWORT. *Blofs.* gaping; furnished with a spur. *Empalement* of two leaves.

*** *Blossoms of one irregular petal; beneath. Seeds naked.*

8. GYPSIE. *Blofs.* nearly equal. *Chives* diftant from each other.

9. CLARY. *Blofs.* gaping. *Threads* very short; each supporting a crofs-thread.

‡ *Jointed Glaffwort.*

**** *Blossoms superior.*

10. CIRCE. *Cup* of two leaves. *Blofs.* two petals; inversely heart-shaped.

Order II. Two Pointals.

11. VERNALGRASS. *Hufks* oblong; one flower in each. *Blofs.* a hufk furnished with an awn.

† *Toadgrafs.*

4. PRIVET.

4 PRIVET. 18 Liguſtrum.

EMPAL. *Cup* of one leaf; tubular; very ſmall; with four upright blunt teeth in its rim.

BLOSS. One funnel ſhaped petal, the *Tube* cylindrical, longer than the empalement. *Border* expanded; divided into four egg ſhaped *Segments*.

CHIVES. *Threads* two; ſimple, oppoſite. *Tips* upright, nearly as long as the Blpſſqm.

POINT. *Seedbud* nearly round. *Shaft* very ſhort. *Summit*, thick, blunt, cloven.

S.VESS. *Berry* globular, ſmooth; of one cell.

SEEDS. Four; convex on one ſide; angular on the other.

Common Vulgare

PRIVET. As there is only one ſpecies known, Linnæus gives no deſcription of it—*Leaves growing in pairs.* Bloſſoms *white.* Berries *black*

Liguſtrum. *Gerard.* 714. *Park.* 1446. *Ray's Syn.* 465. Liguſtrum germanicum. *Bauh. pin.* 475. Prim.

1. The leaves are ſometimes variegated with white or yellow ſtripes.

In the hedges in gravelly ſoils. S. May. June.

It is planted to make hedges—the purple colour upon cards is prepared from the berries—With the addition of alum, the berries dye wool and ſilk of a good and durable green; for this purpoſe they muſt be gathered as ſoon as they are ripe—the leaves are bitter, and ſlightly aſtringent.—Oxen, Goats and Sheep eat it. Horſes refuſe it.

The inſects obſerved to feed upon this plant are the *Privet Hawk Moth*, Sphinx Liguſtri—*Richmond Beauty*, Phalæna Syringaria.

ONE POINTAL. 7

5 SPEEDWELL. 25 Veronica.

EMPAL. *Cup* with four divifions; permanent; the *fegments* fpear-fhaped; fharp.

BLOSS. Wheel-fhaped; of one petal. *Tube* nearly as long as the empalement. The *border* flat, divided into four egg-fhaped *fegments*; the lower fegment narroweft, and that oppofite to it the broadeft.

CHIVES. *Threads* two; thinneft at the bottom; afcending. *Tips* oblong.

POINT. *Seedbud* compreffed. *Shaft* thread-fhaped, declining; as long as the chives. *Summit* fimple.

S. VESS. *Capfule* inverfely heart-fhaped, compreffed at the point; with two cells and four valves.

SEEDS. Several, roundifh.

OBS. *The tube of the bloffom is different in different fpecies; in moft it is very fhort, but in the three firft fpecies it is longer.*

* Flowers growing in Spikes.

SPEEDWELL with a terminating fpike; and oppofite, blunt, fcolloped leaves. Stem afcending; undivided—*Bloffoms blue.* — Spiked Spicata

Veronica Spicata minor. *Bauh. pin.* 247. *Ray's Syn.* 279.
Veronica recta minima. *Gerard.* 627.
Veronica erecta anguftifolia. *Park.* 550.
Upright fpiked male Speedwell, or Fluellin.
In meadows and paftures. P. June.
Cows and Sheep will eat it. Goats and Horfes refufe it.

SPEEDWELL. Spikes terminating; leaves oppofite, bluntly ferrated, rough, ftem upright—*About a fpan long. Spike large and thick. Bloffoms blue. This feems to be the product of the firft fpecies fertilized by the duft of the third fpecies.* — Baftard Hybrida

Veronica Spicata Cambro-Britannica, bugulæ fubhirfuto folio. *Ray's Syn.* 278. tab. 11.
Welfh Speedwell. Bugle leaved Speedwell.—On the fides of mountains in Wales; particularly in Montgomeryfhire.
On Craig Wreidhin. P. July.

B 4

SPEED-

Common Officinalis

SPEEDWELL with fpikes on lateral fruit-ftalks; leaves op-pofite; ftem trailing—Bloffoms *blue: fcored.* Leaves *elliptical, ferrated, hairy. Little fruit-ftalks fhorter than the floral leaves.*
Veronica mas fupina et vulgatiffima. *Bauh. pin.* 246. *Ray's Syn.* 281.
Veronica vera et Major. *Gerard.* 626.
Veronica mas vulgaris et fupina. *Park* 550.
Male Speedwell. Fluellin.
In barren ground and on heaths. P. May.
The leaves have a fmall degree of aftringency, and are fome-what bitter. An infufion of them is recommended by Hoffman, as a fubftitute for tea; but it is more aftringent and lefs grateful than the Indian herb.—It is eaten by Cows, Sheep, Goats and Horfes. Swine refufe it.

* * *Flowers in broad bunches.*

Smooth Serpyllifolia

SPEEDWELL with terminating bunches, nearly refembling a fpike. Leaves fmooth; egg-fhaped; fcolloped—Bloffoms *pale blue.*
Veronica pratenfis ferpyllifolia. *Bauh. pin.* 247.
Veronica pratenfis minor. *Park.* 551. *Ray's Syn.* 279.
Veronica minor. *Gerard.* 627.
Paul's Betony. Little Speedwell.
In meadows and paftures; not uncommon. P. May.
Sheep will eat it.

Brooklime Beccabunga

SPEEDWELL with lateral bunches; a creeping ftem; and flat egg-fhaped leaves—Bloffoms *blue.* Leaves *fitting; ferrated.*
Veronica aquatica rotundifolia Beccabunga dicta minor. *Ray's Syn.* 280.
Anagallis aquatica minor, folio fubrotundo. *Bauh. pin.* 252.
Anagallis aquatica vulgaris, five Beccabunga. *Park.* 1236.
Anagallis five Beccabunga. *Gerard.* 620.
Common Brooklime.
In ditches and rivulets. P. June.
The leaves are mild and fucculent, and are eaten in fallads along with other early fpringing plants.—Cows, Goats and Horfes eat it. Swine refufe it.
This and fome other fpecies of Speedwell afford nourifhment to the *Plantain Fritillary*, Papilio Cinxia—*Black Curculio*, Curcu-lio Beccabungæ *and* Chryfomela Beccabungæ.

ONE POINTAL.

9

SPEEDWELL with lateral bunches an upright stem, and Pimpernel
spear-shaped serrated leaves—Blossoms *pale purple.* Anagallis
Veronica aquatica longifolia media. *Ray's Syn.* 280.
Anagallis aquatica minor, folio oblongo. *Bauh. pin.* 252.
Anagallis aquatica folio oblongo crenato. *Park.* 1237.
Anagallis aquatica minor. *Gerard.* 620.
Long leaved Water Speedwell.
In ditches and shallow ponds. P. July.
Cows, Goats and Sheep eat it. Horses and Swine refuse it.

SPEEDWELL with alternate lateral bunches; little fruit- Narrowleaved
stalks pendant; and very entire strap-shaped leaves—Blossoms Scutellata
white or purplish.
Veronica aquatica angustifolia minor. *Ray's Syn.* 280.
Anagallis aquatica angustifolia scutellata. *Bauh. pin.* 252.
Narrow leaved Water Speedwell or Pimpernel.
In poor swampy soil. P. June.
It is eaten by Cows, Goats, Sheep and Horses. Swine refuse it.

SPEEDWELL with lateral bunches of only few flowers; Stalked
cup rough with hair, leaves wrinkled eggshaped, scolloped, Montana
standing on leaf-stalks. Stem feeble—Blossoms *pale blue.* Leaves
red on the under surface.
Veronica chamædryoides foliis pediculis oblongis infidentibus.
Ray's Syn. 281.
Chamædryi spuriæ affinis, rotundifolia scutellata. *Bauh.
pin.* 249.
Mountain madwort.
In hedges and moist woods. P. May.

SPEEDWELL with lateral bunches; leaves egg-shaped, Germander
wrinkled, toothed; without leaf-stalks. Stem feeble——*the seg-* Chamædrys
ments of the cup unequal, spear-shaped, beset with hairs at the edges.
Bunches of flowers long, opposite, upright. Summit white; Blossoms
blue.
Chamædrys sylvestris. *Gerard.* 657. *Ray's Syn.* 281.
Chamædrys spuria minor rotundifolia. *Bauh. pin.* 249.
Chamædrys spuria sylvestris. *Park.* 107.
Wild Germander.
In pastures. P. May.
The leaves are a better substitute for tea than those of the
common Speedwell, being more grateful and less astringent.
Cows and Goats eat it. Sheep, Horses and Swine refuse it.

SPEED.

*** *Fruitſtalks with one Flower.*

Chickweed Agreſtis
SPEEDWELL with ſolitary flowers; leaves heart-ſhaped; jagged; ſhorter than the fruit-ſtalk—*Segments of the cup, egg-ſhaped; equal.* Seeds *four in each cell of the capſule.* Bloſſoms *blue.*
Veronica floribus ſingularibus, in oblongis pediculis Chamæ-dryfolia. *Ray's Syn* 279.
Alſine foliis Triſsaginis. *Gerard.* 616. *Park.* 760.
Alſine Chamædryfolia, floſculis pediculis oblongis inſidenti-bus. *Bauh. pin.* 250.
Germander leaved Speedwell.
In paſtures and ploughed fields. A. May.
Cows, Goats, Sheep and Horſes eat it.

Wall Arvenſis
SPEEDWELL with ſolitary flowers. Leaves heart-ſhaped; jagged; longer than the fruit-ſtalk—*Segments of the cup, ſpear-ſhaped, unequal.* Bloſſoms *pale blue. The upper leaves have no leaf-ſtalks.*
Veronica floſculis ſingularibus, cauliculis adhærentibus. *Ray's Syn.* 279.
Alſine veronicæ foliis, floſculis cauliculis adhærentibus. *Bauh. pin.* 250.
Alſine foliis veronicæ. *Gerard.* 613.
Alſine foliis ſubrotundis veronicæ. *Park.* 762.
Speedwell or Chickweed.
Upon old walls, amongſt rubbiſh, and in fallow fields. A. May.
Horſes eat it.

Ivy-leaved Hederifolia
SPEEDWELL with ſolitary flowers. Leaves heart-ſhaped; flat; divided into five lobes—*Seeds with a dimple at the top; two in each cell of the capſule.* Segments *of the cup heart-ſhaped.* Bloſſoms *blue.*
Veronica floſculis ſingularibus, hederulæ folio, morſus gal-linæ minor dicta. *Ray's Syn.* 280.
Alſine hederulæ folio. *Bauh. pin.* 250.
Alſine hederulæ folio minor. *Park.* 760.
Alſine hederacea. *Gerard.* 616.
Small Henbit.
Ditch-banks; ploughed fields. A. April. May.

Cloven Triphyllos
SPEEDWELL with ſolitary flowers. Leaves deeply divided into fingers; fruit-ſtalks longer than the cup—*Bloſſoms blue.*
Veronica floſculis ſingularibus, foliis laciniatis, erecta. *Ray's Syn.* 280.
Alſine triphyllos cærulea. *Bauh. pin.* 250.
Alſine triphyllos ſive laciniata. *Park.* 760.
Alſine recta. *Gerard.* 612.
Trifid Speedwell.
In ſandy fields. A. May. June.
Cows, Goats and Sheep eat it.

6. BUTTER-

6 BUTTERWORT. 30 Pinguicula.

EMPAL. *Cup* gaping; fmall; fharp; permanent. *Upper lip* upright; with three clefts; *lower lip* reflected; cloven.

BLOSS. One petal, gaping. The *longer lip* ftraight, blunt, with three clefts; falling back: the *fhorter lip* cloven, more blunt and more expanding, being an expanfion of the lower and hinder part of the petal. *Honey-cup* horn-fhaped.

CHIVES. *Threads* two; cylindrical; crooked; afcending; fhorter than the cup. *Tips* roundifh.

POINT. *Seedbud* globular. *Shaft* very fhort. *Summit* with two lips; the *upper lip* large; flat; reflected; covering the tips; the *lower lip* fhort; very narrow; upright; cloven.

S. VESS. *Capfule* egg-fhaped; of one cell; compreffed and opening at the point.

SEEDS. Many; cylindrical. *The Receptacle* loofe.

BUTTERWORT with a honeycup growing thicker towards the point—*Bloffoms pale red.* Leaves *lying in a circle upon the ground.* — Cornwall, Lufitanica

Pinguicula flore minore carneo. *Ray's Syn.* 281.
On the bogs in Cornwall. P. May.
Refufed by Cows, Goats, Sheep, Horfes and Swine.

BUTTERWORT with a cylindrical honey-cup, the length of the Petal—*Leaves covered with foft, upright, pellucid prickles; fecreting a glutinous liquor.* Bloffoms *pale red, or purple; but fametimes white.* — Common, Vulgaris.

Pinguicula gefneri. *Ray's Syn.* 281.
Pinguicula five fanicula eboracenfis. *Park.* 532. *Gerard.* 788.
Sanicula montana Flore calcari donato. *Bauh. pin.* 243.
Yorkfhire Sanicle.
On bogs. P. May.

If the frefh gathered leaves of this plant are put into the filtre or ftrainer through which the warm milk from the Cow is poured, and the milk is fet by for a day or two to become acefcent, it acquires a confiftence and tenacity, the whey does not feperate, nor does the cream; in this ftate it is an extremely grateful food, and as fuch is ufed by the inhabitants in the North of Sweden. There is no further occafion to have recourfe to the leaves, for half a fpoonful of this prepared milk, mixed with frefh warm milk, will convert it to its own nature; and this again will

change

change another quantity of fresh milk, and so on without end—
The juice of the leaves kills lice; the common people use it to
cure the cracks or chops in Cows elders—The plant is generally
supposed injurious to Sheep: occasioning a disease which the
farmers call the Rot. But it may be made a question whether
the Rot in Sheep is so much owing to the vegetables in marshy
grounds as to a flat insect called a Fluke (*Fasciola hepatica*) which
is found in these wet situations adhering to the stones and plants,
and likewise in the livers and biliary ducts of Sheep that are affected with the Rot.—From experiments made on purpose, and
conducted with accuracy, it appears that neither Sheep, Cows,
Horses, Goats or Swine will feed upon this plant.

7 BLADDERWORT. 31 Utricularia.

Empal. *Cup* of two leaves; the leaves very small; egg-
shaped; concave; deciduous.

Bloss. One petal, gaping *Upper lip* flat, blunt, upright. *Lower lip* large, flat, entire. A heart-shaped
palate standing prominent betwixt the lips. A *Honey-
cup* like a little horn grows from the base of the
petal.

Chives. *Threads* two; very short; bent inwards. *Tips*
small and adhering together.

Point. *Seedbud* globular. *Shaft* thread-shaped; as long
as the cup. *Summit* conical.

S. Vess. *Capsule* large; globular; of one cell.

Seeds. Many.

Obs. *The plants of this genus are very remarkable; the roots being
loaded with small membranaceous bladders.*

Common
Vulgaris.

BLADDERWORT with a conical honeycup, and a stalk bearing but few flowers.—*Honeycup awl aped, as long as the lower lip,
and contiguous thereto. Mouth closed by a prominent palate.* Blossoms *yellow.*

    Lentibularia. *Ray's Syn.* 286.
    Millefolium aquaticum lenticulatum. *Bauh. pin.* 141.
    Millefolium aquaticum flore luteo galericulato. *Park.* 1258.
    Millefolium palustre galericulatum. *Gerard.* 828.
    Common hooded Milfoil.
    In wet ditches and stagnant waters. P. June. July.

BLADDERWORT with a keelſhaped honeycup—*pointing* Leſſer *downwards. Mouth open.* Roots *hair-like, very ſmall, ſwimming,* Minor *beſet with ſmall membranaceous bladders.* Stalk *as long as ones finger, ſimple, very ſlender; dividing towards the top into three fruitſtalks with three floral leaves.* Root-leaves *winged, hair like; little leaves few; equal.* Cup *with the lower leaf reflected.* Bloſſom *of two petals, gaping.* Upper lip *horizontal, heartſhaped, perforated at the baſe, fixed to the receptacle.* Lower lip *larger; heart-ſhaped; reflected at the ſides; hunched at the baſe on the under ſide, prominent and keel ſhaped.* Seedbud *egg-ſhaped.* Shaft *ſimple, ſhort.* Summit *betwixt egg and tongue ſhaped.* Bloſſom *a paler yellow than the preceding ſpecies.*

Lentibularia minor. *Ray's Syn.* 286.
Leſſer hooded Milfoil.
In ditches and muddy ponds. P. June. July.

8 GYPSIE. 33 Lycopus.

EMPAL. *Cup* of one tubular leaf, with five ſhallow clefts; the *ſegments* narrow and ſharp.

BLOSS. One unequal petal. *Tube* cylindrical; as long as the cup. *Border* with four clefts; blunt, open; the *ſegments* nearly equal, but the *lowermoſt* ſomewhat ſmaller, and the *uppermoſt* ſomewhat broader than the others, and imperfect at the margin.

CHIVES. *Threads* two; diſtant; generally longer than the bloſſom, and bending under its upper ſegment. *Tips* ſmall.

POINT. *Seedbud* with four clefts. *Shaft* thread-ſhaped: ſtraight; as long as the chives. *Summit* cloven; reflected.

S. VESS. None. The ſeeds lying at the bottom of the cup.

SEEDS Four: roundiſh.

GYPSIE with indented ſerrated leaves—Bloſſoms *whitiſh*; Water *ſurrounding the ſtem at the joints.* Leaves *oppoſite; ſitting.* Stem Europæus *four cornered.*

Lycopus paluſtris glaber. *Ray's Syn.* 236.
Marrubium paluſtre glabrum. *Bauh. pin.* 230.
Marrabium aquaticum. *Gerard.* 700. vulgare. *Park.* 1230.
Water horehound.

1. There is a variety in which the leaves are very much divided. Upon the banks of rivers and ponds, in ſandy grounds. P. July.

The

The juice gives a permanent colour to linen, wool and, silk, that will not wash out; travelling gypsies stain their faces with it—Sheep and Goats eat it. Cows and Horses refuse it—The Green Tortoise Beetle (*Cassida viridis*), feeds upon it.

9 CLARY. 39 Salvia.

EMPAL. *Cup* of one leaf, tubular, scored; enlarging gradually upwards and compressed at the top; the *Rim* upright; having two lips; the *lower lip* with two teeth.

BLOSS. A single petal; the *Tube* compressed, enlarging gradually upwards. *Border* gaping; *upper lip* concave, compressed; bowed inwards; notched at the end. *Lower lip* broad; with three clefts; the *middle segment* largest, roundish and broken at the margin.

CHIVES *Threads* two; very short supporting two others crofs-ways by the middle, which have *glands* at one end, and *Tips* at the other.

POINT. *Seed-bud* with four clefts. *Shaft* thread-shaped; very long; adjoining the chives. *Summit* cloven.

S. VESS. None; the *cup* closing a little, contains the seeds in its bottom.

SEEDS Four; roundish.

OBS. *The singular cross thread of the chives constitutes the essentia character of this genus. The rudiments of two chives appear in the mouth of the blossom, but they have no tips. The glands in most species are callous, but in a few they appear like tips, and sometimes contain a small quantity of dust.*

Meadow Pratensis

CLARY with oblong heart-shaped, scolloped leaves; the upper leaves embracing the stem; the flowers in whorls with hardly any intermixture of leaves; the helmet of the blossom gummy.—*Blossom four times as large as the cup; bluish purple. Leaves a full green.*

Sclarea pratensis foliis serratis, flore cæruleo. *Ray's Syn.* 237.
Horminum pratense foliis serratis. *Bauh. pin.* 238.
Horminum sylvestre vulgare. *Park.* 55.
Horminum sylvestre Fuchsii. *Gerard.* 760.
In Essex, but not common. P. July.
Sheep and Goats eat it. Cows and Horses refuse it.

CLARY

CLARY with indented ferrated fmoothifh leaves ; the blof- Wild
foms more flender than the cup.—*The lips approach near together.* Verbenaca
The leaves are fometimes indented in a winged manner. Bloffoms *blue*
Horminum fylveftre Lavendulæ flore. *Bauh. pin.* 239. *Park.* 57.
Ray's Syn. 237. *Gerard.* 760.
Common Englifh Wild Clary.
In paftures and meadows. P. May. September.

10 CIRCE. 24 Circæa.

EMPAL. *Cup* of two egg-fhaped, concave leaves; a little
  bent outwards ; deciduous.
BLOSS. *Petals* two ; inverfely heart-fhaped ; expanding ;
  equal ; commonly fhorter than the cup.
CHIVES *Threads* two ; hairlike ; upright ; as long as
  the cup. *Tips* roundifh.
POINT. *Seedbud* turban-fhaped : beneath. *Shaft* thread
  fhaped, as long as the cup. *Summit* blunt ; notched
  at the end.
S. VESS. *Capfule* betwixt egg and turban-fhaped ; covered
  with ftrong hairs ; with two cells, and two valves ;
  opening from the bafe upwards.
SEEDS. Solitary ; oblong, narrow towards the bafe.

CIRCE with upright Stems fupporting feveral bunches of Common
flowers, and egg-fhaped leaves.—*Petals like the cup leaves.* Blof- Lutetiana
foms *whitifh.*
  Circæa lutetiana. *Gerard.* 351. *Ray's Syn.* 289.
  Circæa lutetiana major. *Park.* 351.
  Solanifolia Circæa dicta major. *Bauh. pin.* 168.
  Enchanters Nightfhade.
  In moift hedge bottoms. P. July.
  It is eaten by Sheep.

CIRCE with trailing ftems fupporting a fingle bunch of Mountain
flowers, and heart-fhaped leaves.—*Bloffoms reddifh.* Alpina
  Solanifolia Circæa alpina. *Bauh. pin.* 168.
  Circæa Lutetiana minor. *Park.* 351.
  Mountain Enchanters Nightfhade.
  In Yorkfhire and Weftmoreland. P. Auguft.
  It is eaten by Sheep.

Order

Order II. Two Pointals.

11 VERNALGRASS. 42 Anthoxanthum.

EMPAL. A *Huſk* with two valves containing one flower. The *Valves* concave, eggſh-aped, taper; the innermoſt the largeſt.

BLOSS. A *Huſk* of two valves, the length of the larger valve of the empalement. Each valve ſends out an awn from its back, at the lower part; and one of the awns is jointed. *Honeycup* of two leaves, very ſlender, cylindrical. The leaves nearly egg-ſhaped, and one enfolding the other.

CHIVES *Threads* two; hair-like; very long. *Tips* oblong; forked at each end.

POINT. *Seedbud* oblong. *Shafts* two; thread-ſhaped. *Summits* ſimple.

S. VESS. The *Huſk* of the bloſſom grows to the ſeed.

SEED. Single, cylindrical, tapering at each end.

Odoriferous
Odoratum

VERNALGRASS with an oblong egg-ſhaped ſpike; the florets growing on little fruit-ſtalks, longer than the awns— *Spike yellowiſh green.*

Gramen Vernum ſpica brevi laxa. *Ray's Syn.* 398.
Gramen pratenſe Spica flaveſcente. *Bauh. pin.* 3.
Spring Graſs.
Meadows. P. May.
The delightful ſmell of new mown hay ariſes chiefly from this plant. It is one of the earlieſt Spring Graſſes, and is extremely common in our fertile paſtures.
Cows, Goats, Sheep and Horſes eat it.

CLASS III.

THREE CHIVES.

ORDER I. ONE POINTAL.

12. VALERIAN	-	BLOSSOM with five clefts, hunched at the bafe. *Seed* fingle.
13. SAFFRON	- -	*Blofs* of one petal, but fo deeply divided as to appear like fix nearly upright petals. *Summits coloured*; rolled in a fpiral.
14. FLAG	- - -	*Blofs.* of one petal; but fo deeply divided as to appear like fix alternate reflected petals. *Summits like petals.*

* *Flowers with Valves like Graffes, and hufky. Empalements.*

15 RUSHGRASS	-	*Blofs.* none. *Emp.* chaffy; in bundles. *Seeds* roundifh.
16 GALANGALE	-	*Blofs.* none. *Emp.* chaffy; pointing from two oppofite lines. *Seed* naked.
17 BULLRUSH	- -	*Blofs.* none. *Empal.* chaffy; tiled; *Seed* naked.
18 COTTONGRASS		*Blofs.* none. *Empal.* chaffy; tiled; *Seed* woolly.
19 MATGRASS	-	*Blofs.* two valves. *Empal.* none. *Seeds* covered.

C

Order II. Two Pointals.

* Flowers scattered ; one in each empalement.

20 PANICGRASS. *Empal.* three valves : that upon the back the smallest.

21 FOXTAIL. *Empal.* two valves. *Blofs.* one valve with a simple point.

22 CATSTAIL. *Empal.* two valves ; lopped ; sharp pointed ; sitting.

23 CANARY. - *Empal.* two valves. The *Valves* keel-shaped ; equal ; inclosing the blossom.

24 MILLET. - *Empal.* two valves. The *Valves* distended ; larger than the blossom.

25 BENT. - *Empal.* two valves. The *Valves* sharp ; shorter than the blossom.

26 COCKSFOOT. *Empal.* two valves. The larger and longer valve compressed and keel-shaped.

27 FEATHERGRASS. *Empal.* two valves. *Blofs.* terminated by an awn which is jointed at the base.

† *Small Reed.* † *Branched Reed.* † *Sea Reed.* † *Red Ropegrafs.*

** Flowers scattered ; two in each empalement.

28 HAIRGRASS. *Empal.* two valves, containing two *Florets* without the rudiments of a third.

29 ROPEGRASS. *Empal.* two valves, containing two *Florets*, with the rudiments of a third betwixt them.

‡ *Soft grafs.* ‡ *Tall Oat.* ‡ *Yellow Oat.*

*** *Flowers*

*** *Flowers scattered, several in each empalement.*

30 QUAKEGRASS   *Empal.* two valves. *Bloss.* heart-shaped; with distended valves.

31 MEADOWGRASS. *Empal.* two valves. *Bloss.* eggshaped, with valves somewhat sharp.

32 FESCUE.   *Empal.* two valves. *Bloss.* oblong; with sharp pointed valves.

33 BROOMGRASS. *Empal.* two valves. *Bloss.* oblong, with an awn rising from beneath the point.

34 OAT. - - *Empal.* two valves. *Bloss.* oblong, with a twisted awn upon the back.

35 REED. - - *Empal.* two valves. *Bloss.* woolly at the base, and without an awn.

‡ *Hardgrass.* ‡ *Rough Cocksfoot.*

**** *Flowers on a long toothed seat, without fruit-stalks.*

36 RYE. - - - *Empal.* opposite; of two valves, containing two florets; solitary.

37 WHEAT. - - *Empal.* opposite; of two valves, containing several florets; solitary.

38 BARLEY. - - *Empal.* three together; lateral; with two valves, containing one floret.

39 LIMEGRASS.   *Empal.* two or three together; lateral; with two valves, containing several florets.

40 DARNEL. - *Empal.* solitary; lateral; of one valve, containing several florets.

41 DOGSTAIL. - *Empal.* two valves; containing several florets. *Proper receptacle leafy; Florets* growing from one side.

Order III. Three Pointals.

42 BLINKS. - - *Bloss.* of one petal. *Empalement* of two leaves; *Capsule* of three *Valves* with three *Seeds.*

‡ *Common Chickweed.* † *Heath Crowberry.*

12. VALERIAN. 44. Valeriana.

EMPAL. None, or only a *Rim*, which is superior.

BLOSS. *Tube*, hunched on the under side, and containing honey. *Border* with five clefts; the *Segments* blunt.

CHIVES Three, or fewer than three: awl-shaped, upright; as long as the blossom. *Tips* roundish.

POINT. *Seedbud* beneath. *Shaft* thread-shaped; as long as the chives. *Summit* thick.

S. VESS. A hard substance not opening; deciduous; crowned.

SEEDS Solitary; oblong.

OBS. *There is a wonderful diversity in the parts of the flowers in different species of Valerian, as well in the number, as in the figure of the parts; thus the*

Empal. in some is scarcely discernible; in others there is an evident *Rim* with five clefts.

Bloss. The *Tube* in some oblong, in others very short, and again in another species furnished with a honey cup resembling a spur. The *Border* in some is equal, in others it consists of two *Lips*, and the upper lip cloven.

Chives are for the most part three; in one species two only; in another four, and in some but one. In one species the chives and pointals are found on different plants.

Point. The *Summit* in some species is notched, in others cloven into three; in others globular.

S. Vess. Some species have no seed-vessels; others have a strong thick capsule, and others again a capsule with two cells.

Seeds vary in figure: some are crowned with a feather, whilst others have none.

VALERIAN

VALERIAN with chives and pointals on different plants ; Marſh
leaves winged, very entire.—*The* Bloſſom *of the fertile flowers is* Dioica
much ſmaller than the Bloſſom *of the barren flowers. Root leaves egg-
ſhaped.* Bloſſoms *reddiſh white.*

Valeriana paluſtris minor. *Bauh. pin.* 164.
Valeriana ſylveſtris minor. *Park.* 122. *Ray's Syn.* 200.
Valeriana minor. *Gerard.* 1075.
Small wild Valerian.

1. There is a variety of this which is ſmaller.
In moiſt meadows. P. June.

VALERIAN with three chives in each flower ; all the leaves Wild
winged—*and oppoſite.* Stem *upright, ſcored.* Bloſſoms *reddiſh white.* Officinalis
Valeriana ſylveſtris major. *Bauh. pin.* 164. *Gerard.* 1075.
Park. 122. *Ray's Syn.* 200.
Great Wild Valerian.

1. The Valeriana ſylveſtris major montana, *Bauh. pin.* 164, is
only a variety, and is found in high paſtures. The other is
very common in hedges, woods and marſhes. P. June.

It is the variety, found on heaths and high grounds, which
is in ſo much repute as a medicine. The root hath a
ſtrong and not an agreeable ſmell ; its taſte is warm, bitteriſh,
and ſubacrid ; it communicates its properties to wine, water,
or ſpirit ; but it is beſt in ſubſtance, and may be taken from
half a dram to two drams for a doſe. There is no doubt of its
poſſeſſing antiſpaſmodic virtues in an eminent degree. It is
often preſcribed with advantage in hyſterical caſes ; and inſtances are not wanting where it appears to have removed ſome
obſtinate epilepſies. In habitual coſtiveneſs it is an excellent
medicine, and frequently looſens the bowels when other
ſtronger purgatives have been tried in vain.

Cows eat the leaves ; Sheep are not fond of them. Cats
are delighted with the roots.

VALERIAN with three chives in each flower. Stem forked, Lettuce
leaves ſtrap-ſhaped—Bloſſom *bluiſh white.* Locuſta
Valerianella arvenſis præcox humilis ſemine compreſſo. *Rays
Syn.* 201.
Valeriana campeſtris inodora major. *Bauh. pin.* 165.
Lactuca agnina. *Gerard.* 210. *Park.* 812.
Lambs Lettuce. Corn Sallad.

1. There is a variety with ſerrated Leaves. *Ray's Syn.* 201.
Common in cornfields. A. April. May.

The young leaves in Spring and Autumn are eaten as ſallad,
and are very little inferior to young Lettuce.

Cows, Sheep, and Lambs eat it.

13 SAFFRON. 55 Crocus.

EMPAL. A *Sheath*, of one Leaf.

BLOSS. *Tube* simple, long. *Border* with six divisions; upright. The *Segments* equal; oblong egg-shaped.

CHIVES. *Threads* three; awl-shaped, shorter than the blossom. *Tips* arrow-shaped.

POINT. *Seedbud* beneath; roundish. *Shaft* thread-shaped; as long as the chives. *Summits* three, rolled in a spiral; serrated.

S. VESS. *Capsule* roundish; with three lobes, three cells, and three valves.

SEEDS. Several, round.

Cultivated
Sativus

SAFFRON with a sheath of one valve, arising from the root. The tube of the blossom very long—*Leaves strap-shaped, with a white rib along the middle.* Blossom *purple.*

Crocus. *Gerard.* 151. *Ray's Syn.* 374.

Crocus sativus. *Bauh. pin.* 65.

About Cambridge, and at Saffron-Walden in Essex. A. Aug.

1. Leaves narrower; rolled back at the edges.

Crocus Autumnalis sativus. *Morison. Hist.* 2. p. 335. T. 2. f. 1. The summits of the pointal of the variety (1) carefully collected and moderately dried are the Saffron of the shops. That collected in England is preferred to all other. It affords a beautiful colour to water, wine or spirit, and gives out the whole of its virtues to them. It hath been held in high repute as a cordial, but modern practice pays no great attention to it, since it hath been found to produce no sensible effect, even when given in doses greatly larger than those generally prescribed.

13. FLAG.

14 FLAG. 59 Iris.

EMPAL. *Sheaths* of two valves, feparating the flowers; permanent.

BLOSS. With fix divifions. The *Segments*, which are almoft diftinct petals: oblong, blunt. The three outer ones reflected; the other three upright and fharper. They are all connected together by the claws.

CHIVES. *Threads* three; awl-fhaped, lying upon the reflected Segments. *Tips* oblong; ftraight; depreffed.

POINT. *Seedbud* beneath; oblong. *Shaft* fimple, very fhort. *Summit* very large, confifting of three divifions which refemble petals; broad, reflected, alternately preffing down the chives and fegments; cloven at the end.

S. VESS. *Capfule*, oblong; angular; with three cells and three valves.

SEEDS. Several; large.

OBS. *There is a honeycup of a different kind in different fpecies.*
The capfule in fome fpecies hath three angles, but in others fix.

FLAG with fmooth bloffoms; the innermoft fegments fmaller than the fummit: the leaves fword-fhaped—*The three outer petals* Yellow *have a tooth upon each fide next to the chives.* Seedbud *with three* Pfeud-Acorus *edges; furrowed.* Bloffoms *yellow.*

Iris paluftris lutea. *Gerard.* 50 *Ray's Syn.* 374.
Acorus adulterinus. *Bauh. pin.* 34.
Acorus paluftris, five Pfeudo-Iris, et Iris lutea paluftris. *Park.* 1219.
Yellow water flower de luce.

1. There is a variety with a pale yellow flower. *Ray's Syn.* 375.
On the banks of rivers; in marfhes and wet meadows. P. July.

The juice of the frefh root is very acrid, and hath been found to produce plentiful evacuations from the bowels when other powerful means had failed. *Edin. med. Eff. vol.* 5. *art.* 8. It may be given for this purpofe in dofes of 80 drops, every hour or two; but the degree of its acrimony is fo uncertain, that it can hardly ever come into general ufe. The frefh roots have been mixed with the food of fwine bitten by a mad dog, and they efcaped the difeafe, when others bitten by the fame dog, died raving mad. The root lofes moft of its acrimony by drying.

Goats eat the leaves when frefh, but Cows, Horfes and Swine refufe them. Cows will eat them when dry. The roots are ufed in the ifland of Jura to dye black. *Pennant's* Tour. 1772. p. 214.

Stinking
Fætidiſſima

FLAG with ſmooth bloſſoms; the inner ſegments expanding
very much; the leaves ſword-ſhaped, the ſtem with one angle
*Cylindrical, as long as the leaves, which cover it, and have a very
fætid ſmell.* Seedbud *like the preceding.* Bloſſom *of a diſagreeable purpliſh aſh colour: not ſmelling in the night-time. The
claws of the* outer petals *wrinkled and plaited on the under ſurface.
The* inner petals *larger than the ſummit, expanding.*

Gladiolus fætidus.   *Bauh. pin.* 30.

Xyris. *Gerard.* 60. ſeu ſpatulo fætida. *Park.* 256. *Ray's
Syn.* 375.

1. There is a variety with variegated leaves.

Stinking Gladdon, or Gladwyn.

Near Hornſey and about Charlton Wood in Kent; and near
Braintree in Eſſex. P. June. July.

The juice of the root both of this and the preceding ſpecies,
is ſometimes uſed to excite ſneezing; but it is an unſafe practice;
violent convulſions having been ſometimes the conſequence.

15 RUSHGRASS. 65 Schænus.

EMPAL. A *Huſk* of two valves; large, upright, tapering to a point, permanent: containing ſeveral florets.

BLOSS. *Petals* ſix; ſpear-ſhaped; ſharp; approaching;
permanent: generally tiled; the outermoſt petals
being the ſhorteſt.

CHIVES. *Threads* three; hairlike. *Tips* oblong; upright.

POINT. *Seedbud* egg-ſhaped with three flatted ſides; blunt.
Shaft briſtly; as long as the bloſſom. *Summit* with
three clefts; ſlender.

S. VESS. None; the petals cloſe upon and contain the
ſeed until it is ripe.

SEED. Single, ſhining, nearly egg-ſhaped, but with
three ſides a little flatted; thickeſt towards the
top.

OBS. *In ſome ſpecies very ſmall briſtles ariſing from the receptacle
ſurround the ſeed.*

* Straw Cylindrical.

RUSHGRASS with cylindrical ftraws; the leaves befet Cyperus
with prickles at the edges, and along the back.—   Marifcus
   Cyperus longus inodorus Germanicus. *Bauh. pin.* 14.
   Cyperus longus inodorus Sylveftris. *Gerard.* 29. *Park.* 1263.
Ray's Syn. 426.
   Long-rooted baftard Cyperus.
   In marfhes and bogs.   P July. Auguft.
   It ferves for thatching inftead of ftraw, and often grows in
fuch quantities in pools, as to form floating iflands.

RUSHGRASS with cylindrical, naked ftraws. Flowers in Blackheaded
an eggfhaped head; having a fence of two leaves, one of which Nigricans
is long and awlfhaped—*General fence compofed of two valves, notched
at the end, with a fharp point in the middle; the awn of the outer
Valve much longer than the whole head, and dark brown at the end.*
   Juncus lævis minor, panicula glomerata nigricante. *Ray's
Syn.* 430.
   Round black-headed Bogrufh.
   In bogs and marfhes.   P. June.

RUSHGRASS with cylindrical naked ftraws. Flowers in a Ecwn
double fpike and the larger leaf of the fence juft as long as the Ferrugineus
fpike—*Six times fmaller than the preceding.*
   Juncello accedans graminifolia plantula capitulis Armeriæ
proliferæ. *Ray's Syn.* 430.
   Brown baftard Cyperus.
   In turfy bogs upon mountains. P. July.

** Straw three cornered.

RUSHGRASS with naked ftraws a little three cornered, the Compreffed
flowers in the fpike pointing two oppofite ways. Fence of one Compreffus
leaf.—
   Gramen cyperoides fpica fimplici compreffa difticha. *Ray's
Syn.* 425.
   Compreffed baftard Cyperus.
   In turf bogs. P. July.

RUSHGRASS with leafy ftraws, a little three cornered. White
Flowers growing in bundles, leaves briftly.   Albus
   Cyperus minor paluftris hirfutus, paniculis albis, paleaceis.
Ray's Syn. 427.
   Gramen junceum leucanthemum. *Gerard.* 30.
   White flowered Rufhgrafs.
   In marfhes. P. July.
   Goats eat it.

16 GALANGALE. 66 Cyperus.

EMPAL. A tiled *Spike*; the flowers pointing from two opposite lines, separated by egg-shaped *Scales* which are keel-shaped on the back, and bent inwards at the edges.

BLOSS. None.

CHIVES. *Threads* three; very short. *Tips* oblong, furrowed.

POINT. *Seedbud*, very small. *Shaft* thread-shaped; very long. *Summits* three, hairlike.

S. VESS. None.

SEED. Single, three cornered, pointed; not hairy.

English
Longus

GALANGALE with leafy three cornered straws. The flowers in a trebly compound leafy bundle: fruit-stalks naked; spikes alternate—*Root long, odoriferous.*

Cyperus longus. *Gerard.* 30. Odoratus. *Park.* 145. *Ray's Syn.* 425.

Cyperus odoratus radice longa seu Cyperus Officinalis. *Bauh. pin.* 14.

Sweet Cyperus.

In fens and marshes. P. July.

The root is agreeably aromatic to the smell, and warm and bitter to the taste. The modern practice disregards it, but perhaps it is not inferior to some of the more costly medicines brought from abroad.

17 BULLRUSH. 67 Scirpus.

EMPAL. A tiled *Spike*; the flowers growing equally from every side and separated by egg shaped *Scales* bent inwards at the Edges.

BLOSS. None.

CHIVES *Threads* three; which continue growing longer. *Tips* oblong.

POINT. *Seedbud*, very small. *Shaft* thread-shaped, long. *Summits* three; hairlike.

S. VESS. None.

SEED. Single, three cornered, pointed; furnished with soft hairs, which are shorter than the empalement.

OBS. *The hairs in some species grow to the point, in others to the base of the seed.*

BULL-

* Spike single.

BULLRUSH with a cylindrical, naked straw; spike nearly Club
egg-shaped; terminating.— Paluftris
   Scirpus equiseti capitulo majori. *Ray's Syn.* 428.
   Juncus equiseti capitulis, major. *Bauh. pin.* 12.
   Juncus aquaticus capitulis equiseti. *Park.* 1196.
   Juncus minor capitulis equiseti. *Gerard.* 34.
   Clubrush.
   On the banks of rivers, ponds and ditches : frequent. P. July.
   Swine devour the roots greedily when fresh, but will not
touch them when dry.—Goats, Horses and Hogs eat it. Cows
and Sheep refuse it.

BULLRUSH with the straw scored, and naked; spikes ter- Dwarf
minating, with two valves as long as the empalement. The roots Cæspitosus
separated by scales.—
   Scirpus montanus capitulo breviori. *Ray's Syn.* 429.
   Dwarf Clubrush.
   In turf bogs and upon dry heaths. P. July.
   Goats eat it.

BULLRUSH with cylindrical, naked, bristle-shaped straws. Sharp
Spike egg-shaped, consisting of two valves. Seeds naked—*Leaves* Acicularis
crooked, stiff, cylindrical.
   Scirpus minimus capitulis equiseti. *Ray's Syn.* 429.
   The least upright Club-rush.
   In marshes and bogs. P. August.

BULLRUSH with cylindrical, naked, alternate straws. Floating
Stem leafy and limber — Fluitans
   Scripus equiseti capitulo minor. *Ray's Syn.* 431.
   Juncellus, capitulis equiseti minor fluitans. *Bauh. pin.* 12.
   Floating Clubrush.
   In ponds and ditches. P. July. August.

* * Straw cylindrical, having several Spikes.

BULLRUSH with cylindrical, naked, straws; terminated Lake
by several eggshaped spikes supported upon fruitstalks.— Lacuftris
   Scirpus palustris altissimus. *Ray's Syn.* 428.
   Juncus maximus seu Scirpus major *Bauh. pin.* 12.
   Juncus aquaticus maximus. *Gerard.* 35.
   In rivers, pools and fens. P. July. August.
   When fodder is exhausted, cattle will live upon it. Cotta-
ges are sometimes thatched, and pack-saddles stuffed with it.
Bottoms of chairs are very commonly made of this rush : if it is
cut at one year old it makes the fine bottoms. Coarse bottoms
are made of it at two years old; and those that are still older
mixed with the leaves of the Yellow Flag, make the coarsest bot-
                                    tom

toms of all. Mats are likewife made either of the Lake Bullrufh alone, or mixed with the aforefaid leaves.—Goats and Swine eat it. Cows and Sheep refufe it.

Roundheaded
Holofchænus

BULLRUSH with cylindrical, naked ftraws; fupporting nearly globular, congregated fpikes, ftanding on fruit-ftalks; with a fence, confifting of two unequal fharp pointed leaves.—
Scirpus maritimus, capitulis rotundioribus glomeratis. *Ray's Syn.* 429.
Juncus acutus maritimus, capitulis rotundis. *Bauh. pin.* 11.
Juncus acutus maritimus alter. *Park.* 1106.
On the fea coaft P. July.

Leaft
Setaceus

BULLRUSH with naked, briftle-fhaped ftraws. Spikes one or two, lateral; fitting—*Leaves numerous, flender cylindrical.*
Scirpus foliaceus humilis. *Ray's Syn.* 430.
Juncellus inutilis, five Chamæfchænos. *Bauh. pin.* 12. *pr.* 22.
Gramen junceum minimum capitulo fquamofo. *Bauh. pin.* 6.
Gramen junceum maritimum exile Plimmothii. *Park.* 1270.
Leaft rufh, or fmall Plymouth Rufhgrafs.
Wet fandy ground. A. July. Auguft.

*** *Straw three cornered, fpike naked.*

Pointed
Mucronatus

BULLRUSH with three cornered, taper, naked ftraws. The fpikes lateral; fitting; clofe.—*The fharp points much longer than the flowers, and bent fideways.*
Juncus acutus maritimus, caule triangulo. *Bauh. pin.* 11.
Juncus acutus maritimus, caule triquetro, rigido, mucrone pungente. *Ray's Syn.* 429
On the fea fhore and the banks of large rivers. P. July. Auguft.

**** *Straw three cornered, panicle leafy.*

Marine
Maritimus

BULLRUSH with three cornered ftraws. Panicle clofe, and leafy; the fcales of the little fpikes cloven into three fegments; the middle fegment awl-fhaped—*Leaves ftiff, and fharp at the edges.*
Gramen Cyperoides panicula fparfa majus. *Bauh. pin.* 6.
Gramen Cyperoides paluftris, panicula fparfa majus. *Park.* 1266. *Ray's Syn.* 425.
Round rooted baftard Cyperus.
1. There is a variety that hath no aromatic fmell. *Bauh. pin.* 14.
On the fea coaft and the banks of rivers, P. Auguft.
The roots dried and ground to powder, have been ufed inftead of flour, in times of fcarcity.—Cows eat it.

BULLRUSH with three cornered, leafy straws. Flowers in Millet leafy bundles. Fruit-stalks naked; trebly compound. Spikes Sylvaticus crowded. —

Gramen Cyperoides miliaceum. *Bauh. pin.* 6.
Pseudo-cyperus miliaceus. *Park.* 1171.
Cyperus gramineus. *Ray's Syn.* 426.
Cyperus gramineus miliaceus. *Gerard.* 30.
Millet Cyperus grass.
In wet shady places. P. July.
Cows, Sheep, Horses and Goats eat it. Swine refuse it.

18. COTTONGRASS 68 Eriophorum.

EMPAL. A tiled *Spike*; the flowers growing equally from every side, and seperated by oblong egg-shaped membranaceous, flexible, tapering *Scales*, which are flat but turned in at the edges.

BLOSS. None.

CHIVES *Threads* three; hairlike. *Tips* upright, oblong.

POINT. *Seedbud*, very small. *Shaft* thread shaped; as long as the scales of the empalement. *Summits* three; longer than the *shaft*; reflected.

S. VESS. None.

SEED Three cornered, tapering to a point; furnished with soft hairs, which are longer than the spike.

COTTONGRASS with cylindrical sheathed straws, and Hares-tail a skinny spike —— *Perennial*; Root leaves *imperfectly three cor-* Vaginatum *nered, sharp; with two of the sides scored; Stalk twice as long as the leaves; scored; cylindrical, but flatted on one side; Stem-leaves not sharp pointed; purple at the base; Spike egg-shaped, tiled on every side with brown membranes; the lower spikes barren; the upper ones woolly and fertile.*

Juncus capitulo lanuginoso, seu schænolaguros. *Bauh. pin.* 12.
Gramen juncoides lanatum alterum danicum. *Park.* 1471.
Ray's Syn. 436.
Hares-tail Rush.
On bogs. February. March. P.

Common
Polyſtachion

COTTONGRASS with cylindrical ſtraws; flat leaves, and ſpikes on fruitſtalks—*pendant.*

Gramen pratenſe tomentoſum panicula ſparſa. *Bauh. pin.* 4.
Gramen tomentarium. *Gerard.* 29.
Gramen junceum lanatum, vel juncus bombycinus vulgaris.
Park. 1271.
Linagroſtis. *Ray's Syn.* 435.
Moorgraſs.
In marſhes and bogs. P. June.

This plant is uſeful in the Iſle of Skie to ſupport cattle in the earlier part of the ſpring, before the other graſſes are ſufficiently grown. *Pennant's Tour.* 1774. p. 308. Poor people ſtuff their pillows with the down, and make wicks of candles with it.

19 MATTGRASS 69 Nardus.

EMPAL. None.

BLOSS. Two valves; the *outer Valve* long, betwixt ſtrap and ſpear-ſhaped; pointed at the end, and incloſing the *leſſer Valve,* which is ſtrap-ſhaped and ſharp pointed.

CHIVES *Threads* three; hair-like; ſhorter than the bloſſom. *Tips* oblong.

POINT. *Seedbud* oblong. *Shaft* ſimple, long, downy. *Summit* ſimple.

S. VESS. The bloſſom adheres to the ſeed, without opening.

SEED. Single, incloſed in the Bloſſom; long and narrow, tapering to a point at each end, the upper part narroweſt.

Small
Stricta

MATGRASS with a briſtly, ſtraight ſpike; and the florets all pointing in one direction—*this graſs is ſtiff and hard to the touch, but being ſhort it eludes the ſtroke of the ſcythe.*

Gramen ſparteum juncifolium. *Bauh. pin.* 5. *Ray's Syn.* 393.
Spartum batavum et anglicum. *Park.* 1199.
Spartum noſtras parvum lobelio. *Gerard.* 43.
Small Matweed.
In dry heaths; fens and marſhes. P. April. July.

Goats and Horſes eat it. Cows and Sheep are not fond of it. Crows ſtock it up, for the ſake of the larva of inſects which they find at the root.

OBS. *The laſt five genera are nearly allied to the* GRASSES, *and a ſuperficial obſerver would be tempted to conſider them as ſuch, but an attentive peruſal of the generic characters will afford ſufficient diſtinctions.* (See the Plate of Graſſes.)

Order

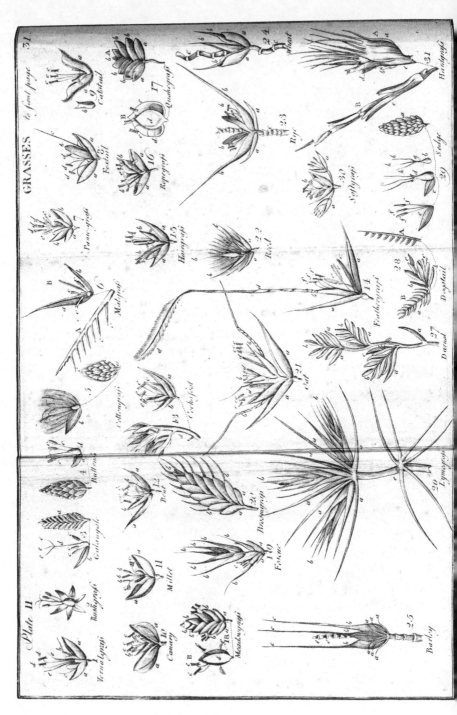

Order II. Two Pointals.

G R A S S E S.

THIS order comprehends the greater part of thofe
vegetables, commonly called Graffes ; and although
the flowers are generally difregarded, they will not to an
attentive obferver, appear lefs curioufly coinftructed, than
thofe which boaft of gayer colours and more confpicuous
parts.

NATURAL CHARACTER OF GRASSES.

EMPAL. A hufk ; generally compofed of two valves : the
larger valve hollow ; the *smaller*, flat.

BLOSS. Strictly fpeaking, none : but inftead thereof, a
Hufk of two valves, dry and fkinny. *Honeycup*
of two leaves ; oblong ; very fmall ; fuperior.

CHIVES *Threads* three ; hair-like. *Tips* oblong ; with
two cells.

POINT. *Shafts* two ; downy ; reflected. *Summits* downy.

S. VESS. None.

SEED. Single ; inclofed either by the bloffom, or the
empalement ; oblong ; tapering at each end.

STEM fimple, ftraight, hollow ; with knots or joints : it
is commonly called a *Straw*.

LEAVES entire ; narrow ; tapering to a point : one placed
at each joint of the ftraw.

Befides the plants which fall under this order, there
are others of the grafs kind that differ in fome of their
characters, and are arranged accordingly. Thus the
VERNALGRASS hath only two chives, and the SOFT-
GRASS and HARDGRASS have their chives and pointals
varioufly difpofed upon one plant ; *(See the figures in the
plate of* Graffes.)

The great folicitude of nature for the prefervation
of graffes is evident from this ; that the more the
leaves are confumed, the more the roots increafe.
The great author of nature defigned, that the delightful verdure of thefe plants fhould cover the furface

of

of the earth, and that they should afford nourishment to an almost infinite number of animals. But what increases our admiration most is, that although the Grasses constitutes the principal food of herbivorous animals, yet, whilst they are left at liberty in the pasture, they leave untouched the straws which support the flowers; that the seeds may ripen and sow themselves. Add to this, that many of the seemingly dry and dead leaves of Grasses, revive, and renew their verdure in the spring. And on lofty mountains where the summer heats are hardly sufficient to ripen the seeds, the most common Grasses are the SHEEPS FESCUE, the MOUNTAIN MEADOWGRASS, and the TURFFY HAIRGRASS, all which are viviparous.

In general, the leaves furnish pasturage for cattle; the smaller seeds are food for birds, and the larger for men. But some are preferred to others; as the FESCUE for Sheep; the MEADOWGRASS for Cows; the CANARY for Canary Birds; the OAT for Horses; the RYEGRASS, BARLEY and WHEAT for Men.

A Variety of insects too derive their nourishment from grasses; as the Great Argus Butterfly, *Papilio mæra*, the Wood Argus, *Papilio Ægeria*; the Marble Butterfly, *Papilio Galathea*; the Meadow brown Butterfly, *Papilio Jurtina*; the Plaintain Fritilary, *Papilio Cinxia*; the Lappit Moth, *Phalæna quercifolia*: the Drinker Moth, *Phalæna Potatoria*; the small Pasture Moth, *Phalæna culmella*; the *Chrysomela Graminis*, and several others that will be mentioned under the different species.

No part of botany appeared to me more difficult than the study of Grasses, but the method of accurate dissection and observation once adopted, nothing was more certain or more easy. However, when the great importance of the subject is considered, we cannot labour too much to fix the public attention to it, by rendering it as easy as possible: for which reason the exceptions are carefully noted under each subdivision of the orders, and in the following plate an example is selected from each genus. To gain a clear idea of the structure of the flowers, they must be examined just before the *Tips* discharge their dust; and by comparing them in that state with the figures in the plate, and with the generic description, every difficulty will soon be surmounted. The Botanic Microscope will be found extremely useful in dissecting the minuter parts.

Explanation

EXPLANATION of the PLATE.

Fig. 1. VERNALGRASS. *a a* hufks of the empalement. *b* the awn of the inner valve of the bloffom, twifted and jointed. *c* the ftraight awn of the outer valve of the bloffom. *d d* the two tips. *e e* the two fhafts.

Fig. 2. RUSHGRASS. The fix petals, the three chives, and tips; the feedbud, the fhaft, and the fummit cloven into three parts.

Fig. 3. GALANGALE. *a* the tiled fpike pointing from two oppofite lines. *b* the fcale of the empalement. *c c c* the tips. *d* the fhaft. *e e e* the fummits.

Fig. 4. BULLRUSH. *b* the tiled fpike. *a* the fcale of the empalement. *c c c* the chives and tips. *d* the feedbud, a little woolly.

Fig. 5. COTTONGRASS. *a* the woolly tiled fpike. *b* the fcale of the empalement including the hairy feedbud, the chives and the pointal.

Fig. 6. MATGRASS. A the fpike pointing one way. *c c c* the bloffoms. B one of the flowers a little magnified. *a* the lower and larger valve that embraces the fmaller valve *b*, which is here drawn out of its natural fituation. *c c c* the tips.

Fig. 7. PANICGRASS. *b b* the two equal valves of the empalement. *a* the third fmaller and outer valve. *c c* the valves of the bloffoms. *d d d* the tips. *e e* the downy fummits of the fhafts.

Fig. 8. FOXTAIL. *a a* the valves of the empalement. *b* the fingle valve of the bloffom, with the awn *c* proceeding from its bafe. *d d d* the tips.

Fig. 9. CATSTAIL. *a a* the hufks of the empalement opened and magnified to fhew the bloffom. *b* the floret in its natural ftate to fhew the two points at the top of it. *c c c* the tips.

Fig. 10. CANARY. *a a* the keeled hufks of the empalement. *b b* the hufks of the bloffom. *c c c* the tips.

Fig. 11. MILLET. *a a* the hufks of the empalement. *b b b* the tips. *c c* pencil-fhaped fummits.

Fig. 12. BENT. *a a* the two pointed valves of the empalement. *b b* the two valves of the bloffom. *c c c* the tips.

D   13. Fig.

Fig. 13. COCKSFOOT. *a* the outer and larger valve of the empalement. *b* the shorter valve. *c* the keel-shaped valve of the blossom. *c c c* the tips. *d.* the panicle pointing one way.

Fig. 14. FEATHERGRASS. *a a* the valves of the empalement. *b* the outer valve of the blossom, with the awn jointed at the base and twisted. *c* the inner valve of the blossom. *d d* the feathered awn. *e e* the hairy shafts and summits. *f f f* the tips.

Fig. 15. HAIRGRASS. *a a* the empalement, *b b* the blossoms, without the rudiment of a third betwixt them.

Fig. 16. ROPEGRASS. *a a* the empalement, *b b* the fertile blossoms with *e* the rudiment, of a third blossom betwixt them.

Fig. 17. QUAKEGRASS. *a a* the valves of the empalement, *b b b b b* the blossoms, of which the outer valves only are visible, B one of the blossoms taken out of the little spike, *c c* the outer heart-shaped valve of the blossom, *d d* the inner valve inversely egg-shaped.

Fig. 18. MEADOWGRASS. A, an entire little spike. *a a* the two husks of the empalement. *b b b b b* the blossoms. B one of the florets separated from the little spike. *c* the outer valve, *d* the inner valve of the blossom. *e e e* the forked tips. *f f* the woolly summits.

Fig. 19. FESCUE. *a a* the valves of the empalement. *b b b b b b b* the blossoms of the little spike terminating in sharp points. *c* the inner valve of one of the blossoms.

Fig. 20. BROOMGRASS. *a a* the empalement. *b b b* the blossoms, the outward valves only of which are visible, with the awns growing from beneath the point.

Fig. 21. OAT. *a a* the valves of the empalement. *b b b* the florets, the outer valves of which are furnished with a twisted jointed awn, growing from the back. *d d d* the inner valves. *c c c c c c* the tips.

Fig. 22. REED. *a a* the valves of the empalement. *b b b* the woolly blossoms.

Fig. 23. RYE. *a a* the valves of the empalement. *b b b b* the blossoms; the inner valve of which is flat, but the outer concave and furnished with an awn. *c c* the spike-stalk with its little teeth.

Fig. 24. WHEAT. *a a* the blunt valves of the empalement, embracing the three blossoms *b b b.* the outer valve only of which is seen, furnished with an awn. *c c* the spike-stalk.

Fig. 25. BARLEY. *a a a a a a* the six valves of the empalement, two of which belong to each of the blossoms *b b b. c c c* the long awns of the outer valves of the blossoms. *e e* the naked spike-stalk as it appears after the florets are pulled off.

Fig. 26.

Fig. 26. LYMEGRASS. *a a a a a a* the valves of the empalement, two of which belong to each little fpike *b b b.* *e* the empalement as it appears after the little fpikes are taken away.

Fig. 27. DARNEL. *a a a* the empalements of one vaive. *b b b* the little fpikes confifting of feveral florets. *c* one of the florets opened to fhew the two valves of the bloffom.

Fig. 28. DOGSTAIL. A the fpike pointing all one way, compofed of the florets B, in which *a* reprefents the fence with many clefts ; *b b* the valves of the empalement containing feveral florets, and *c c* the florets.

Fig. 29. SEDGE. *a* the tiled Cat-kin. *c* the fcaly empalement of the fertile floret. *d* the honeycup cloven at the top. *b* the feedbud, and *g* the fhaft, taken out of the honey-cup : *b b b* the fummits. *e* the fcaly empalement of the barren floret, with the three chives *f f f.*

Fig. 30. SOFTGRASS. *a a* the barren florets on fhort fruit ftalks. *b* the fertile floret furnifhed with chives and pointals.

Fig. 31. HARDGRASS. A, fhews a fingle floret taken from a fpike. *a* and *b* the hufks of the empalement embracing the fpike-ftalk. *c d* three bloffoms, in which the chives and pointals are varioufly difpofed. B the *St.i Hardgrafs* with the valves of the empalement undivided.

19. PANIC.

19 PANICGRASS. 76 Panicum.

EMPAL. *Hufk* with three valves, containing one Floret.
The *Valves* nearly egg-fhaped. The fmalleft ftanding behind the other two.

BLOSS. Two valves. The *Valves* nearly egg-fhaped,
the fmaller valve the flatteft.

CHIVES. *Threads* three ; hair-like ; fhort. *Tips* oblong.

POINT. *Seedbud* roundifh. *Shafts* two ; hair-like ; *Summits* downy.

S. VESS. The bloffom adheres to the Seed without opening.

SEED. Single, inclofed in the bloffom ; roundifh, but
a little flatted on one fide.

Green
Viiide

PANICGRASS with a cylindrical Spike. Partial fence in
cluding two Florets ; hairy and bundled. Seeds ftringy.—
Gramen paniceum fpica fimplici lævi. *Ray's Syn.* 393. *Bauh.
pin.* 8.
Gramen panici effigie fpica fimplici. *Gerard.* 17.
Panicgrafs with a fingle fmooth Ear.
1. There is a variety with a rough Spike.
In Paftures but not very common. A. July.

Loofe
Crus Galli

PANICGRASS with the fpikes alternate and in pairs. The
hufks have Awns : and are rough with hair. Spike ftalk with
five angles.—
Gramen paniceum fpica divifa. *Bauh. pin.* 8. *Ray's Syn.* 394.
Panicum vulgare. *Gerard* 85.
Panicum fylve re herbariorum. *Park.* 1154.
In paftures. A. Auguft.

Cocks-foot
Sanguinale

PANICGRASS with fingered Spikes ; knotty on the infide
the bafe. Flowers in pairs ; without Awns. The Sheath of the
leaves dotted—
Gramen dactylon latiore folio. *Bauh. pin.* 8.
Ifchæmon fylveftre latiore folio. *Park.* 1178.
Ifchæmon vulgare. *Gerard.* 27.
Not uncommon. A. Auguft.

Creeping
Dactylon

PANICGRASS with figured expanding Spikes, and foft
hairs on the infide the bafe. Flowers folitary. The plant fends
out creeping runners.—
Gramen dactylon folio arundinaceo majus aculeatum. *Bauh.
pin* 7.
Gramen dactyloides radice repente. *Gerard.* 28.
Gramen canarium ifchæmi paniculis. *Park.* 1179.
In fandy ground and on the fea-fhore. P. July.

20. FOX

TWO POINTALS. 37

20. FOXTAIL. 78. Alopecurus.

Empal. A *Hufk* of two valves containing one floret. The *Valves* equal, betwixt egg and fpear-fhaped; concave, comprefled.

Bloss. One concave *Valve* as long as the empalement, with a long *Awn* upon the back fixed towards the bafe.

Chives. *Threads* three; hair-like. *Tips* forked at each end.

Point. *Seedbud* roundifh. *Shafts* two, like tendrils; reflected; longer than the empalement. *Summits* fimple.

S. Vess. The bloffom inclofes the feed.

Seed. Single; roundifh; covered by the bloffom.

FOXTAIL with upright ftraws; cylindrical fpikes; and bulbous roots.— **Bulbous** **Bulbofus**
Gramen myofuroides nodofum. *Rays Syn.* 39 7.
Bulbous Fox-tail Grafs.
In paftures. P. June.

FOXTAIL with the fpiked ftraw upright; hufks woolly; **Meadow** bloffoms without awns.—*The fpike cylindrical, fomewhat divided* **Pratenfis** *like a panicle. The bloffoms are not always without awns.*
Gramen alopecuroides majus. *Gerard* 10.
Gramen Phalaroides fpica molli, five Germanicum. *Bauh. pin.* 4.
Gramen Phalaroides majus, five Italicum. *Bauh. pin.* 4. *Park.* 1164.
Meadow Foxtail Grafs.
In meadows very common. P. June.
This is the beft grafs to fow in low meadow grounds, or in boggy places that have been drained.—Sheep, Horfes and Goats eat it. Cows and Swine are not fond of it.

FOXTAIL with the fpiked ftraw upright. Hufks fmooth— **Field** *Spike cylindrical, very long. Straw not quite upright.* **Agreftis**
Alopecurus fpica cylindrica longiffima, glumis glabris; culmo fub-erecto. *Hudfon* 23.
Gramen myofuroides majus; fpica longiore, ariftis rectis. *Ray's Syn.* 397.
Gramen typhoides fpica anguftiore. *Bauh. pin.* 4.
Gramen alopecuroides minus *Gerard* 10.
1. There is a variety with crooked awns mentioned by Ray. *Syn.* 397.
Field Foxtail Grafs.
In fields and road-fides. P. July. Auguft.

D 3 FOXTAIL

THREE CHIVES.

Floating
Geniculatus
FOXTAIL with the spiked straw knee-jointed. Blossoms without awns—*Spike long and slender. In some situations the blossoms have awns.*
Gramen aquaticum geniculatum spicatum. *Bauh. pin.* 3. *Rays Syn.* 396.
Flote Foxtail Grass.
The Leaves lie upon the surface of pools and wet ditches.
P. June.---August.
Cows, Horses, Sheep and Goats eat it. Swine refuse it.

Bearded
Monspeliensis
FOXTAIL with a panicle not unlike a spike; the Empalement rough; the blossoms furnished with awns—*resembling the next species but three times as large. There is a little lump or pimple under the empalement. The* Blossom *is very short.* Straw and Leaves *inflexible.*
Gramen Alopecuroides anglo-britanicum maximum. *Bauh. pin.* 4.
Bearded Dog-tail Grass.
In Marshes. A. June. July.

Hairy
Paniceus
FOXTAIL with a panicle not unlike a spike; the empalement woolly; the blossoms furnished with awns.—
Gramen alopecurum minus; spica longiore. *Eauh. pin.* 4.
Bearded Foxtail Grass.
In dry soil. A. July.

21 CATSTAIL. 77 Phleum.

EMPAL. A *Husk,* of two valves. including a single floret: the husk oblong, strap-shaped, compressed; open at the end, and furnished with two spit.points. The *Valves* equal, straight, concave, compressed; one embracing the other; lopped; with a sharp point at the end of the keel.

BLOSS. Two valves, shorter than the empalement; the *outer Valve* embracing the *inner Valve,* which is smaller.

CHIVES *Threads* three; hair-like; longer than the empalement. *Tips* oblong, forked at each end.

POINT. *Seedbud* roundish. *Shafts* two; hair-like; reflected *Summits* downy.

S. VESS. None. The empal. and the bloss. inclosing the seed.

SEED. Single; roundish.

1 CATSTAIL

CATSTAIL. The fpike cylindrical; long; fringed. Straw Meadow upright.— Pratenfe
Gramen typhoides maximum, fpica longiffima. *Dauh. pin.* 4.
Gramen typhinum majus, feu prinum. *Gerard.* 11. *Ray's Syn.* 398.
Gramen typhinum medium, feu vulgatiffimum. *Park* 1170.
1. There is a variety which is fmaller. *Gerard.* 11. *Park.* 1170.
Timothy Grafs.
In paftures; common. P. July.
Cows, Horfes and Goats eat it. Swine refufe it.

CATSTAIL. The fpike cylindrical; ftraw afcending; leaves Bulbous oblique; root bulbous—*ftraw fheathed.* Spikes *fmooth.* Tips Nodofum *white.* Leaves *pointing from oppofite fides of the ftraw; fmooth; except the edges which are rough.*
Gramen nodofum fpica parva. *Bauh. pin.* 2.
Very common. P. July. Auguft.

CATSTAIL with the fpikes egg-fhaped and fringed. Straw Marine branching.—*The fpikes feldom rife entirely out of the fheaths of the* Arenarium *leaves.*
Gramen typhinum maritimum minus. *Ray's Syn.* 398.
Gramen typhynum danicum minus. *Park.* 1170.
Danifh Cats-tail Grafs, or Sea Canary Grafs.
On the fea-fhore. A. July.

22 CANARY. 74 Phalaris.

EMPAL. A *Hufk* of two valves including a fingle floret; compreffed; blunt. The *Valves* boat-fhaped, compreffed; keel-bottomed, but more blunt upwards; the edges ftraight; parallel; approaching.
BLOSS. Two valves fmaller than the empalement. The *outer Valve* oblong; rolled in a fpiral; tapering to a point. The *inner Valve* fmaller than the other.
CHIVES. *Threads* three; hair-like; fhorter than the empalement. *Tips* oblong.
POINT. *Seedbud* roundifh; *Shafts* two; hair-like. *Summits* woolly.
S. VESS. The bloffom clofely furrounds the feed, without opening.
SEED. Single; fmooth; nearly cylindrical in the middle, but tapering towards each end.

Manured
Canariensis

CANARY with a panicled spike nearly eggshaped; and keel-shaped husks.—

Phalaris major semine albo. *Bauh. pin.* 28.

Phaloris. *Gerard.* 86.

Road sides and uncultivated ground.　A. June. September.

It is ·often cultivated for the sake of the seeds, which are found to be the best food for the Canary, and other small birds.

It nourishes the Canary Bug, *Coccus Phalaridis.*

Cats-tail
Phleoides

CANARY with a cylindrical panicled spike, smooth and viviparous.—*The spike is pale, divisible into lobes, with here and there a viviparous husk. This plant so exactly resembles some of the* Cats-tails *that it may easily be mistaken for one of that genus; but when you examine the spike and press it with your fingers, it separates, and proves to be a panicle; and the husks are not notched at the end.*

Gramen typhoides asperum primum. *Bauh. pin.* 4.

Branched Cats-tail Grass.

In the fields below Kings-Weston near Bristol.　P. July.

Sheep and Goats eat it.　Swine refuse it.

Reed
Arundinacea

CANARY with a large oblong panicle, swelling out in the middle.—*Leaves broad, scored. Panicle nearly egg-shaped.*

Gramen arundinaceum acerosa gluma nostras. *Park* 1273.

Ray's Syn. 400.

Gramen aquaticum paniculatum latifolium. *Bauh. pin.* 6.

Banks of rivers.　P July.

It is used to thatch ricks or cottages, and endures much longer than straw.　In Scandinavia they mow it twice a year, and their cattle eat it.

Horses, Cows, Sheep and Goats eat it.　Swine refuse it.

There is a cultivated variety of this in our gardens with beautifully striped leaves.　The stripes are generally green and white; but sometimes they have a purplish cast.　This is commonly called *Painted Lady-grass,* or *Ladies Traces.*

23 MILLET. 79 Milium.

EMPAL. *Hufk* with two valves inclofing a fingle floret. The
Valves egg-fhaped, tapering to a point.

BLOSS. Two valves, fmaller than the empalement. *Valves*
egg fhaped ; one larger than the other.

CHIVES. *Threads* three ; hair-like ; very fhort. *Tips* oblong.

POINT. *Seedbud* roundifh. *Shafts* two ; hair-like. *Summits*
pencil-fhaped.

S. VESS. The bloffom, which is very fmooth, inclofes
the feed.

SEED Single, roundifh.

MILLET with the flowers without awns ; in fcattered Soft
panicles.—   Effufum
Gramen miliaceum. *Gerard.* 6. *Ray's Syn.* 402. vulgare.
Park. 1153.
Gramen fylvaticum panicula miliacea fparfa. *Bauh. pin.* 8.
Millet Grafs.
In wet woods, common. A. June. July.
Horfes, Cows, Sheep and Goats eat it.

24 BENT. 80 Agroftis.

EMPAL. *Hufk* of two valves, inclofing one floret, tapering
to a point.

BLOSS. Two valves, tapering to a point, *one Valve* larger
than the other ; hardly fo long as the empalement.

CHIVES. *Threads* three ; hair-like ; longer than the bloffom. *Tips* forked.

POINT. *Seedbud* roundifh. *Shafts* two ; reflected, woolly ;
Summits woolly or hairy.

S. VESS. The bloffom adheres to the feed without opening.

SEED Single ; cylindrical, but tapering towards each end.

BENT

THREE CHIVES.

With Awns.

Silky
Spicaventi
BENT with a ſtraight, ſtiff, very long awn, fixed to the out-
ward petal. Panicle expanding.—
Gramen ſegetum panicula arundinacea. *Bauh. pin.* 3.
Gramen agrorum venti ſpica. *Park* 1151.
Gramen arundinaceum. *Gerard.* 5.
Gramen miliaceum majus, glamus ariſtatis, ſpadiceis et pal-
lidis. *Ray's Syn.* 405.
In fields very common. A. July.
Horſes and Goats eat it. Sheep refuſe it.

Red
Rubra
BENT with the part of the panicle that is in bloſſom very
much expanded. The outer petal ſmooth. The terminating
awn twiſted and bent back.—*Before it is in flower it is cloſe and
contracted like a ſpike ; when in flower the bloſſoms ſtand out in ho-
rizontal whorls ; and after flowering it becomes entirely red.*
Panicum ſerotinum arvenſe, ſpica pyramidata. *Ray's Syn.* 394.
In meadows. A. July.
Horſes eat it. Sheep are not fond of it.

Brown
Canina
BENT with very long empalements. The awn upon the
back of the petals much curved : the ſtraw proſtrate ; ſomewhat
branched.—*The panicle is often very long ; its branches compact,
and of a ſhining purple : the awn like a briſtle, double the length of
the flower ; white, ſtraight, marked in the middle by a brown joint.*
Gramen ſupinum caninum paniculatum, folio varians. *Bauh.
pin.* 1.
In wet meadows, frequent: P. July. Auguſt.
Cows and Horſes eat it.

**Without Awns.*

Creeping
Stolonifera
BENT without awns ; the branches of the panicle expand-
ing ; the ſtraws creeping ; the empalements equal.—
Gramen montanum miliaceum minus, radice repente. *Ray's
Syn.* 402.
Gramen caninum ſupinum minus. *Bauh. pin.* 1.
1. There are two varieties ; one with a very narrow leaf and
another,
2. With extremely ſmall empalements.
In meadows. Not very common. P. Auguſt.
Cows, Horſes and Sheep eat it.

BENT

BENT with a panicle fine like hair; expanding: empale- Fine
ments awl-shaped, equal; a little rough with hair; coloured. Capillaris
blossoms without awns.—*The very slender panicle and fruit-stalks
cannot escape observation.*
    Gramen montanum panicula spadicea delicatiore. *Bauh. pin.* 3.
    Gramen miliaceum locustis minimis, panicula fere arundi-
nacea. *Ray's Syn.* 402.
    Very common. P. August.

BENT with a compact panicle without awns. Empalements Wood
equal; shorter than the blossom before flowering; but afterwards Sylvatica
twice as long.—
    Gramen Miliaceum sylvestre, glumis oblongis. *Ray's Syn.* 404.
    In Bishops Wood near Hampstead. P. August.

BENT with a flexible panicle. Empalaments without awns, Marsh
equal—*Stem creeping and striking root at the joints.* Alba
    Agrostis panicula coarctata mutica, calycibus æqualibus his-
pidiusculis coloratis; culmo repente. *Hudson's Flor. Ang.* 27.
    Gramen miliaceum majus, panicula spadicea. *Ray's Syn.* 464.
1. There is a variety with a green panicle.
    In ditches and marshes, frequent. P. July.

BENT with a thread-shaped panicle without awns.—*Straw* Small
slender, and very short. Minima
    Gramen minimum paniculis elegantissimis. *Bauh. pin.* 2.
    Gramen minimum anglo-britanicum. *Ray's Syn. Pl.. dub.*
    In high dry pastures. P. May.

25 COCKSFOOT. 86 Dactylis.

EMPAL. *Husks* compressed; keel-shaped, sharp; pointing
    one way. One *Valve* longer than the blossom, the
    other shorter.
BLOSS. A *Husk*; compressed; oblong; sharp. The longest
    Valve keel-shaped; lying within the longest valve of
    the empalement.
CHIVES. *Threads* three; hair-like; as long as the blossom.
    Tips forked at each end.
POINT. *Seedbud* turban-shaped. *Shafts* two; hair-like;
    expanding, woolly. *Summits* simple.
S. VESS. None. The blossom incloses the seed until it is
    ripe.
SEEDS Solitary; naked; depressed on one side, convex
    on the other.

    OBS. *The* Rough Cocks-foot *hath several florets in each empale-
ment*

Smooth
Cynoſuroides   COCKSFOOT with numerous rough ſcattered ſpikes, pointing one way.—*Six leaves upon each ſtraw; broad, and longer
than the ſtraw; very ſmooth; rough at the edges; ſea green on the
inner ſurface.* Spikes *ſix or more; chaffy.* Pointals *long, woolly.*
Spartum Eſſexianum, ſpica germina clauſa. *Ray's Syn.* 393.
In marſhes in Eſſex. A. Auguſt. September.

Rough
Glomerata   COCKSFOOT with a congregated panicle, pointing one
way. Each empalement contains four florets,—*The bundles of
flowers but little expanded. Little ſpikes compreſſed; in bundles, ſitting;
ſpreading wide when in flower.* Tips *reddiſh.*
Gramen ſpicatum folio aſpero. *Bauh. pin.* 3.
Gramen pratenſe ſpica multiplici rubra. *Park.* 1161.
Calamagroſtis toroſa panicula. *Park.* 1182.
In meadows and paſtures, common. P. June. Auguſt.
Horſes, Sheep and Goats eat it. Cows refuſe it.

26 FEATHERGRASS. 90 Stipa.

EMPAL.   A *huſk* of two valves, taper, flexible; incloſing
a ſingle floret.
BLOSS.   Two valves. The *outer Valve* terminated by a
very long, ſtraight, twiſted awn, jointed at the
baſe. The *inner Valve* ſtrap-ſhaped, without any
awn: as long as the outer valve.
CHIVES. *Threads* three; hair like. *Tips* ſtrap-ſhaped.
POINT. *Seedbud* oblong. *Shafts* two; rough with hair;
united at the baſe. *Summit* downy.
S. VESS. An adhering huſk.
SEED. Single; oblong.

Downy
Pennata   FEATHERGRASS. The awns covered with wool.
Gramen Sparteum pennatum. *Bauh. pin.* 5. *Ray's Syn.* 393.
Spartum Auſtriacum pennatum. *Gerard.* 42.
On mountains. P. July. Auguſt.

EMPAL.

27 HAIRGRASS. 81 Aira.

EMPAL. A *Husk* of two valves, containing two florets.
The *Valves* betwixt egg and spear-shaped, nearly
equal; sharp.
BLOSS. Two valves, resembling those of the empale-
ment.
CHIVES. *Threads* three; hair-like; as long as the blos-
som. *Tips* oblong; forked at each end.
POINT. *Seedbud* egg-shaped. *Shafts* two; bristly; ex-
panding. *Summits* downy.
S. VESS. None. The blossom incloses and adheres to
the seed.
SEED. Nearly egg-shaped.

* Without awns.

HAIRGRASS with a compact panicle. Flowers without Purple
awns, on fruit stalks, awl-shaped; and the edges turned in. Cærulca
Leaves flat.— *In very fertile soils there are sometimes four florets in
each empalement.* Roots *bulbous.* Tips *purplish blue.* Pointals
purple. Straw *long and without joints. There is sometimes a shaft
betwixt the florets but without any summit; or else it must have been
arranged under the genus* ROFEGRASS.
Gramen pratense serotinum panicula longa purpurascente.
Rai's Syn. 404.
On bogs, turfy, barren and marshy places frequent. P.
August.
Horses, Sheep and Goats eat it.
Upon this and the other species is found the *Chermes Graminis.*

HAIRGRASS with an expanded panicle: flowers smooth, Water
without awns; longer than the empalement. Leaves flat—*In* Aquatica
*dry situations sometimes there are five florets in each empalement,
and the flowers stand far asunder.*
Gramen miliaceum aquaticum. *Ray's Syn.* 402.
Gramen caninum supinum paniculatum dulce. *Bauh. pin.* 2.
Gramen exile tenuifolium canario simile, seu Gramen dulce.
Park. 1174.
Banks of rivers. P. June. July.
It hath a sweet taste. Cows are very fond of it. Horses
and Sheep eat it.

HAIR

** *With awns.*

Turty
Cæfpitofa

HAIRGRASS with flat leaves; expanding panicles; petals woolly at the bafe and furnifhed wth ftraight fhort awns—*Outer fides of the valves tinged with purple.*
Gramen fegetale. *Gerard.* 5.
Gramen fegetum altiffimum panicula fparfa. *Bauh. pin.* 3.
Gramen fegetum panicula fpeciofa. *Park.* 1158.
Gramen miliaceum fegetale majus. *Ray's Syn.* 403.
Moift meadows and woods. P. June. Auguft.
It is very apt to grow in tufts and occafion irregularities in the furface of meadows.—Cows, Goats and Swine eat it. Horfes are not fond of it.

Twifted
Flexuofa

HAIRGRASS with leaves like briftles; ftraw almoft naked; panicle ftraddling; fruitftalks zigzag—*Each floret hath a twifted awn as long as the bloffom, fixed to the outer fide of the bafe of the petal.*
Gramen paniculatum locuftis parvis purpuro-argenteis majus et perenne. *Ray's Syn.* 407.
Gramen avenaceum capillaceum minoribus glumis. *Bauh. pin.* 10.
Rocky and barren ground. P. July. Auguft.
Horfes, Cows and Sheep eat it.

Mountain
Montana

HAIRGRASS with leaves like briftles; panicle clofe; florets hairy at the bafe, awn rather long and twifted—*Hudfon makes this only a variety of the foregoing, but it differs in the florets being hairy at the bafe.*
Gramen nemorofum paniculis albis, capillaceo folio. *Bauh. pin.* 7. *Ray's Syn.* 407.
High barren ground. P. July. Auguft.
Sheep are extremely fond of it.

Grey
Canefcens

HAIRGRASS with leaves like briftles; the upper leaf fheathing and inclofing the lower part of the panicle.—*The awns are encompaffed with teeth in the middle part, brown and thick below, but white, flender and club-fhaped above. After flowering the panicle rifes higher out of the fheath; before flowering it refembles a fpike. This is paler than moft other graffes, and from that circumftance may be diftinguifhed at firft fight.*
Gramen foliis junceis, radice jubata. *Bauh. pin.* 5.
Gramen miliaceum maritimum molle. *Ray's Syn.* 405.
In fand on the fea fhore. P. July.
Cows and Goats eat it.

HAIRGRASS with leaves like briftles; angular fheaths; pa- *Early*
nicle like a fpike; florets with an awn at the bafe—*Nearly allied* Praecox
to the foregoing, but fmaller; being amongft the fmalleft of all the
graffes.
  Gramen parvum praecox panicula (potius fpica) laxa canef-
cente. *Ray's Syn.* 407.
  Common in barren fands. A. June. July.

  HAIRGRASS with leaves like briftles; panicle ftraddling; Silver
florets diftant; furnifhed with awns.— Caryophyllea
  Gramen paniculatum, locuftis purpureo-argenteis annuum.
Ray's Syn. 407.
  Caryophyllus arvenfis glaber minimus. *Bauh. pin.* 105.
  In fandy paftures, frequent. A. July.

28 ROPEGRASS. 82 Melica.

EMPAL.. A *Hufk* of two valves, containing two florets.
    The *Valves* egg-fhaped, concave; nearly equal.
BLOSS. Two valves. *Valves* egg-fhaped; without awns;
    one concave, the other flat. Betwixt the two
    florets is a rudiment of a third, and fometimes of
    a fourth.
CHIVES. *Threads* three; hairlike as long as the bloffom.
    Tips oblong, forked at each end.
POINT. *Seedbud* betwixt egg and turban-fhaped. *Shafts*
    two; like briftles; expanding. *Summits* oblong;
    woolly.
S.VESS. None, the bloffom inclofes the feed until it
    ripens.
SEED. Single, eggfhaped.

  OBS. *The rudiment of a third floret ftanding upon a little fruit-ftalk*
betwixt the other two florets, gives the effential charaƈter of this genus.
It confifts of two rudiments, or florets; lopped; alternate. The
hufks rolled fpirally inwards and pellucid.

Red
Nutans

ROPEGRASS with a simple nodding panicle, and the blossoms not fringed—*Panicle red. In this species there is often only one perfect flower in each empalement.*

Gramen avenaceum nemorense, glumis rarioribus exfusco xerampelinis. *Ray's Syn.* 403.

Gramen avenaceum locustis rarioribus. *Bauh. pin.* 10.

Gramen locustis rubris. *Park.* 1151.

Gramen avenaceum locustis rubris montanum. *Bauh. pin.* 10.

Purple melic grass.

In moist woods, frequent. P. June. July.

In the Isle of Rasa they make this grass into ropes for fishing nets, which are remarkable for lasting long without rotting. *Pennant's Tour.* 1774. p. 297.—Cows, Horses and Goats eat it.

29 QUAKEGRASS. 84 Briza.

EMPAL. A *Husk* of two valves, expanding; containing several florets pointing from two opposite lines, collected into a heart-shaped spike. The *Valves* blunt, heart-shaped, concave, equal.

BLOSS. Two valves. The *lower Valve* the size and figure of the empalement. The *superior Valve* small, flat, roundish, closing the hollow of the other.

CHIVES. *Threads* three; hairlike. *Tips* oblong.

POINT. *Seedbud* roundish. *Shafts* two; hairlike; much curved. *Summits* downy.

S. VESS. The blossom unchanged, contains the seed until it is ripe.

SEED. Single; very small; roundish, compressed.

Small
Minor

QUAKEGRASS. The little spikes triangular; and the florets shorter than the empalement—*Seven florets in each empalement.*

Gramen tremulum minus, panicula parva. *Bauh. pin.* 2.

Gramen tremulum minus, panicula ample, locustis parvis triangulis. *Ray's Syn.* 412.

In pastures. P. July.

QUAKEGRASS. The little fpikes egg-fhaped and the florets Common longer than the empalement—*The hufks and flowers are fometimes* Media *white. Seven florets in each empalement.*

Gramen tremulum majus. *Bauh. pin.* 2. *Ray's Syn.* 412.

Gramen tremulum, feu phalaris media Anglica prima an fecunda. *Park.* 1165.

Phalaris Pratenfe. *Gerard.* 86.

Middle quaking grafs, cow-quakes, ladies hair.

In fields and paftures. P. July.

Cows, Sheep and Goats eat it.

If a feed is carefully diffeated in a microfcope, with a fine lancet, the young plant will be found with its root and leaves pretty perfectly formed.

30 MEADOWGRASS. 83 Poa.

EMPAL. A *Hufk* of two valves without awns; containing feveral florets pointing from two oppofite lines and collected into an oblong egg-fhaped fpike. The *Valves* egg-fhaped, tapering.

BLOSS. Two valves. The *Valves* egg-fhaped, tapering, concave, compreffed; fomewhat longer than the empalement; fkinny at the edges.

CHIVES. *Threads* three; hair-like. *Tips* forked at each end.

POINT. *Seedbud* roundifh. *Shafts* two; reflected; woolly. *Summits* like the fhafts.

S.VESS. The bloffoms adheres to the feed without opening.

SEED. Single; oblong, compreffed, tapering to a point at each end: covered by the bloffom.

MEADOWGRASS with a fpreading panicle. Little fpikes Reed ftrap-fhaped, containing fix florets— *Which are generally purplifh.* Aquatica Leaves *channeled, fmooth, broad. The number of florets in each empalement varies from four to ten.*

Gramen paniculatum aquaticum latifolium. *Bauh. pin.* 3.

Gramen aquaticum majus. *Gerard.* 6. 411.

Gramen paluftre paniculatum altiffimum. *Bauh. pin.* 3.

In marfhes and on the banks of rivers. P. July. Auguft.

Sheep eat it. Horfes and Cows are not fond of it.

It is an extremely ufeful grafs to fow upon the banks of rivers or brooks.

E   MEA-

Common
Trivialis

MEADOWGRASS with a spreading panicle. Little spikes of three florets; downy at the base. Straw cylindrical, upright —*Leaves smooth along the back. Florets of a yellower colour at the ends than in the other species.*
Gramen pratense paniculatum medium. *Bauh. pin.* 2. *Ray's Syn.* 409.
Gramen pratense minus. *Park.* 1156. *Gerard.* 2.
In most meadows and pastures. P. June. August.

Narrowleaved
Angustifolia

MEADOWGRASS with spreading panicles. Little spikes with two, three, or four florets; downy. Straw cylindrical, upright—*Tips yellow. Panicle oblong; crowded, so as to appear almost tiled.*
Gramen pratense paniculatum majus angustiore folio. *Bauh. pin.* 2.
In woods and hedges. A. June. July.
Horses, Cows, Sheep, Goats and Swine eat it.

Great
Pratensis

MEADOWGRASS with a spreading panicle. Little spikes with five florets; smooth. Straw cylindrical, upright.—*Tips blue.*
Gramen pratense paniculatum majus latiore folio,—Poa Theophrasti. *Bauh. pin.* 2. *Ray's Syn.* 409.
Gramen Pratense. *Gerard.* 2.
Gramen pratense vulgatius. *Park.* 1156.
In fields and pastures. P. June. July.
Horses, Cows, Sheep and Swine eat it. Goats are not fond of it.

Annua
Annua

MEADOWGRASS with the panicle spreading horizontally from the straw. Little spikes blunt. Straw oblique, compressed.—
Gramen pratense minus seu vulgatissimum. *Ray's Syn.* 408.
Gramen pratense minimum album et rubrum. *Park.* 1156. *Gerard.* 3.
Gramen pratense paniculatum minus. *Bauh. pin.* 3.
Suffolk grafs.
In most pastures. A. April. September.
Horses, Cows, Sheep, Goats and Swine eat it.

Hard
Rigida

MEADOWGRASS with a ·spear-shaped panicle; a little branched; the florets all pointing one way. The branches alternate—*Straw inflexible, very short. Panicles spear-shaped, inflexible, doubly compounded; branches alternate, with alternate little spikes on inflexible fruitstalks which are shorter than the spike they support. The little Spikes are strap-shaped, sharp, and contain six or eight florets which are a little sharp and skinny at the point. Empalement keel-shaped.*
Gramen panicula multiplici. *Bauh. pin.* 3. *Park.* 1157.
Gramen exile duriusculum in muris et aridis proveniens. *Ray's Syn.* 410.
On walls, roofs, and dry sandy soil, frequent. A. July.
MEA-

MEADOWGRASS with a compact panicle, in which the Creeping florets point all one way. Straw oblique, compressed—*Generally* Compress'a *six florets in each empalement.*

Gramen pratense medium culmo compresso. *Ray's Syn.* 409.
Gramen murorum radice repente. *Bauh. pin* 2.
On walls, house tops and other very dry places. A. June.
Horses, Cows, Sheep and Goats eat it.

MEADOWGRASS with a panicle growing gradually smaller Wood towards the point; each little spike contains about two sharp- Nemoralis pointed rough florets. Straw bowed inwards.—
In woods and shady places. A. June.

MEADOWGRASS with a panicle a little expanding; the Bulbous florets all pointing one way; each little spike containing about Bulbola four florets—*Straw with bulbs or knobs at the bottom.*
Gramen arvense panicula crispa. *Bauh. pin.* 3.
In pasture ground near Clapham in Surry. P. June.

MEADOWGRASS with a spiked panicle. About four florets. Crested in each empalement, which is longer than the fruit-stalk and a Cristata little hairy. Petals with awns.—
Gramen spica cristata subhirsutum. *Bauh. pin.* 2.
Gramen pumilum spica purpuro-argentea molli. *Ray's Syn.* 396.
Crested Hairgrass. *Hudson's Flor. Angl.* 28.
In high barren pastures. P. July. August.

31 FESCUE. 88 Festuca.

EMPAL. A *Husk* of two valves; upright; containing several florets collected into a slender spike. The *Valves* awl-shaped, tapering. The *inferior Valve* the smallest.

BLOSS. Two valves. The *inferior Valve* the figure of the empalement but larger; rather cylindrical but tapering, and ending in a sharp point.

CHIVES. *Threads* three; hair-like; shorter than the blossom. *Tips* oblong.

POINT. *Seedbud* turban-shaped. *Shafts* two, short, reflected. *Summits* simple.

S. VESS. The blossom shuts close upon the seed.

SEED. Single; slender; oblong; very sharp pointed at each end, with a furrow running lengthways.

Panicle with the florets all pointing one way.

Sheep
Ovina

FESCUE with a compact panicle; florets with awns, all pointing one way. Straw four cornered, almost naked. Leaves. like bristles—*Hairy. The lower little spikes on fruit-stalks; four florets in each; with the rudiment of a fifth.*

Gramen capillaceum, locustellis pennatis non aristatis. *Ray's Syn.* 410.

Gramen foliis junceis brevibus majus, radice nigra. *Bauh. pin.*5.
1. There is a variety which is viviparous. *Ray* in his *Syn.* p. 410, calls it Gramen sparteum montanum, spica foliacea graminea, majus et minus. Tab. 22. fig. 1.

In high and dry situations. P. June. July.

It flourishes best in a dry sandy soil: Cows, Horses, and Goats will eat it, but it is the favourite food of Sheep: they prefer it before all other grasses, and soonest grow fat upon it; for though small, it is succulent. The Tartars who lead a wandering life, tending their flocks and herds, always choose those spots where this grass abounds. Is not the superiority of the Spanish and English wool owing to the abundance of this grass in in the hilly pastures where the Sheep are kept?

Hard
Duriuscula

FESCUE with an oblong panicle. Florets all pointing one way. Little spikes oblong and smooth. Leaves like bristles—*Smooth. Branches of the panicle alternate. The leaves which arise from the root, thread-shaped and channelled; those upon the straw flat; the Husks smooth. Little Spikes on fruit-stalks, five or six florets in each, with short awns. One valve of the husk twice as large as the other.*

Gramen pratense. panicula duriore laxa unam partem spectante. *Ray's Syn.* 413. Tab. 19. fig. 1.

In dry pastures. P. June.

Purple
Rubra

FESCUE with a rough panicle and the florets all pointing one way. The little spikes contain six florets, all of which have awns except that at the end. Straw semi-cylindrical---*The size, the red colour when ripe; and the semi-cylindrical straw distinguish this from the Sheep's Fescue. The upper surface of the leaves soft and covered with a very fine wool.*

In dry pastures. P June.
Horses and Goats eat it. Sheep refuse it.

FESCUE with fpiked nodding panicles. Empalement very Wall
minute; without awns. Bloffoms rough and furnifhed with Myuros
awns---*Leaves awl fhaped, fcored, fheathing the ftraw. Five florets
in each little fpike.* Panicle *branched but not fpreading ; very flender
and long ; bent a little, but not nodding.*
   Gramen murorum fpica longiffima. *Gerard* 31. *Ray's Syn.* 415.
   Gramen fpica nutante longiffima. *Park.* 1162.
   Capons-tail grafs.
   On walls and in very dry and barren foils. A. June.

FESCUE with a panicle in which the florets all point one way: Barren
little fpikes upright. One valve of the empalement entire, the Bromoides
other tapering to a point.---*The panicle fomewhat refembling a fpike
and the hufks not being fringed, diftinguifh this from the* Wall Fefcue.
   Gramen paniculatum bromoides minus, paniculis ariftatis
unam partem fpectantibus. *Ray's Syn.* 415.
   In fandy fields. A. May. June.

FESCUE with an upright panicle, in which the florets all Tall
point one way. Little fpikes with a few awns ; the outer ones Elatior
cylindrical.--- *Meadow*
1. Panicle upright, little fpikes narrow, without awns: leaves
flat. *Hudfon.* 37.
   Gramen paniculatum nemorofum, latiore folio glabrum, panicula nutante non ariftata. *Ray's Syn.* 411.
   Gramen paniculatum elatius, fpicis longis muticis et fquamofis.
Ray's Syn. 411.
2. There is another variety, which Ray calls Gramen arundinaceum aquaticum panicula avenacea. *Syn.* 411.
   In paftures ; but not common. P. July.
   It makes an excellent pafture, but it requires a rich foil.
   Horfes, Cows, Sheep and Goats eat it.

   * * *Panicle with the florets pointing various ways.*

FESCUE with an upright panicle : little fpikes fomewhat egg- Small
fhaped ; without awns. Florets fhorter than the empalement. Decumbens
Straw drooping--*Panicle fimple.* Empalement *as long as the little fpikes.*
   Gramen avenaceum parvum procumbens, paniculis non
fariftatis. *Ray's Syn.* 408.
   In barren, but rather moift ground. P. Auguft.
   Sheep refufe it.

FESCUE with an upright branched panicle. Little fpikes Floating
cylindrical ; without awns ; with very fhort fruit-ftalks.--- Fluitans
   Gramen aquaticum cum longiffima panicula. *Ray's Syn.* 412.
   Gramen fluviatile. *Gerard.* 14. *Park.* 1275.
   Gramen aquaticum fluitans, multiplici fpica. *Bauh. pin.* 2.
   Flote Fefcue grafs.
   In wet ditches and ponds, very common. P. June. July.
                                                            The

The feeds are fmall, but very fweet and nourifhing. They are collected in feveral parts of Germany and Poland, under the name of *Manna Seeds*; and are ufed at the tables of the great in foups and gruels, upon account of their nutritious quality, and grateful flavour. When ground to meal, they make bread very little inferior to that in common ufe. The bran feparated in preparing the meal, is given to Horfes that have the worms; but they muft be kept from water for fome hours afterwards. Geefe are very fond of the feeds, and well know where to look for them. The plant affords nourifhment to the gold fpotted Moth, *Phalæna Feftucæ*.

Wood
Sylvatica

FESCUE with flowers forming a fpike; little fpikes alternate, nearly all pointing in two directions; fitting: furnifhed with awns.—

Feftuca graminea nemoratis latifolia mollis. *Bauh. pin.* 10.
Gramen avenaceum dumetorum fpicatum. *Ray's Syn.* 394.
In woods and hedges. P. July.

32 BROOMGRASS. 89 Bromus.

EMPAL.. A *Hufk* of two valves, expanding; containing feveral florets collected into a fpike. The *Valves* oblong egg-fhaped, taper, without awns. The *inferior Valve* fmaller than the other.

BLOSS. Two valves. The *inferior Valve* large; the fize and figure of the empalement; concave, blunt, cloven; fending out a ftraight *Awn* from beneath the ends. The *fuperior Valve* fpear-fhaped, fmall; without an awn.

CHIVES. *Threads* three; hair-like; fhorter than the bloffom. *Tips* oblong.

POINT. *Seedbud* turban-fhaped. *Shafts* two; fhort; woolly; reflected. *Summits* fimple.

S. VESS. The bloffom fhuts clofe upon and adheres to the feed.

SEED. Single, oblong, convex on one fide, furrowed on the other.

Field
Secalinus

BROOMGRASS. The panicle expanding. Little fpikes egg-fhaped; awns ftraight. Seeds feparate—*Leaves rough on one fide.*

BROOM-

1. BROOMGRASS. Panicle upright and compact—*This variety* Compact
is occasioned by a dry stiff soil, for when sown in a garden it again Hordeaceus
becomes the BROMUS SECALINUS.
2. Spikes thick, empalements smooth.
Festuca avenacea spicis habitioribus glumis glabris. *Ray's
Syn.* 414.
3. Empalements smooth and slender.
In meadows and pastures. 1. 2. in ploughed fields. A. May.
June.

The seeds of this species mixed with better corn, may be used
to make bread; but when mixed in too large a proportion they
render the bread brown and bitter, and those that eat it experience a temporary giddiness.

The panicles are used by the common people in Sweden for
dying green.—This grass is eaten by Horses, Cows; Goats and
Sheep.

BROOMGRASS. The panicle nearly upright. Little spikes Soft
egg-shaped, downy; awns straight; leaves very soft and woolly.— Mollis
Festuca avenacea hirsuta, paniculis minus sparsis. *Ray's
Syn.* 413.
On walls and ditch-banks. B. May. June.

BROOMGRASS. The panicle open. Little spikes oblong; Barren
the florets pointing from two opposite lines. Husks awl-shaped, Sterilis
with awns—*six or seven florets in each.* Leaves *smooth.*
Festuca avenacea sterilis elatior seu bromos dioscoridis. *Bauh.
pin.* 9. *Ray's Syn.* 412.
Bromos herba, sive avena sterilis. *Park.* 1147.
Bromos sterilis. *Gerard* 16
In woods and hedge rows, frequent. A. June. July.

BROOMGRASS with a nodding panicle. Little spikes ob- Corn
long egg-shaped—*Straw upright; often as thick as a goose-quill.* Arvensis
*The sheaths and the upper surface of the leaves downy with white
hairs. Ten florets in each little spike.* Husks *membranaceous at the
edges.*
Festuca graminea effusa juba. *Bauh. pin.* 192.
Festuca elatior paniculis minus sparsis, locustis oblongis
strigosis, aristis purpureis splendentibus. *Ray's Syn.* 414.
In cornfields. Plentiful about Gravesend. P. July.
Cows, Horses, Sheep and Goats eat it.

Wall
Tectorum

BROOMGRASS with a nodding panicle. Little spikes strap-shaped—*Straw with five joints.* Sheaths *scored.* Leaves *soft on the upper surface; fringed at the edges.* Florets *in the panicle all pointing one way; and when the seeds are ripe it hangs down to the ground. The five lowermost fruit-stalks hair-like; limber; rough.* Florets *five in each little spike.* Husks *awl-shaped, with upright awns, as long as the husks.*

     Festuca avenacea sterilis, spicis erectis. *Ray's Syn.* 413.
     Bromus ciliatus. *Hudson's Flor. Angl.* 40.
     Upon walls, roofs, and very dry pastures. A. May.
     Cows, Horses, Goats and Sheep eat it.

Tall
Giganteus

BROOMGRASS with a nodding panicle. Little spikes with four florets. Awns shorter than the little spikes—*Straw about five feet high.* Leaves *the breadth of a man's finger.*

     Gramen avenaceum glabrum, panicula a spicis raris strigosis composita, aristis tenuissimis. *Ray's Syn.* 415.
     In woods and moist hedges. P. August.
     Cows, Horses, Goats and Sheep eat it.

Cluster
Racemosus

BROOMGRASS. Each empalement standing on a fruit-stalk, and including six smooth blossoms with awns: the whole forming an undivided bunch.

     Festuca avenacea spicis strigosioribus e glumis glabris compactis. *Ray's Syn.* 414.
     Hudson makes this only a variety of the BROMUS SECALINUS. *Flor. Angl.* 39.
     Common in woods and hedges. A. July. August.

Wood
Ramosus

BROOMGRASS with a branched nodding, rough panicle. About ten florets in each little spike, longer than the awn. Leaves rough—

     Festuca graminea nemoralis latifolia mollis. *Bauh. pin.* 9.
     Gramen avenaceum dumetorum panicula sparsa. *Ray's Syn.* 415.
     In woods and hedges, frequent. A. July August.

Spiked
Pinnatus

BROOMGRASS. The straw undivided. Little spikes cylindrical; alternate; with awns; supported upon very short fruit-stalks.—*The flatter sides of the little spikes are turned towards the straw. The awns terminating. During the time of flowering, the little spikes stand out horizontally, but before, and afterwards, they lye close to the straw.*

     Gramen spica brizæ majus. *Bauh. pin.* 9. *Ray's Syn.* 392.
     In dry fields and commons. P. June.
     Horses and Goats eat it. Sheep are not fond of it.

33 OAT. 91 Avena.

EMPAL. A *Husk* of two valves; frequently containing several florets loosely collected. The *Valves* large, spear shaped distended; sharp; without awns.

BLOSS. Two valves. The *inferior Valve* of the size of the husk but harder; somewhat cylindrical, distended, tapering towards each end. sending out from its back an awn, spirally twisted and bent back as if it was jointed.

CHIVES. *Threads* three; hair-like. *Tips* oblong; forked at each end.

POINT. *Seedbud* blunt. *Shafts* two; reflected; hairy. *Summits* simple.

S. VESS. The *Blossom* shuts close upon and adheres to the seed.

SEED. Single; slender; oblong, tapering at each end; marked with a furrow lengthways.

OBS. *The essential character of this genus consists in the jointed, twisted, Awn, growing from the back of the blossom. The number of the florets in each empalement is from one to four or more.*

OAT with flowers forming a panicle; each empalement in-Tall closing two florets. One floret containing both chives and Elatior pointals and hardly any awn; the other having only chives and bearing the awn.—
Gramen nodosum avenacea panicula. *Bauh. pin.* 2. *Ray's Syn.* 406.
Gramen caninum nodosum. *Gerard.* 23.
Gramen caninum nodosum sive bulbosum vulgare. *Park.* 1175.
Tall Oatgrass.
1. There is a variety with long shining flowers.
Gramen avenaceum elatius, juba longa splendente. *Ray's Meth. Em.* 179.
In fields and pastures.
Cows, Sheep and Goats eat it.

OAT with flowers forming a panicle. Each empalement Naked containing three blossoms. The receptacle is longer than the Nuda empalement. The blossoms have awns upon their backs—*The seeds when ripe fall out of the husks.*
Avena nuda. *Bauh. pin.* 23. *Gerard.* 77. *Park.* 1134. *Ray's Syn.* 389.
Pilcorn.
In ploughed fields, but not common. A. July.
This is nearly as good as the cultivated Oat; it will make gruel or oat cake, and feed cattle the same as that.

OAT

Bearded
Fatua

OAT with flowers forming a panicle. Each empalement containing three florets which are all hairy at the base. The awns entirely smooth—

Festuca utriculis lanugine flavescentibus. *Bauh. pin.* 10.

Ægilops quibusdam aristis recurvis, seu avena pilosa. *Ray's Syn.* 389.

Ægilops bromoides. *Gerard.* 77.

Ægilops bromoides belgarum. *Park.* 1148.

In the fields about Stockwell in Surry. P. August.

Horses, Sheep and Goats eat it.

The awns are used in making hygrometers.

Rough
Pubescens

OAT with flowers almost forming a spike. Each empalement generally containing three florets; hairy at the base; leaves flat, downy—*Spiked panicle of a purplish shining white.*

Gramen avenaceum hirsutum, s. glabrum, panicula purpuro-argentea splendente. *Ray's Syn.* 406. Tab. 21. fig. 2.

In dry soils. Plentiful upon Banstead Downs. P. June.

Yellow
Flavescens

OAT with a loose panicle. Each empalement short and containing two or three florets; all of them furnished with awns—*Panicle yellowish.*

Gramen avenaceum pratense elatius, panicula flavescente, locustis parvis. *Ray's Syn.* 407.

Very common. P. July.

Cattle are not fond of it.

Meadow
Pratensis

OAT with flowers almost forming a spike. Each empalement containing five florets—*Leaves channelled, smooth; panicle compact. Florets smooth; the upper one in each little spike barren.*

Gramen avenaceum montanum spica simplici, aristis recurvis. *Ray's Syn.* 405. Tab. 21. fig. 1.

Upon heaths and chalk hills. P. July.

Horses, Cows, Sheep and Goats eat it.

34 REED. 93 Arundo.

EMPAL. A *Husk* of two upright valves, containing one or more florets. The *Valves* oblong, tapering; without awns. One shorter than the other.

BLOSS. Two valves. The *Valves* as long as the empalement; oblong, tapering; with soft and tender hairs arising from the base and nearly as long as the blossom.

CHIVES. *Threads* three; hair-like. *Tips* forked at each end.

POINT. *Seedbud* oblong. *Shafts* two; hair-like; reflected; woolly. *Summits* simple.

S.VESS. The blossom adheres to the seed without opening.

SEED. Single; oblong, tapering towards each end, furnished with long feathers at the base.

REED with a flexible panicle: each empalement containing five florets— *Common Phragmitis*

Arundo vulgaris five phragmites diofocoridis. *Bauh. pin.* 17.
Arundo vallatoria. *Gerard.* 36. *Ray's Syn.* 401.
Arundo vulgaris five vallatoria. *Park.* 1208.
Rivers, lakes, ditches; very common. P. July.
Horfes, Cows and Goats eat it. Sheep refuse it.
The panicles are used by the ruftics in Sweden to dye green. The Reeds are much more durable than ftraw for thatching. Skreens to keep off the cold winds in gardens, are made of them; and they are laid acrofs the frame of wood-work as the foundation for plaifter floors.

REED with an upright panicle; each empalement containing a single floret. Leaves fmooth on the under fide—*Straw two feet high.* Leaves *the breadth of a finger.* Panicle *denfe green, with a tinge of red.* *Small Epigejos*

Calamagroftis minor glumis rufis et viridibus. *Roy's Syn.* 401.
In woods P. July.
It very much refembles the *rough Cocksfoot* in its external appearance.

REED

Tranched
Calamagroftis

REED with one floret in each empalement. The empalement
fmooth; the bloffoms furnifhed with long foft hairs; the ftraw
branched—
Gramen arundinaceum panicula molli fpadicea majus. *Bauh.
pin.* 7. *Ray's Syn.* 401.
Gramen tomentorium arundinaceum. *Gerard.* 9.
Calamagroftis feu gramen tomentofum. *Park.* 1182.
In moift woods and hedges. P. June. July.
Goats eat it.

Sea
Arenaria

REED with one floret in each empalement. Leaves rolled
inwards at the edges and furnifhed with very fharp points—*This
plant probably originated from the* Small Reed *impregnated by the
duft of the* Sea Lymegrafs.
Gramen fparteum fpicatum foliis mucronatis longioribus vel
fpica fecalina. *Bauh pin.* 5. *Ray's Syn.* 393.
Spartum marinum noftras. *Park.* 1198.
Spartum anglicanum. *Gerard* 42.
Sea matweed.
On the fea fhore. P. June. July.
It grows only on the very drieft fand upon the fea fhore; and
it prevents the wind carrying the fand from the fhore and difperf-
ing it over the adjoining fields: which is not unfrequently the
cafe where this Reed is wanting. Many a fertile acre hath been
covered with unprofitable fand and rendered entirely ufelefs;
which might have been prevented by fowing the feeds of this
plant upon the fhore.
The country people cut, and bleach it for making matts.
Where it is plentiful, houfes are thatched with it.

35 RYE. 97 Secale.

Common *Receptacle* lengthened into a fpike.

EMPAL. A *Hufk* of two leaves, containing two florets.
The *Leaves* of the hufk oppofite; diftant; upright;
ftrap fhaped, tapering, fmaller than the bloffom.
Florets fitting.

BLOSS. Two valves. The *outer Valve* inflexible. diftended,
tapering, compreffed; the keel fringed and terminating in a long awn. The *inner Valve* flat, fpear-
fhaped.

CHIVES. *Threads* three; hair-like; hanging out of the
bloffom. *Tips* oblong, forked.

POINT. *Seedbud* turban-fhaped. *Shafts* two; reflected;
woolly. *Summits* fimple.

S. VESS. None. The bloffom contains the feed until it
is ripe.

SEED. Single; oblong; naked; fomewhat cylindrical,
but tapering at one end.

OBS. *There is frequently a third floret upon a fruit-ftalk betwixt
the other two larger ones which have no fruit-ftalk.*

RYE. The hufks fringed with foft hairs, and the fcales of the Woolly
empalement wedge-fhaped.— Villofum
Gramen fecalinum majus fylvaticum. *Ray's Syn.* 393.
Wood Rye-grafs.
In woods and hedge-rows. A. July.
It nourifhes the following infects. Brown Moth, *Phalæna
granella*—Rye Moth, *Phalæna fecalis*—Weevel, *Curculio gra-
narius*—and *Trips Phyfapus.*
The little fpikes appear very beautiful in the microfcope.

36 WHEAT

36 WHEAT. 99 Triticum.

Common *Receptacle* lengthened into a spike.

EMPAL. A *Husk* of two valves, containing several florets. The *Valves* egg-shaped, blunt, concave.

BLOSS. Two valves, nearly equal; the size of the empalement. The *outer Valve* distended, blunt but tapering. The *inner Valve* flat.

CHIVES. *Threads* three; hair-like. *Tips* oblong, forked at each end.

POINT. *Seedbud.* turban shaped. *Shafts* two; hair-like, reflected. *Summits* downy.

S.VESS. None. The blossom contains the seed until it is ripe.

SEED. Single; oblong egg-shaped; blunt at each end, convex on one side, furrowed on the other.

OBS. *The outer valve of the blossom in some species is furnished with an awn; in others not. The middle floret is frequently without pointals.*

Rush
Junceum

WHEAT with lopped empalements, each containing five florets. Leaves rolled inwards—*and sharp pointed.*
1. There is one variety with a thick spike and a trailing straw, the Gramen loliaceum maritimum supinum spica crassiore. *Ray's Syn.* 391.
And another
2. With a leafy spike, the Gramen caninum maritimum spica foliacea. *Bauh. pin.* 2.
Sea Wheat-grass.
On the sea shore; very common. P. June. July.

Couch
Repens

WHEAT with awl-shaped tapering empalements; three or four florets in each. Leaves flat—*There is generally only three florets in each empalement.*
Gramen spica triticea repens vulgare, caninum dictum. *Ray's Syn.* 390.
Gramen caninum. *Gerard.* 23.—Vulgatius. *Park.* 1173.
Gramen caninum arvense, seu gramen dioscoridis. *Bauh. pin.* 1.
CommonWheat-grass.Dogs-grass,Squitch grass,orCouch-grass.
In every field. P. June. August.
The roots dried and ground to meal, have been used to make bread in years of scarcity: and the juice of them drank liberally is recommended by Boerhaave in obstructions of the viscera; particularly in cases of schirrhus liver and jaundice. Cattle are frequently found to have schirrhus livers in the winter, and they soon get cured when turned out to grass in the spring—Dogs eat the leaves, to excite vomiting—Horses, Cows, Sheep and Goats eat them. WHEAT

WHEAT with empalements containing feveral florets, each Sea
having a fharp point. Spike branched.—*Leaves as long as the* Maritimum
ftraw; the fheathing part purple. Eight or ten florets in each em-
palement; alternate, compreffed; thickeft at the bafe; fitting; in the
fubdivifions fometimes folitary.

Poa fpicata, fpiculis alternis feffilibus, fub fex floris. *Hudfon's*
Flor. Angl. 35.

Gramen pumilum loliaceofimile. *Ray's Syn.* 395.

Spiked Meadow-grafs.

In the fand on the fea fhore, frequent. ·P· May. June.

37 BARLEY. 98 Hordeum.

Common *Receptacle* lengthened into a fpike.

EMPAL. A *Hufk* of two narrow taper valves, containing
one floret, flowers fitting, three in a place.

BLOSS. Two valves. The *lower Valve* longer than the
empalement; diftended; angular; egg-fhaped; tapering; ending in a long awn. The *interior Valve*
fmaller; flat, fpear fhaped.

CHIVES. *Threads* three; hair-like, fhorter than the bloffom. *Tips* oblong.

POINT. *Seedbud* betwixt egg and turban-fhaped. *Shafts*
two; woolly; reflected. *Summits* like the fhafts.

S. VESS. The bloffom rows round the feed without opening.

SEED. Single; oblong, diftended, angular, tapering at
each end; furrowed on one fide.

OBS. *In fome fpecies all the three florets that grow together have*
both chives and pointals; but in others the middle *floret alone is fur-*
nifhed with chives and pointals, the lateral florets having only
two chives.

BARLEY. The lateral florets without awns and without Knotted
pointals. The bufks briftle-fhaped and fmooth. *Root bulbous,* Nodefum
large. The bufks have purple awns, longer than the bloffoms.

Alopecurus fpica cylindrica, culmo erecta, radice bulbofa.
Hudfon's Flor. Angl. 24.

Gramen myofuroides nodofum. *Ray's Syn.* 397.

Bulbofe Fox-tail-grafs.

In meadows and paftures. P. June.

BARLIY

Wall
Murinum

BARLEY. The lateral florets have awns, and chives. The intermediate hufks are fringed—*There is often the appearance of pointals in the lateral florets as well as in the middle floret; but the lateral ones do not produce perfect feeds.*

Hordeum fpurium vulgare. *Park.* 1147.
Gramen fecalinum et fecale fylveftre. *Gerard.* 73. *Ray's Syn.* 391.
Gramen hordeaceum minus et vulgare. *Bauh. pin.* 9.
There are two varieties, viz.

1. Lateral florets very indiftinct.
Gramen fecalinum. *Gerard.* 29. and
2. Marine wall barley.
Gramen fecalinum paluftre et maritimum. *Ray's Syn.* 392.
Wall Barley-grafs. Way Bennet.

Upon walls and road-fides. The firft variety is found in wet paftures; the fecond near the fea coaft. A. April. Auguft.

Sheep and Horfes eat it.

It feeds the Brown Moth *Phalæna granella*—and the Barley Fly, *Mufca frit*.

38 LYMEGRASS. 96 Elymus.

Common *Receptacle* lengthened into a fpike.

EMPAL. A *Hufk* of four leaves, pointing from two oppofite lines, two of the leaves which are awl-fhaped, belonging to each little fpike.

BLOSS. Two valves; the *outer Valve* large, tapering; furnifhed with an awn. The *interior Valve* flat.

CHIVE. *Threads* three; hair-like; very fhort. *Tips* oblong, forked at the bafe.

POINT. *Seedbud* turban-fhaped. *Shafts* two; ftraddling, hairy, bent inwards. *Summits* fimple

S. VESS The bloffom inclofes the feed.

SEED. Single; ftrap-fhaped; convex on one fide.

OBS. *The empalement may be confidered as a hufk of two leaves, and two of thefe empalements growing together.*

LYMEGRASS with an upright close and long spike. Em- Sea palement furnished with down; longer than the blossom.—*Leaves* Arenarius *like those of reeds; bluish green or whitish; rolled inwards and sharp-pointed.*

Gramen caninum maritimum spica triticea nostras. *Ray's Syn.* 390.

Gramen caninum maritimum spicatum. *Bauh. pin.* 1.

Gramen caninum geniculatum maritimum spicatum.*Park.*1277.

Common on the sea coast. P. May. June.

It resists the spreading of loose sand on the sea shore. Is it not capable of being formed into ropes as one species of the Feathergrass (Stipa tenacissima) is in Spain?—Cows, Horses and Goats eat it. Sheep refuse it.

LYMEGRASS with a nodding; close and long spike. The Dogs little spikes straight, without any fence; the lower ones grow- Caninus ing in pairs—

Gramen caninum aristatum, radice non repente, sylvaticum *Ray's Syn.* 390.

Triticum caninum. *Hudson's Flor. Angl.* 45.

Bearded Wheat-grass.

In woods and hedges. P. June. July.

LYMEGRASS with an upright spike. Two florets in each Wood little spike. Empalement equal.— Europæus

Gramen hordeaceum montanum, seu majus. *Bauh. pin.* 9. *Ray's Syn.* 392.

In woods. P.

F   39 DARNEL

39 DARNEL. 95 Lolium.

Common *Receptacle* lengthened into a spike. The florets pointing from two oppolite lines and each preffed clofe to a bend in the receptacle.

EMPAL. A *Hufk* of one valve, awl-fhaped, permanent; ftanding oppofite to a bend in the receptacle.

BLOSS. Two valves. The *inferior Valve* narrow; fpear-fhaped; rolled inwards; tapering; as long as the empalement. The *fuperior Valve* fhorter, more blunt, ftrap fhaped, concave in the upper part.

CHIVES. *Threads* three; hair-like; fhorter than the blof-fom. *Tips* oblong.

POINT. *Seedbud*, turban-fhaped. *Shafts* two; hair-like; reflected. *Summits* downy.

S. VESS. None. The bloffom inclofes the feed until it is ripe.

SEED. Single; oblong; compreffed; convex on one fide, flat and furrowed on the other.

OBS. *The angles in the fpike-ftalk lying in the fame plane with the florets fupply the defect of inner valves to the hufks.*

Red
Perenne

DARNEL. The fpike without awns. The little fpikes contain feveral florets and are compreffed.—*Three or four florets in each little fpike. The larger valve of the bloffom cloven; fometimes furnifhed with a fhort, foft fort of an awn.*

Gramen loliaceum anguftiore folio et fpica. *Bauh. pin.* 9. *Ray's Syn.* 395.

Lolium rubrum. Gerard-Phænix. *Park* 114².

1. There is a variety in which the fpike refembles a panicle; viz. the gramen loliaceum paniculatum. *Ray's Syn.* 395.

In oads and dry paftures, very common. P. June.

Ray Grafs, Rye Grafs.

It makes excellent hay upon dry chalky or fandy foils. It is cultivated with advantage along with Clover, and fprings earlier than the other graffes; thereby fupplying food for cattle, at a feafon when it is moft difficult to be obtained. Cows, Horfes and Sheep eat it. Goats are not fond of it.

DARNEL.

DARNEL. The ſpike with awns. The little ſpikes con- White
tain ſeveral florets and are compreſſed.—— Temulentum
  Lolium album. *Gerard.* 78. *Park* 1146. *Ray's Syn.* 396.
  Gramen loliaceum ſpica longiore. *Bauh. pin.* 9.
  Annual Darnel Graſs.
  In ploughed fields. A. July. Auguſt.
  The ſeeds mixed with bread-corn produce but little effect, un-
leſs the bread is eaten hot; but if malted with Barley the ale
ſoon occaſions drunkenneſs.
  Sheep are not fond of it.

40 DOGSTAIL. 87 Cynoſurus.

EMPAL. Partial *Fence* large; lateral; generally conſiſting
  of three leaves. A *Huſk* of two valves containing
  ſeveral florets. The *Valves* ſtrap-ſhaped; tapering;
  equal.
BLOSS. Two valves. The *outer Valve* concave and longer
  than the other. The *inner Valve* flat, without an awn.
CHIVES. Threads three; hair-like. *Tips* oblong.
POINT. *Seedbud* turban-ſhaped. *Shafts* two; woolly; reflected. *Summits* ſimple.
S. VESS. None. The bloſſom cloſely wrapping round
  the ſeed.
SEED. Single; oblong, tapering at each end.

  OBS. *In moſt ſpecies the fence is like a comb.*

  DOGSTAIL. The floral leaves with winged clefts.——*Little* Creſted
ſpikes compoſed of three florets. Valves of the huſk without awns. Criſtatus
Petals unequal; the larger one terminated by a ſhort awn.
  Gramen criſtatum. *Gerard* 29. *Ray's Syn.* 398.
  Gramen criſtatum anglicum. *Park* 1159.
  Gramen pratenſe criſtatum, ſive gramen ſpica criſtata læve.
Bauh. pin. 3.
1. There is a variety in which the ear is four cornered, viz. the
  Gramen criſtatum quadratum, ſeu quatuor criſtatum gluma-
rum verſibus. *Ray's Syn.* 399.
  In paſtures. P. Auguſt.
  Sheep eat it.

Rough
Echinatus

DOGSTAIL. The floral leaves winged; chaffy; furnished with short awns.—*The florets are collected into a congregated bunch. The outer florets only have floral leaves, alternately winged; each containing a single floret, and ending in an awn. The* Husk *hath two valves and contains two florets; is membranaceous, and grows very fine at the point. The* Blossom *of two valves with an awn upon the outermost point.* Shaft *cloven.*

Gramen alopecuroides spica aspera. *Bauh. pin.* 4.

Gramen alopecuroides spica aspera brevi. *Park.* 1168. *Ray's Syn.* 39 .

In sandy soils. A. July.

Blue
Cæruleus

DOGSTAIL with the floral leaves entire.—*Leaves bluish green. Straws oblique. The spike is sometimes white.*

Gramen glumis variis. *Bauh. pin.* 10.

Gramen parvum montanum spica crassiore purpuro-cærulea brevi. *Ray's Syn.* 399.

In marshes sometimes, but mostly on the tops of mountains in the poorest soil. P. July.

Horses, Sheep and Goats eat it. Swine refuse it.

Order III. Three Pointals.
41 BLINKS. 101 Montia.

EMPAL. *Cup* of two leaves; egg-shaped, concave, blunt; upright; permanent.

BLOSS. One petal, deeply divided into five parts. The three alternate segments smaller than the rest, and supporting the chives.

CHIVES. *Threads* three; hair-like; as long as the blossom, into which they are inserted. *Tips* small.

POINT. *Seedbud* turban-shaped. *Shafts* three; woolly; expanding. *Summits* simple.

S. VESS. *Capsule* turban-shaped, blunt; of one cell and three valves.

SEEDS Three; roundish.

OBS. *The cup hath frequently three leaves and then the flower produces five chives.*

Vater
Fontana

BLINKS. As there is only one species known Linnæus gives no description of it.— *Leaves opposite:* Flowers *on long fruit-stalks.* Blossoms *white. This is one of the smallest plants we are acquainted with.*

Alsine parva palustris tricoccos, portulacæ aquaticæ similis. *Ray's Syn.* 352.

Portulaca arvensis. *Bauh. pin.* 288.

About springs and watery lanes. A. April.

CLASS

C L A S S   IV.

THE Chives in this Clafs are four, and all of the fame length ; whereas in the fourteenth Clafs, which is likewife compofed of flowers with four chives, the chives are unequal in lengh, two of them being long and two of them fhort.

The fourth divifion of the firft ORDER including the STARRY plants, admits of the following natural character.

EMPAL. *Cup* fmall ; with four teeth ; permanent : fu- perior.

BLOSS. One petal ; tubular. *Border* expanding ; with four divifions.

CHIVES. *Threads* four. *Tips* fimple.

POINT. *Seedbud* beneath : double. *Shaft* thread-fhaped ; cloven.

SEIDS, Two ; fomewhat globular.

OBS. *Stem four cornered. Leaves furrounding the ftem in form of a ftar.*

The plants correfponding with this natural character are aftringent and diuretic.

CLASS IV.

FOUR CHIVES.

ORDER I. ONE POINTAL.

* *Flowers of one Petal, and one Seed—superior.* INCORPORATED.

42 TEASEL.   -   Common *Empalement* leafy. *Receptacle* conical, chaffy. *Seeds* like little pillars.

43 DEVILSBIT. - *Empalement* common to several florets. *Receptacle* raised, a little chaffy. *Seeds* crowned; rolled in a cover.

** *Flowers of one Petal—beneath; and one Seed-vessel.*

44 CHAFFWEED.   *Blossom* wheel-shaped. *Empalement* deeply divided into four parts. *Capsule* of one cell; cut round.

45 PLANTAIN.   *Blossom* bent back as if broken. *Empalement* divided into four parts. *Capsule* of two cells; cut round.

† *Dwarf Gentian.*   † *Marsh Gentian.*

*** *Flowers of one Petal—superior; and one Seed-vessel.*

46 BURNET.   *Blossom* flat. *Cup* two leaves. *Capsule* four cornered; placed betwixt the cup and the blossom.

47 MAD-

**** *Flowers of one Petal.—superior; and two Berries.* STARRY.

47 MADDER.   - *Blossom* bell-shaped. *Fruit,* berries.
48 GOOSEGRASS.   *Blossom* flat. *Fruit* nearly globular.

         † *Yellow Cross-wort.*

49 WOODROOF.   *Blossom* tubular. *Fruit* nearly globular.

50 REDWORT.   *Blossom* tubular. *Fruit* crowned. *Seeds* with three teeth.

***** *Flowers of four Petals—beneath.*

† *Hairy Ladys-smock.* † *Gatteridge Spindle.* † *Wall Pellitory.*

****** *Flowers of four Petals—superior.*

51 CORNEL.   *Empalement* with four teeth.; deciduous. *Seed-vessel* pulpy, including a stone with two cells.

******* *Flowers imperfect—beneath.*

52 LADIESMANTLE. *Empalement* with eight clefts. *Seed* one, inclosed in the empalement.

F 4   Order

Order II.   Two Pointals.

53 Toadgrass.   *Bloss.* four petals. *Empal.* four leaves. *Capsule* with one cell; two valves; and two feeds.

54 Dodder.   *Bloss.* with four clefts; egg-shaped. *Empal.* with four clefts. *Capsule* with two cells; cut round.

55 Parsleypiert.   *Bloss.* none. *Empal.* with four clefts and four intermediate teeth. *Seeds* one or two.

    † *Dwarf Gentian.*   † *Marsh Gentian.*

Order III.   Four Pointals.

56 Holly.   - *Bloss.* one petal, *Empal.* with four teeth. *Berry* with four feeds.

57 Pearlwort.   *Bloss.* four petals. *Empal.* four leaves. *Capsule* with four cells; *Seeds* many.

58 Pondweed.   *Bloss.* none. *Empal.* four leaves. *Seeds* four; fitting.

59 Seagrass.   *Bloss.* none. *Empal.* none. *Seeds* four; on foot-stalks.

    † *Little Flax.*

42 TEASEL.

42 TEASEL. 114 Dipfacus.

EMTAL. *Common Cup* of many leaves containing feveral florets The *little Leaves* which form the cup, longer than the florets; flexible; permanent. *Proper Cup*, fuperior; fcarcely perceptible.

BLOSS. *General*, regular. *Individuals* of one petal, tubular. *Border* with four clefts; upright. The outer *Segment* larger and fharper than the reft.

CHIVES. *Threads* four; hair-like; longer than the bloffom. *Tips* fixed fide-ways to the threads.

POINT. *Seedbud* beneath, *Shaft* thread-fhaped; as long as the bloffom. *Summit* fimple.

S.VESS. None.

SEED. Solitary; refembling fquare pillars; crowned with the entire margin of the proper cup. *Receptacle Common* conical; the florets feparated by long chaffy leaves.

TEASEL. The leaves fitting and ferrated—*Awns bent back like hooks.* Clothiers Fullonum

Dipfacus.fativus. *Bauh. pin.* 385. *Gerard.* 1167. *Park.* 983. *Ray's Syn.* 192.
Manured Teafel.

1. There is a variety in which the awns are not bent backwards; it is the dipfacus fylveftris, aut virga paftoris major. *Bauh. pin.* 385.
Dipfacus fylveftris. *Gerard.* 1167. *Park.* 984. *Ray's Syn.* 192. *Hudfon.* 49.
Wild Teafel.

In hedges and uncultivated places. B. July.

The clothiers employ the heads with crooked awns to raife the knap upon the woollen cloths. For this purpofe they are fixed round the circumference of a large broad wheel, which is made to turn round, and the cloth is held againft them.

TEASEL. The leaves fupported upon leaf-ftalks with little appendages.— Smaller Pilolus

Dipfacus fylveftris capitulo minore. vel virga paftoris minor. *Bauh. pin.* 385.
Dipfacus minor, feu virga paftoris. *Gerard.* 1168. *Ray's Syn.* 192.
Virga paftoris. *Park.* 984.
Small wild Teafel.

In hedges and amongft old ruins. P. Auguft.

43 DEVILSBIT.

43 DEVILSBIT 115 Scabiofa.

EMPAL. *Common Cup* of many leaves, expanding; containing many florets. The *Leaves* fit upon and furround the receptacle in feveral rows, the inner ones of which become gradually fmaller and fmaller.
Proper Cup double; fuperior.
Outer Cup fhort; membranaceous; plaited; permanent.
Inner Cup with five divifions; the *Segments* bewtixt awl and hair-fhaped.
BLOSS. *General* regular; but moftly compofed of irregular florets.
Individuals of one petal; tubular; with four or five clefts; equal, or unequal.
CHIVES. *Threads* four; betwixt awl and hair-fhaped; feeble. *Tips* oblong; fixed fideways to the threads.
POINT. *Seedbud* beneath; rolled in a proper fheath, like a little cup. *Shaft* thread-fhaped, as long as the bloffom. *Summit* blunt; obliquely notched at the end.
S.VESS. None.
SEED. Solitary; oblong egg-fhaped; rolled in a cover; varioufly crowned by the proper cup.
Receptacle Common, convex befet with chaffy leaves.

OBS. *The outer bloffoms are often the largeft and the moft unequal. The crown of the feed varies in different fpecies. In fome few inftances the receptacle is naked.*

Common
Succifa

DEVILSBIT. The bloffoms with four nearly equal clefts. Stem fimple. Branches near together. Leaves betwixt egg and fpear-fhaped—*Proper cup four cornered; hairy; with four fringed teeth. Crown of the feedbud adorned with four red briftles fringed at the bafe.* Bloffoms blue.
Scabiofa radiea fuccifa, flore globofo. *Ray's Syn.* 191.
Succifa glabra et hirfuta. *Bauh. pin.* 269.
Morfus diaboli. *Gerard.* 726.
Morfus diaboli vulgaris, flore purpureo. *Park.* 491.
In fields and paftures. P. June. Auguft.
The dried leaves are ufed to dye wool yellow or green. *It. Oel.* 97. 101. *It. Scan.* 277.

DEVILSBIT. The bloſſoms with four clefts; radiate. Stem Field
covered with ſtrong hair. Leaves with winged clefts and Arvenſi‑
jagged—*Lower leaves egg-ſhaped.* Seed *woolly; compreſſed.*
Crown of the ſeed toothed. Receptacle *hairy but not chaffy.*
Bloſſoms *blue.*

Scabioſa pratenſis hirſuta. *Bauh. pin.* 269.
Scabioſa major vulgaris *Gerard.* 719.
Scabioſa vulgaris pratenſis. *Park.* 484.
Common field ſcabious.
In paſtures and cornfields. P. Auguſt.
Sheep and Goats eat it. Horſes and Cows are not fond of it.
It is ſlightly aſtringent, bitter and ſaponaceous.

DEVILSBIT. The bloſſoms with five clefts; radiate. Feathered
Root leaves egg-ſhaped, ſcollopped. Stem leaves briſtly; wing‑ Columbaria
ed—Bloſſoms *pale blue, convex in the center.* Seed *crowned by
five feathers as long as the bloſſom.*

Scabioſa capitulo globoſo major et minor. *Bauh. pin.* 270.
Scabioſa minor campeſtris. *Park.* 487.
Scabioſa minor five columbaria. *Gerard.* 719.
Leſſer Field Scabious.
In dry hilly paſtures. P. June. July.
Horſes, Sheep and Goats eat it.
The Heath Fritillary, *Papilio Maturna* finds its nouriſhment
in all the ſpecies of Devilſbit.

44 CHAFFWEED. 145 Centunculus.

EMPAL. *Cup* with four diviſions; expanding: permanent,
the *Segments* ſharp, ſpear-ſhaped; longer than the
bloſſom.
BLOSS. One petal; wheel ſhaped. *Tube* ſomewhat glo‑
bular. *Border* flat, with four diviſions; the *Segments*
nearly egg-ſhaped.
CHIVES. *Threads* four; nearly as long as the bloſſom.
Tips ſimple.
POINT. *Seedbud* roundiſh; within the tube of the bloſſom.
Shaft thread-ſhaped; as long as the bloſſom: perma‑
nent. *Summit* ſimple.
S. VESS. *Capſule* globular; of one cell; cut round.
SEEDS. Several; roundiſh; very ſmall.

CHAFFWEED. As there is only one ſpecies known Linnæus Pimpernel
gives no deſcription of it.—*Bloſſoms minute; white; at the baſe* Minimus
of the leaves.
Centunculus. *Ray's Syn.* 1.
Baſtard Pimpernel.
In moiſt ſandy ground. A. June.

45 PLANTAIN

45 PLANTAIN. 142 Plantago.

EMPAL. *Cup* with four divisions; short; upright; permanent.
BLOSS. One petal, permanent: shrivelling. *Tube* cylindrical but somewhat globular. *Border* with four divisions; depressed. *Segments* egg-shaped, sharp.
CHIVES. *Threads* four; hair-like; upright: exceedingly long. *Tips* rather long; compressed; fixed sideways to the threads.
POINT. *Seedbud* egg-shaped. *Shaft* thread-shaped; longer than the blossom. *Summit* simple.
S. VESS. *Capsule* egg-shaped; with two cells; cut round. *Partition* loose,
SEEDS. Several oblong.

OBS. *The empalement in some species is equal; in others unequal. In the grass-leaved Plantain the chives and pointals are in separate flowers.*

Great
Major

PLANTAIN with smooth egg-shaped, leaves. Stalk cylindrical; spike tiled with florets---
Plantago latifolia vulgaris. *Park.* 493.
Plantago latifolia. *Gerard.* 419. sinuata. *Bauh. pin.* 189.
Way-bread.
Near roads and foot-paths, common. A. June. July.
Sheep, Goats and Swine eat it. Cows and Horses refuse it.
The common people apply the green leaves to cuts.

Hoary
Media

PLANTAIN. The leaves betwixt spear and egg-shaped; downy. Spike and stalk cylindrical---*Leaves not toothed.* Cups *smooth.* Threads *purplish.* Blossoms *white.*
Plantago major incana. *Park.* 493. *Ray's Syn.* 314.
Plantago incana. *Gerard.* 419.
Plantago latifolia incana. *Bauh. pin.* 189.
Gravelly soil and road sides. P. July August.
Sheep, Goats and Swine eat it. Cows and Horses refuse it.

Ribwort
Lanceolata

PLANTAIN with spear-shaped leaves. Spike somewhat egg-shaped, naked. Stalks angular---*The Root appears as if bitten off.* Tips *white. Five ribs on each leaf.* Cup unequal. Floral *leaf green, and woolly on the under surface.*
Plantago angustifolia major. *Bauh. pin.* 189.
Plantago quinquenervia. *Gerard.* 422. major. *Park.* 495.
In pastures, very common. P. June. July.
Horses, Sheep and Goats eat it. Cows refuse it.

PLANTAIN

PLANTAIN with very entire femi-cylindrical leaves; woolly Marine
at the bafe. Stalk cylindrical.----*Tips yellow.*   Maritima
Coronopus. *Gerard.* 425. maritima major. *Bauh. pin.* 190.
Narrow-leaved Plantain.
On the fea coaft. P. June. July.
Mr. Pennant in the Britifh Zool. vol. 1. p. 13, fays it is cul-
tivated and fown with clover in North Wales, and that it is
greedily eaten by Horfes and Cows; but Linnæus fays that
Sheep and Goats eat it but Cows are not fond of it.

PLANTAIN with ftrap-fhaped, toothed leaves; and cylin- Buckfhorn
drical ftalks—*Spike pendant before it flowers.* Leaves *lying upon* Coronopus
the ground in form of a ftar.
Plantago foliis laciniatis, coronopus dicta. *Ray's Syn.* 315.
Coronopus vulgaris, five cornu cervinum. *Park.* 502.
Coronopus fylveftris hirfutior. *Bauh. pin.* 190.
Cornu cervinum. *Gerard.* 427.
1. There is a variety which is fmaller, viz.
Plantago gramineo folio hirfuto minor, capitulo rotundo brevi.
Ray's Syn. 316.
Star of the earth.
In gravelly foil. A. July. Auguft.
This was formerly in repute as an antidote againft the bite of
a mad dog, but it is now partly fallen into difufe. Sheep and
Goats eat it.

PLANTAIN with ftrap-fhaped leaves, a little toothed; Sea
ftalk cylindrical; fpike egg-fhaped, floral leaves membranaceous; Læflingii
keel-fhaped—*It is a fmaller and an earlier plant than the preceding
fpecies.* Flowering *ftalk hairy.*
Plantago marina. *Gerard.* 423. *Ray's Syn.* 315. vulgaris.
Park. 498.
In falt marfhes. P. July.

PLANTAIN with awl-fhaped leaves; and a ftalk fupporting Grafs-leaved
only one flower—*The flower fupported upon a tall fruit-ftalk is barren.* Uniflora
The fertile *flower is fituated at the root and fends out a fhaft as long
as the ftalk of the other flower.*
Plantago paluftris gramines folio, monanthos parifienfis.
Ray's Syn. 316.
In marfhes and wet fandy places. P. July.
The plantain fritillary, *Papilio Cinxia,* and Wood Tyger
Moth, *Phalœna Plantaginis,* feed upon the different fpecies of

46 BURNET. 146 Sanguiforba.

EMPAL. *Cup* two leaves, little *Leaves* oppofite, very fhort; fhedding.

BLOSS. One petal; wheel-fhaped; with four divifions. *Segments* blunt egg-fhaped; united by the claws.

CHIVES. *Threads* four; broadeft in the upper part; as long as the bloffom. *Tips* fmall; roundifh.

POINT. *Seedbud* four cornered; fituated betwixt the cup and the bloffom. *Shaft* thread-fhaped; very fhort. *Summit* blunt.

S. VESS. *Capfule* fmall; with two cells.

SEEDS. Small.

Common   BURNET with egg-fhaped fpikes.—*Stem cylindrical, fcored,*
Officinalis   *fmooth.* Leaves *winged; fmooth; alternate.* Spikes *brown.*
    Sanguiforba major, flore fpadiceo. *Ra.'s Syn.* 203.
    Pimpinella fanguiforba major. *Bauh. pin.* 160.
    Pimpinella major vulgaris. *Park.* 582.
    Pimpinella fylveftris. *Gerard.* 1045.
1. There is a variety which is fmaller.
    In moift paftures. The variety in high ground. P. June. July.
The root ftands recommended againft the bite of a mad dog, but without any great foundation. The whole plant is aftringent. The green leaves are fometimes put into wine, to give it a grateful flavour; and the very young fhoots are agreeable in fallads; but its principal ufe is as food for cattle early in the fpring, and it grows fo luxuriantly as to allow of three mowings in one fummer. Cows, Horfes, Sheep and Goats eat it.

47 MADDER. 127 Rubia.

EMPAL. *Cup* with four teeth; very fmall; fuperior.

BLOSS. One petal; bell-fhaped; with four divifions; without a tube.

CHIVES. *Threads* four; awl-fhaped; fhorter than the bloffom. *Tips* fimple.

POINT. *Seedbud* beneath; double. *Shaft* thread-fhaped; cloven at the top. *Summits* nearly globular.

S. VESS. Two fmooth *Berries,* united.

SEED. Solitary; roundifh; with a hollow dot.

OBS. *The bloffom hath fometimes five divifions and five chives.*

MADDER with oval, perennial leaves; smooth on the Wild upper surface—*four at each joint of the stem.* Blossom *on the tops* Peregrina *of the branches;* yellow.

Rubia anglica. *Hudson's Flor. Angl.* 54.

Rubia sylvestris. *Park.* 274.

Upon St. Vincent's Rock near Bristol, and in hedges in Devonshire. P. July. August.

48 GOOSEGRASS. 125 Galium.

EMPAL. *Cup* very small; with four teeth; superior.

BLOSS. One petal; wheel shaped; with four sharp segments, without any tube.

CHIVES. *Threads* four; awl-shaped; shorter than the blossom. *Tips* simple.

POINT. *Seedbud* beneath; double. *Shaft* thread-shaped; cloven half way down: as long as the Chives. *Summits* globular.

S. VESS. Two dry globular *Berries,* united.

SEED. Solitary; large; kidney-shaped.

OBS. *In several species the seeds grow two together and one roundish.*

* Fruit smooth.

GOOSEGRASS. The leaves growing by fours; inversely White egg-shaped; unequal. Stems spreading—*Blossoms numerous;* Palustre *terminating;* white.

Gallium palustre album. *Bauh. pin.* 335.

Gallium album. *Gerard.* 1126.

Molluginis vulgatiores varietas minor. *Park.* 565. *Ray's Syn.* 224.

White Ladies Bed-straw.

Cows, Sheep and Horses eat it. Goats and Swine refuse it.

GOOSEGRASS. The leaves generally growing by fours; Mountain strap-shaped; smooth. Stem rough, feeble; seeds smooth— Montanun *Stem leaves growing by fives, reflected. Branch leaves growing by fours. Flowers in broad-topped spikes, cloven into three parts; white; but purple before they unfold themselves.* Tips *brown.*

Mollugo montana minor gallio albo similis. *Ray's Syn.* 224.

Mountain Ladies Bed-straw.

Upon Hills. P. July.

Marsh
Uliginosum

GOOSEGRASS. The leaves growing by sixes; spear-shaped, inversely serrated with inflexible sharp pointed prickles. Blossom larger than the fruit—*white*.

Aparine palustris minor, parisiensis, flore albo. *Ray's Syn.* 225.
On heathy marshes and wet pastures. P. July. August.
Horses, Cows, Sheep, Goats and Swine eat it.

Smoothseeded
Spurium

GOOSEGRASS. The leaves spear-shaped; growing by sixes; a little keel-shaped on the back; rough, beset with prickles pointing backwards. Stems simple, with crooked joints; fruit smooth—*Blossoms white*.

Aparine semine læviore. *Ray's Syn.* 225.
Aparine lævis. *Park* 567.

Little
Pusillum

GOOSEGRASS. The leaves growing by eights; rough with strong hairs, strap-shaped, tapering at the end, somewhat tiled. Fruit stalks forked—*Stems numerous, angular. Leaves six or eight in a whorl, narrow, sharp; rough with expanding hairs on all sides, and so is the stem. Branches few, alternate. Flowers terminating; forming a very loose panicle upon doubly forked fruit stalks.* Blossoms *white*.

Rubeola saxatilis. *Bauh. pin.* 334.
Least Ladies Bed straw.
In mountains in Westmoreland. P. August.

Yellow
Verum

GOOSEGRASS. The leaves growing by eights; strap-shaped; furrowed. The flowering branches short.—*Blossoms yellow. The Tips after shedding the dust become brown; not only in this but in other species.*

Gallium luteum. *Bauh. pin.* 335. *Gerard.* 1126. *Park.* 564. *Ray's Syn.* 224.
Yellow Ladies Bed-straw, or Cheese-renning: or Pettymuguet. In dry ground and road-sides. P. July. August.
The flowers will coagulate boiling milk; and the best Cheshire-cheese is said to be prepared with them. According to an experiment of Borrchius, they yield an acid by distillation. The French prescribe them in Hysteric and Epileptic cases.—Boiled in alum-water they tinge wool yellow.—The roots dye a very fine red, not inferior to madder, and are used for this purpose in the island of Jura. *Pennant's Tour.* 1772. *p.* 214.
Sheep and Goats eat it. Horses and Swine refuse it. Cows are not fond of it.

GOOSEGRASS. The leaves growing by eights; betwixt Madder egg and ftrap-fhaped, expanding; fomewhat ferrated, and fharp Mollugo pointed. Stem limber, branches expanding—*Bloffoms yellowifh white, numerous.*

Mollugo vulgatior. *Park.* 565. *Ray's Syn.* 223.
Rubia fylveftris lævis. *Bauh. pin.* 333.
Rubia fylveftris. *Gerard.* 1118.
Wild Madder. Great Baftard Madder.
Hedges, heaths, frequent. P. June. July,

* * *Fruit rough with ftrong hairs.*

GOOSEGRASS. The leaves growing by fours; fpear-fhaped; Crofs-wort fmooth; three fibred. Stem upright—*fmooth.* Bloffoms *white.* Boreale Seedbud *covered with a white woolly fubftance.*

Mollugo montana erecta quadrifolia. *Ray's Syn.* 224.
Rubia pratenfis lævis, acuto folio. *Bauh. pin.* 333.
Crofs-wort Madder.
On the hills in Yorkfhire and Weftmoreland. P. June. Aug.
The roots afford a red dye for woollen cloths. *Flor. Lappon.* 60.
Horfes, Sheep and Goats eat it. Cows are not fond of it. Swine refufe it.

GOOSEGRASS. The leaves growing by eights; keel- Cleavers fhaped upon the back; rough; fpear-fhaped; fet with prickles Aparine pointing backwards. Joints woolly.—*Stem four cornered;* Leaves *eight or ten in a whorl.* Branches *oppofite.* Bloffoms *white, on flender fruit-ftalks; not numerous.*

Aparine vulgaris. *Bauh. pin.* 334. *Park.* 567.
Aparine. *Gerard.* 1122. *Ray's Syn.* 225.
Catchweed.
In hedges, frequent. A. May, Auguft.
The branches are fometimes ufed inftead of a foi to ftrain milk. Young geefe are very fond of them. The feeds may be ufed inftead of coffee. The plant is eaten by Horfes, Cows, Sheep and Goats. Swine refufe it.

GOOSEGRASS. The leaves growing in whorls; ftrap- Small fhaped; each fruit-ftalk fupporting two flowers.—*Stems feeble;* Parifienfe *four cornered, rough when ftroked upwards. Seven leaves in each whorl, fharp pointed; particularly rough at the edges.* Flowering branches *oppofite, fhort.* Fruit-ftalks *naked; with two or three fmall yellow flowers upon each.*

Aparine minima. *Ray's Syn.* 225. *Tab.* 9. fig. 1.
Leaft Goofe grafs.
On walls and in moift barren places. P. July.
The large Bee-moth, *Sphinx ftellatarum* and the Spurge Moth, *Spinx Euphorbiæ,* feed upon the different fpecies of Goofe-grafs.

49 WOODROOF. 121 Afperula.

EMPAL. *Cups* fmall; with four teeth; fuperior.

BLOSS. One petal, funnel-fhaped. *Tube* long; cylindrical. *Border* with four divifions; the fegments oblong, blunt, reflected.

CHIVES. *Threads* four; fituated at the top of the tube.
Tips fimple.

POINT. *Seedbud* beneath: double; roundifh. *Shaft* thread-
fhaped; cloven at the top. *Summits* knobbed.

S. VESS. Two dry globular *Berries* adhering together.

SEED. Solitary; roundifh; large.

Sweet
Odorata

WOODROOF. The leaves growing by eights; fpear-
fhaped. Flowers in bundles, on fruit-ftalks.—*Bloſſoms white,
odoriferous.*
   Afperula. *Gerard.* 1224. *Ray's Syn.* 224. odorata *Park.* 563.
   Afperula, feu rubeola montana odorata. *Bauh. pin.* 334.
   In woods and fhady places. P. May.
   The fmell of it is faid to drive away Ticks, and other infects.
Iter Ocland. 60. *Weſtrogoth.* 33.—It gives a grateful flavour to
wine.
   Cows, Horfes, Sheep and Goats eat it.

Squinancy
Cynanchica

WOODROOF. The leaves growing by fours; ftrap-fhaped.
The upper leaves oppofite. Stem upright. Bloffoms with four
divifions.—*Stem rough*; Seedbuds red. Bloffoms *a little wrinkled
and rough on the outer part; reddiſh, or white.*
   Rubeola vulgaris quadrifolia lævis, floribus purpurantibus.
Ray's Syn. 225.
   Rubia cynanchica. *Bauh. pin.* 333.
   Synanchica. *Gerard.* 1120.
   Afperula repens gefneri, feu faxifraga altera cæfalpini. *Park.*
453.
   Squinancy Wort.
   On chalky hills. P. July.
   The roots are ufed in Sweden to dye red.

ONE POINTAL. 83

50 REDWORT. 120 Sherardia.

EMPAL. *Cup* fmall; with four teeth; fuperior; permanent.
BLOSS. One petal; funnel-fhaped. *Tube* cylindrical, long.
Border with four divifions. The *Segments* flat and
fharp.
CHIVES. *Threads* four; fituated at the top of the tube.
Tips fimple.
POINT. *Seedbud* beneath: double; oblong. *Shaft* thread-
fhaped; cloven at the top. *Summits* nearly globular.
S. VESS. None. *Fruit* oblong; crowned; feparable
length-ways into two feeds.
SEEDS. Two; oblong; convex on one fide; flat on the
other; with three fharp points at one end.

REDWORT with the leaves all growing in whorls, and the Wild
flowers terminating.—*Bloffoms blue.*   Arvenfis
Rubeola arvenfis repens cærulea. *Bauh. pin.* 334. *Ray's Syn.*
225.
Rubeola minor pratenfis cærulea. *Park.* 276.
Little Field Madder.
In corn-fields. A. May. June.
Goats are extremely fond of it. Horfes eat it, Sheep are in-
different to it.

51 CORNEL. 149 Cornus.

EMPAL. *Fence* generally of four leaves; including feveral
florets. *Leaves* egg-fhaped; coloured; deciduous;
two, oppofite, fmaller. *Cup* very fmall, with four
teeth; fuperior; fhedding.
BLOSS. *Petals* four; oblong; fharp; flat; fmaller than
the fence.
CHIVES. *Threads* four; awl-fhaped; upright; longer than
the bloffom. *Tips* roundifh; fixed fide-ways to the
threads.
POINT. *Seedbud* beneath: roundifh. *Shaft* thread-fhaped;
as long as the bloffom. *Summit* blunt.
S. VESS. Pulpy; including a nut or ftone; nearly globular, and dimpled.
SEED. A heart-fhaped, or oblong nut; with two cells.

G 2   CORNEL

Dogberry
Sanguinea

CORNEL. A tree, with ſtraight branches, and flowers in naked tuſts.—*Leaves oppoſite, egg-ſhaped, very entire.* Seed-veſſel *crowned by the ſhaft and the cup.* Branches *of a reddiſh colour.* Bloſſoms *white.*

Cornus fœmina. *Bauh. pin.* 447. *Gerard.* 1467. *Park.* 1521. *Ray's Syn.* 460.

Female Cornel. Dogberry-tree. Gatter tree. Prick-wood.

1. There is a variety with variegated leaves.

In woods and hedges. S. June.

The wood is very hard and ſmooth, fit for the purpoſes of the turner. The berries are bitter and ſtyptic: they dye purple. Horſes, Sheep and Goats eat it. Swine and Cows refuſe it.

Dwarf
Suecica

CORNEL. Herbaceous; with two branches.—*Bloſſom white, placed betwixt the branches.* Fruit *red.*

Cornus herbacea. *Hudſon's Flor. Angl.* 58.

Chamæpericlymenum. *Park.* 1461. *Gerard.* 1296. *Ray's Syn.* 261.

Periclymenum humile. *Bauh. pin.* 302.

Dwarf Honeyſuckle.

On hills in the north. P. June.

The pulpy berries which have a ſweet and not unpleaſant flavour, are acceptable to children.—

Horſes, Sheep, Goats and Swine eat it. Cows refuſe it.

52 LADIES-MANTLE. 165 Alchemilla.

EMPAL. *Cup* of one leaf; tubular; permanent. *Rim* flat, with eight diviſions: every other *Segment* ſmaller.

BLOSS. None.

CHIVES. *Threads* four; awl-ſhaped; upright; ſmall; ſtanding on the rim of the empalement. *Tips* roundiſh.

POINT. *Seedbud* egg-ſhaped. *Shaft* thread-ſhaped; as long as the chives; ſtanding on the baſe of the ſeedbud. *Summit* globular.

S. VESS. None. The neck of the empalement cloſes upon the ſeed.

SEED. Solitary; oval; compreſſed.

LADIES-MANTLE. The leaves gashed—*plaited like a fan*; Common
pale green; *hairy*; *on long leaf-stalks.* Blossoms *yellowish green.* Vulgaris
Alchemilla. *Geraid.* 949. *Ray's Syn.* 158. vulgaris. *Bauh.*
pin. 319.
Alchemilla major vulgaris. 538.
1. Empalement white.
In meadows, frequent. P. June. August.
The whole plant is astringent. In the province of Smolandia
in Gothland they make a tincture of the leaves, and give it in
spasmodic or convulsive diseases. Horses, Sheep, and Goats
eat it. Swine refuse it. Cows are not fond of it.

LADIES-MANTLE. The leaves fingered and serrated.— Cinquefoil
Leaves cloven into seven little leaves, only serrated at the ends, and Alpina
of a shining white on the under surface. Blossoms *greenish.*
Alchemilla alpina pentaphyllos. *Ray's Syn.* 158.
Pentaphyllum petrosum, heptaphyllum clusii. *Gerard.* 988.
Tormentilla alpina folio sericeo. *Bauh. pin.* 326.
Tormentilla argentea. *Park.* 393.
1. There is a variety, viz. the Alchemilla minor, *Hudson's* Bastard
Flor. Anglic. 59, which is probably the product of this species Hybrida
impregnated by the dust of the first species. The leaves are
gashed, plaited; sharply serrated and silky.
Least Ladies-mantle.
On mountains in the north. P. July.
Goats and Cows eat it. Horses, Sheep and Swine refuse it.

Order II. Two Pointals.

53 TOADGRASS. 168 Bufonia.

EMPAL. *Cup* of four awl-shaped leaves, keel-shaped on the
back; with membranaceous edges; upright and
permanent.
BLOSS. Petals four; oval; upright; equal; notched at
the ends; shorter than the empalement.
CHIVES. *Threads* four; equal; as long as the seedbud.
Tips double.
POINT. *Seedbud* egg-shaped; compressed. *Shafts* two;
as long as the chives. *Summits* simple.
S. VESS. *Capsule* oval; compressed; of one cell, and two
valves.
SEEDS. Two; oval, compressed; but marked with a little
protuberance. Concave on one side.

OBS. *It hath generally four chives; sometimes only two, rarely*
three.

Chickweed
Tenuifolia

TOADGRASS. As there is only one species known Linnæus gives no description of it.—*Stem cylindrical, upright, jointed. Branches alternate. Leaves two at each joint of the stem. Bloſſoms at the baſe of the leaves; white.*

Alſinoides. *Ray's Syn.* 346.
Baſtard Chickweed.
On Hounſlow Heath. On the ſea-coaſt near Boſton in Lincolnſhire. P. May.

54 DODDER. 170 Cuſcuta.

EMPAL: *Cup* of one leaf; ſhaped like a drinking glaſs; divided halfway down into four *Segments*, which are blunt, and fleſhy at the baſe.

BLOSS: One petal; egg-ſhaped; a little longer than the empalement. The *Rim* with four clefts; the *Segments* blunt.

CHIVES. *Threads* four; awl-ſhaped; as long as the empalement. *Tips* roundiſh.

POINT: *Seedbud* roundiſh. *Shafts* two; upright; ſhort. *Summits* ſimple.

S. VESS: Roundiſh; fleſhy; with two cells cut round.

SEEDS. Two.

OBS. *Sometimes the empalement and bloſſom are divided into five ſegments and then there are five chives.*

Common
Europæa

DODDER. The flowers ſitting.—*This is a very ſingular plant, almoſt deſtitute of leaves; paraſitical; creeping; fixing itſelf to whatever is next to it; it decays at the root and afterwards is nouriſhed by the plant which ſupports it. Hops, flax, and nettles, are its moſt common ſupport; but principally the common nettle.* Bloſſoms white.

Cuſcuta major. *Bauh. pin.* 219. *Ray's Syn.* 281. Caſſutha. *Gerard,* 577. *Park.* 10.
Hellweed. Devil's Guts.
In cornfields and heaths, very common. A. July.

As ſoon as the ſhoots have twined about an adjoining plant, they ſend out from their inner ſurface a number of little veſicles or papillæ, which attach themſelves to the bark, or rind of the plant. By degrees, the longitudinal veſſels of the ſtalk, which appear to have accompanied the veſicles, ſhoot forth from their extremities, and make their way into the foſter plant, by dividing the veſſels and inſinuating themſelves into the tendereſt part of the ſtalk; and ſo intimately are they united with it, that it is eaſier to break than to diſengage them from it.

The whole plant is bitter. It affords a pale reddiſh colour.
Cows, Sheep and Swine eat it. Horſes refuſe it. Goats are not fond it.

55 PARSLEY.

55 PARSLEYPIERT. 166 Aphanes.

EMPAL: *Cup* of one leaf; tubular; permanent. *Rim* flat; with four divisions, and a little tooth betwixt each division.

BLOSS: None.

CHIVES. *Threads* four; upright; awl-shaped; very small; standing upon the the rim of the empalement. *Tips* large; oblong

POINT: *Seedbud* double; egg-shaped. *Shafts* thread-shaped; as long as the Chives; growing from the base of the seedbud. *Summits* somewhat globular.

S. VESS: None; the *Rim* of the empalement closing, confines the seeds.

SEEDS. One or two; egg shaped; tapering; compressed, as long as the shafts.

OBS. *This genus is nearly allied to the Ladies-mantle. In some specimens only one chive and one pointal are to be found; and when that is the case, there is only one seed.*

PARSLEYPIERT. As there is only one species known Linnæus gives no description of it.—*The* Stems *trailing and leafy.* Leaves *divided into three lobes*; jagged. Blossoms *small*; *greenish white.* Common Arvensis

Percepier Anglorum. *Gerard.* 1594. *Ray's Syn.* 159.
Polygonum selinoides. *Park.* 449.
Chærophyllo non nihil similis. *Bauh. pin.* 152.
In cornfields and dry gravelly soils, common. A. May.

Order III. Four Pointals.

56 HOLLY. 172 Ilex.

EMPAL: *Cup* with four teeth: very fmall: permanent.

BLOSS: One petal; flat; with four divifions. *Segments*
roundifh, concave, expanding, rather large; connected together by the claws.

CHIVES. *Threads* four; awl-fhaped; fhorter than the
bloffom. *Tips* fmall.

POINT: *Seedbud* roundifh. *Shafts* none. *Summits* four;
blunt

S VESS: A roundifh *Berry*, with four cells.

SEEDS. Solitary; hard as bone; oblong; blunt; huncn-
ed on one fide, angular on the other.

OBS. *Great variations take place in the flowers of the Holly; fome-*
times the chives and pointals are found on diftinct plants; fome-
times on the fame plant, but in different flowers; fometimes again
the flowers have five chives, and frequently the difpofition of the
chives and pointals is fuch that it claims a place in the fecond Order
of the twenty-third Clafs.

Tree   HOLLY with egg-fhaped leaves, befet with fharp thorns.—
Aquifolium *Bloffoms fmall; whitifh.* Berries red.

Ilex aculeata baccifera. *Taub. pin.* 425.

Agrifolium. *Gerard.* 1338. five aquifolium. *Park.* 1486.
Ray's Syn. 466.

In woods, hedges and heaths. S. April.

All the varieties which gardiners reckon to the amount of
forty or fifty, are derived from this one fpecies, and depend
upon the variegations of the leaves or thorns, and the colour
of the berries.

Sheep are fed in the winter with the croppings. *Pennant's*
Tour. 1772. p. 32. Birds eat the berries. The bark fermented
and afterwards wafhed from the woody fibres makes the common
Birdlime. It makes an impenetrable fence, and bears cropping;
nor is its verdure or the beauty of its fcarlet berries ever obferved
to fuffer from the feverest of our winters. The wood is ufed
in finceing, and is fometimes ftained black, to imitate Ebony.
Handles for knives, and cogs for mill-wheels are made of it.

57 PEARL-

57 PEARLWORT. 176 Sagina.

BLOSS: *Cup* with four leaves, permanent. *Leaves* egg-shaped; concave; greatly expanded.

BLOSS: *Petals* four; egg-shaped; blunt; expanding; shorter than the empalement.

CHIVES. *Threads* four; hair-like. *Tips* roundish.

POINT: *Seedbud* somewhat globular. *Shafts* four; awl-shaped; bent backward; downy. *Summits* simple.

S. VESS: *Capsule* egg-shaped; straight; with four cells, and four valves.

SEEDS. Numerous; small; fixed to the receptacle.

OBS. *In the* Chickweed Pearlwort *the capsule hath only one cell, and the cup sometimes consists of five leaves.*

PEARLWORT with trailing branches.—*Blossoms greenish* Chickweed *white. The petals are not always to be found; but the four valves of* Procumbens *the capsule after it opens have very much the appearance of petals.*

Alsinella muscoso flore repens. *Ray's Syn.* 345.

Saxifraga anglicana alsine folio. *Gerard.* 568.

Caryophyllus minimus muscosus nostras. *Park.* 1340.

1. There is a variety noticed by Plot in his Natural History of Oxfordshire, tab. 9. fig. 7. and

2. Ray mentions another variety with shorter, thicker and more succulent leaves. *Ray's Syn.* 345.

Chickweed Breakstone. Mofs-like Pink.

On walls; roofs and dry sandy places, common. A. June.

PEARLWORT. The stem upright; generally with only one Upright flower. Leaves betwixt strap and spear-shaped—*Sitting; smooth.* Erecta Blossom *white. Leaves of the cup spear-shaped tapering to a point.*

Alsinella foliis caryophylleis. *Ray's Syn.* 344. tab. 15. fig. 4.

Least Stitchwort.

In gravelly pastures. A. April. May.

59 PONDWEED.

FOUR CHIVES.

58 PONDWEED. 174 Potamogeton.

EMPAL: None.

BLOSS: *Petals* four; nearly circular; blunt; concave; upright: furnished with a little claw; deciduous.

CHIVES. *Threads* four; flat; blunt; very short. *Tips* double; short,

POINT: *Seedbuds* four; egg-shaped; tapering. *Shafts* none. *Summits* blunt.

S. VESS: None.

SEEDS. Four; roundish, tapering; hunched on one side, compressed on the other, and angular.

Broadleaved **PONDWEED.** The leaves oblong egg-shaped, on leaf-stalks;
Natans floating—*Leaf-stalks for the most part longer than the leaves.* Blossoms *in spikes, yellowish.*
Potamogiton rotundifolium. *Bauh. pin.* 193. *Ray's Syn.* 148.
Potamogiton latifolium. *Gerard.* 821.
Fontalis major latifolia vulgaris. *Park.* 1254.
In ponds and rivers, common. P. August.
The leaves floating upon the surface of the water afford an agreeable shade to fish.
This plant is the habitation and food of the Pondweed Moth, *Phalæna potamogeton.*

Perforated **PONDWEED.** The leaves heart-shaped, embracing the
Perfoliatum stem—*Mostly alternate, but at the divisions of the flowering branches opposite.* Blossoms *yellowish.*
Potamogiton perfoliatum. *Ray's Syn.* 149.
Potamogiton foliis latis splendentibus. *Bauh. pin.* 193.
Potamogiton 3 dodonæi. *Gerard.* 822.
Perfoliated Pondweed.
In rivers and ponds, frequent. P. June. July.
Goats and Cows eat it. Horses, Sheep and Swine refuse it.

Longleaved **PONDWEED.** The leaves flat and spear-shaped; ending at
Lucens the base in leaf-stalks—*Surface covered with a network of pellucid, veins.* Blossoms *in a reddish spike.*
Potamogiton aquis immersum folio pellucido, lato, oblongo, acuto. *Ray's Syn.* 148.
Potamogiton foliis angustis splendentibus. *Bauh. pin.* 193.
Potamogiton longis acutis foliis. *Gerard.* 822.
Rivers and ponds. P. June.

PONDWEED. The leaves alternate; fpear-fhaped; ferrated Caltrop
and waved at the edge—*Bloffoms on fruit-ftalks*; *white*, *or* Crifpum
reddifh.
Potamogiton, feu fontinalis crifpa. *Ray's Syn.* 149.
Potamogiton foliis crifpis, feu lactuca ranarum. *Bauh. pin.* 193.
Tribulus aquaticus minor quercus floribus. *Gerard.* 824.
Tribulus aquaticus minor prior. *Park.* 1248.
Greater water caltrops.
In ponds and flow rivers, very common. P. May. June.

PONDWEED. The leaves oppofite; fpear-fhaped; a little Serrated
waved at the edges—*Bloffoms greenifh yellow on fruit-ftalks.* *Is* Serratum
not this a variety only of the former?
Potamogiton feu fontalis media lucens. *Ray's Syn.* 149.
Potamogiton longo ferrato folio. *Bauh. pin.* 193.
Tribulus aquaticus minor, mofcatellæ floribus. *Gerard.* 823.
Tribulus aquaticus minor alter. *Park.* 1248.
1. There is a variety with a ftrap-fhaped leaf. *Ray's Syn.* 148.
Leffer water caltrops, or frogs lettuce.
In flow rivers. P. June.

PONDWEED. The leaves ftrap-fhaped; blunt; ftem com- Branched
preffed—*Bloffoms greenifh*; *on fhort fpikes.* Compreffum
Potamogiton caule compreffo, folio graminis canini. *Ray's
Syn.* 149.
Small branched Pondweed with a flat ftalk.
In rivers. P. June. July.

PONDWEED. The leaves like briftles; parallel; ftanding Fennel-leaved
near together, and pointing from two oppofite lines—*Bloffoms* Pectinatum
whitifh; Tips *yellow*.
Potamogiton millefolium, feu foliis gramineis ramofum.
Ray's Syn. 150.
Potamogiton gramineum ramofum. *Bauh. pin.* 193.
Millefolium tenuifolium. *Gerard.* 828.
In rivers and ponds. P. June.

PONDWEED. The leaves betwixt ftrap and fpear-fhaped; Grafs-leaved
alternate; fitting; broader than the props.— Gramineum
Potamogiton gramineum latiufculum, foliis et ramificationi-
bus denfe ftipatis. *Ray's Syn.* 149. Tab. 4. Fig. 3.
In ditches. P.

Marine
Marinum

PONDWEED. The leaves ſtrap-ſhaped ; alternate; diſtinct; the baſe ſheathing the ſtem—*The props are not diſtinct in this as in the other ſpecies, but the leaves are fixed to them.*

Potamogiton maritimum ramoſiſſimum, grandiuſculis capitulis, capillaceo folio noſtras. *Ray's Syn.* 150.

Sea Pondweed.

In ſalt water ditches. P. Auguſt.

Small
Puſillum

PONDWEED. The leaves ſtrap-ſhaped ; oppoſite and alternate ; expanding at the baſe. Stem cylindrical.—

Potamogiton puſillum, gramineo folio, caule tereti. *Ray's Syn.* 150.

Potamogiton minimum capillaceo folio. *Bauh. ʃin.* 193.

Small graſs-leaved Pondweed.

Pools and ditches. P. June. Auguſt.

59 SEAGRASS. 175 Ruppia.

EMPAL: *Sheath*, hardly any but what is formed by the baſe of the leaves. *Sheathed fruit-ſtalk,* awl ſhaped ; undivided ; ſtraight ; but becoming crooked when the fruit ripens. Beſet with flowers which point in two oppoſite directions. *Cup* none.

BLOSS: None.

CHIVES: Threads none. *Tips* four ; ſitting ; equal ; ſomewhat roundiſh ; rather double.

POINT : *Seedbuds* four or five ; ſomewhat eggſhaped ; approaching. *Shafts* none. *Summits* blunt.

S. VESS : None. The ſeeds are ſupported upon little foot-ſtalks, thread-ſhaped, and as long as the fruit.

SEEDS. Four or five ; egg-ſhaped ; oblique ; terminated by a flat circular ſummit.

Longleaved
Maritima

SEAGRASS. As there is only one ſpecies known Linnæus gives no deſcription of it---*Stem undivided.* Leaves *alternate.* Flowers *on fruit-ſtalks.*

Gramen maritimum fluitans cornutum. *Bauh. pin.* 3.

Potamogiton maritimum gramineis longioribus ſoliis, fructu ſere umbellato. *Ray's Syn.* 134.

Fucus folliculaceus, fæniculi folio longiore. *Bauh. pin.* 363.

In ſalt water ditches. P. Auguſt.

The

C L A S S V.

THE first division of the first ORDER of this class, includes the plants with ROUGH LEAVES; which admit of the following natural character.

EMPAL. *Cup* one leaf; with five clefts, or five divisions: permanent
BLOSS. One petal; with five clefts.
CHIVES. *Threads* five, fixed to the tube of the blossom.
POINT. *Seedbuds* four. *Shaft* single; thread-shaped. *Summit* blunt.
S. VESS. None.
SEEDS. Four; inclosed by the cup.

OBS. Leaves *rough and hairy; without leaf-stalks. As there is no seed-vessel, the cup does not fall off but remains after the blossom decays, and contains the seeds.*

In the second division of this order, those plants which bear berries and have a blossom composed of one petal, are generally poisonous.

The third division of the SECOND ORDER consists of plants whose flowers are disposed in RUNDLES; or the *Umbelliferous* plants of many authors. They admit of the following natural character.

Rundle composed of several *Rundlets.*
Fence, general, inclosing the whole rundle. or *partial,* inclosing only the rundlets.

EMPAL.

Empal. *Cup* hardly difcernible.

Bloss. Five petals; ftanding on the feedbud; fhed-ding. *Petals* generally heart-fhaped and bent inwards.

Chives. *Threads* five; fimple, hair-like. *Tips* fimple, or roundifh.

Point. *Seedbud* beneath. *Shafts* two; generally diftant. *Summits* fimple.

S. Vess. None.

Seeds. Two; fcored: convex on one fide, flat on the other.

Obs. *Stems hollow, or pithy.* Leaves *alternate.*

In dry fituations thefe plants are aromatic and carminative: in moift ones, often poifonous. The greateft virtues are contained in the feeds and roots. Many of them are employed in our kitchens, as the roots of Carrot and Parsnep; the leaves of Celery, and the feeds of Coriander.

CLASS V.

FIVE CHIVES.

ORDER I. ONE POINTAL.

* *Flowers of one petal: beneath. Seeds four; naked.* ROUGH LEAVES.

60 VIPERGRASS. *Bloff.* mouthnaked; irregular; bell-shaped.

61 LUNGWORT. *B'off.* mouth naked; funnel shaped. *Cup* prism-shaped.

62 GROMWELL. *Bloff.* mouth naked.; funnel-shaped. *Cup* with five divisions.

63 COMFREY. - *Bloff* mouth toothed; distended.

64 BORRAGE. - *Bloff.* mouth toothed ; wheel-shaped.

65 BUGLOSS. - *Bloff.* mouth vaulted; funnel-shaped. *Tube* crooked.

66 MADWORT. - *Bloff.* mouth vaulted; funnel shaped; *Fruit* compressed.

67 HOUNDSTONGUE. *Bloff.* mouth vaulted; funnel shaped. *Seeds* depressed and fixed to the side of the shaft.

68 ALKANET. - *Bloff.* mouth vaulted; funnel-shaped. *Tube* prism-shaped at the base.

69 SCORPIONGRASS. *Bloff.* mouth vaulted; salver-shaped: *Segments* notched at the end.

* * *Flowers of one petal: beneath. Seeds in a veffel.*

70 PIMPERNEL. *Capfule* of one cell: cut round. *Bloff.* wheel-shaped. *Summit* a knob.

71 LOOSESTRIFE. *Capf.* of one cell and ten valves. *Bloff.* wheel-shaped. *Summit* blunt.

72 Primrose

72 PRIMROSE. -   *Capſ.* of one cell. *Bloſſ.* funnel-ſhaped: mouth open. *Summit* globular.

73 FEATHERFOIL.   *Capſ.* of one cell. *Bloſſ.* the tube beneath the chives ! *Summit* globular.

74 BUCKBEAN. -   *Capſ.* of one cell. *Bloſſom* woolly ! *Summit* cloven.

75 BINDWEED. -   *Capſ.* with two cells and two ſeeds. *Bloſſ.* bell-ſhaped. *Summit* cloven.

76 THORNAPPLE.   *Capſ.* with two cells, and four valves ! *Bloſſ.* funnel-ſhaped. *Cup* deciduous.

77 HENBANE. -   *Capſ.* with two cells : covered with a lid ! *Bloſſ.* funnel-ſhaped. *Summit* a knob.

78 MULLEIN. -   *Capſ.* with two cells. *Bloſſ.* wheel-ſhaped. *Summit* blunt. *Chives* declining.

79 JACOBSLADDER.   *Capſ.* with three cells. *Bloſſ.* with five diviſions. *Chives* ſtanding upon the valves of the tube.

80 ROSEBAY. -   *Capſ.* with five cells. *Bloſſ.* bell-ſhaped. *Summit* blun .

81 PERRIWINKLE.   *S. Veſſ.* two upright little bags. *Bloſſ.* ſalver-ſhaped. *Seeds* not winged.

82 NIGHTSHADE.   *Berry* with two cells. *Tips* with two holes in each !

83 DWALE. -   *Berry* with two cells. *Chives* diſtant : bowed inwards.

† *Centaury Gentian.* † *Peloria* ; ſee *Common Toad-flax.*

* * * *Flowers of one Petal; ſuperior.*

84 MARSHWORT.   *Capſ.* with one cell : and five valves at the top. *Bloſſ.* ſalver-ſhaped : *Summit* a knob.

85 RAMPION. -   *Capſ.* with two or three cells : perforated. *Bloſſ* : with five diviſions. *Summit* with two or three clefts.

86 BELLFLOWER.   *Capſ.* with three or five cells ; perforated. *Bloſs* : bell-ſhaped. *Summit* with three clefts,

87 HONEYSUCKLE.   *Berry,* with two cells ; roundiſh. *Bloſſ.* unequal. *Summit* a knob.

88 BUCKTHORN.

* * * * *Flowers of five Petals; beneath.*

88 BUCKTHORN.   *Berry* with three cells; globular. *Cup* tubular, refembling a bloffom; with five converging fcales at the mouth of the tube.

89 SPINDLE.   *Berry* refembling a capfule; lobed. *Cup* expanding. *Seed* like a berry; covered with an outer coat.

† *Violet.*

* * * * * *Flowers of five Petals; fuperior.*

90 CURRANT. - *Berry* with feveral feeds. *Petals* ftanding on the cup. *Shaft* cloven.

91 IVY. - - *Berry* with five feeds. *Cup* binding round the fruit. *Summit* fimple.

* * * * * * *Flowers imperfect; beneath.*

92 KNOTGRASS.   *Capf.* with one feed and five valves. *Cup* fimple; rough and inelegant.

93 SALTWORT. - *Capf.* with five feeds; and five valves. *Cup* fimple; bell-fhaped; very rough and inelegant.

† *Perennial Snake-weed.* † *Orache.*

* * * * * * * *Flowers imperfect; fuperior.*

94 FLUELLIN. - *Seed* fingle; crowned. *Cup* bearing the chives.

Order II. Two Pointals.

* *Flowers of one Petal: beneath.*

95 FELWORT - *Capf.* with one cell and two valves. *Bloff.* wheel-fhaped, with ten honeycup pores.

96 GENTIAN. - *Capf.* with one cell and two valves. *Bloff.* tubular; varioufly fhaped.

H   * * *Flowers*

FIVE CHIVES.

** *Flowers of five Petals; beneath.*

† *Tree Bladdernut.*

*** *Flowers imperfect.*

97 KELPWORT. - *Seed* one; resembling a snail-shell, covered. *Cup* five leaves.

98 BLITE. - - *Seed* one; round and flat. *Cup* five concave leaves.

99 BEET. - - *Seed* one; kidney-shaped. *Cup* five leaves; seed at the bottom of the cup.

100 RUPTUREWORT. *Seed* one; egg-shaped; covered. *Cup* with five divisions. Five *Threads* with, and five without tips.

101 ELM. - - *Berry* not juicy; compressed. *Cup* one leaf; soon shrivelling.

† *Common Dodder.*

**** *Flowers of five Petals: superior and two seeds. In* RUNDLES.

A. FENCE *both general and partial.*

102 ERYNGO. - *Florets* growing in globular heads. *The receptacle* chaffy.

103 PENNYWORT. *Florets* in a sort of rundle; fertile. *Seeds* compressed.

104 SANICLE. - *Florets* in a sort of rundle: those in the centre barren. *Seeds* covered with sharp points.

105 MADNEP. - *Florets* unequal; almost all fertile. *Fence* deciduous. *Seeds* membranaceous.

106 DROPWORT. *Florets* unequal; those in the circumference barren. *Fence* simple. *Seeds* crowned; sitting.

107 PRICKLENEP. *Florets* unequal; several barren. *Fence* simple. *Seeds* sitting.

108 HENSFOOT. *Florets* unequal; those in the centre barren. *Fence* simple. *Seeds* covered with sharp points.

109 CARROT. *Florets* unequal; those in the centre barren. *Fence* winged. *Seeds* covered with prickles.

110 HARTWORT.

110 HARTWORT. *Florets* unequal; all fertile. *Fence* simple. *Seeds* with a scolloped border.

111 HARESPONG. *Florets* equal; those in the centre barren. *Fence* simple, *Seeds* scored; covered.

112 HEMLOCK. *Florets* equal; all fertile. *Petals* heartshaped. *Partial fence* extending but half way round. *Seeds* hunched, ribbed and furrowed.

113 PIGNUT. - *Florets* equal; all fertile. *Petals* heartshaped. *Partial fence* bristly.

114 SPIGNAL. - *Florets* equal; all fertile. *Petals* heartshaped. *Seeds* convex, scored.

115 THOROUGHWAX. *Florets* equal; all fertile. *Petals* rolled inwards. (The leaves of the partialfence often resemble petals.)

116 KEX. - *Florets* equal; all fertile. *Petals* heartshaped. *Seeds* nearly egg-shaped, scored.

117 SAMPHIRE. *Florets* equal; all fertile. *Petals* rather flat. *Fence* horizontal.

118 LOVAGE. - *Florets* equal; all fertile. *Petals* rolled inwards. *Fence* membranaceous.

119 ANGELICA. *Florets* equal; all fertile. *Petals* rather flat. *Rundlets* globular.

120 HONEWORT. *Florets* equal; all fertile. *Petals* rather flat. *Rundlets* of few florets.

† *Coriander.* ‡ *Celery Smallage.*

B. *Fence only partial.*

121 CICELY. - *Florets* rather unequal; all fertile. *Partialfence* extending but half way round.

122 CORIANDER. *Florets* unequal; some of them barren. *Fruit* nearly globular.

123 SHEPHERDSNEEDLE. *Florets* unequal; those in the centre barren. *Fruit* oblong.

124 COW-WEED. *Florets* unequal; those in the centre generally barren. *Partialfence* of five leaves.

125 WATERWORT. *Florets* nearly equal; all fertile. *Fruit* crowned.

H 2   126 COWBANE.

126 COWBANE.   *Florets* nearly equal; all fertile. *Petals* rather flat.

      † *Round-leaved Thorough-wax.*
      † *Cows Madnep.*
      † *Hemlock Dropwort.*
      † *Fine-leaved Hens-foot.*
      † *Creeping Flax.*
      † *Wild Angelica.*

      C. *No Fence.*

127 ALEXANDERS.   *Florets* equal; thofe in the centre barren. *Seeds* kidney-fhaped: angular.
128 CARAWAY.   *Florets* nearly equal; thofe in the centre barren. *Seeds* hunched; fcored.
129 PARSNEP. - *Florets* nearly equal; all fertile. *Seeds* depreffed and flat.
130 FENNEL. - *Florets* nearly equal; all fertile. *Seeds* bordered; fcored.
131 GOUTWEED.   *Florets* nearly equal; all fertile. *Seeds* hunched; fcored. *Petals* heart-fhaped.
132 SMALLAGE.   *Florets* equal; almoft all fertile. *Seeds* fmall; fcored. *Petals* bent inwards.
133 BREAKSTONE. *Florets* nearly equal; all fertile. *Petals* heart-fhaped. Before flowering the *Rundles* nodding.

Order III. Three Pointals.

      * *Bloffoms* fuperior.

134 MEALTREE.   *Bloff.* with five clefts. *Berry* with one feed.
135 ELDER. - *Bloff.* with five clefts. *Berry* with three feeds.

      * * *Bloffoms beneath.*

136 BLADDERNUT. *Bloff.* five petals. *Capf.* with two or three clefts: bladder-fhaped.
137 CHICKWEED.   *Bloff.* five petals. *Capf.* with one cell. *Cup* five leaves. *Petals* cloven.

   † *Water Elinks.* † *Sea Sandwort..*

              Order

Order IV. Four Pointals.

138 PARNASSUS. *Bloff.* five petals. *Capf.* with four valves. *Honey-cups* five. Fringed with glands.

Order Five. Five Pointals.

139 FLAX. - - *Bloff.* five petals. *Capf.* with ten cells.
140 SUNDEW. - *Bloff.* five petals. *Capf.* with one cell; opening at the top.
141 SIBBALD. - *Bloff.* five petals. Seeds five. *Cups* with ten clefts.
142 THRIFT. - *Bloff* with five divifions. *Seed* fingle; inclofed in the funnel-fhaped Cup.

† *Five-chived Moufe-ear.* † *Five-chived Spurrey.* † *Cranes-bill.*

Order VI. Many Pointals.

143 MOUSETAIL. *Cup* with five leaves. *Honeycups* five; tongue-fhaped. *Seeds* numerous.

† *Ivy-leaved Crowfoot.*

60 VIPERGRASS. 191 Echium.

EMPAL. *Cup* with five divifions, upright, permanent. *Segments* awl-fhaped, upright.
BLOSS. One petal; bell-fhaped. *Tube* very fhort. *Border* gradually widening; with five clefts, blunt; upright. *Segments* often unequal, the two upper ones being the longeft; the *lower,* fmaller; fharp; reflected. *Mouth* open.
CHIVES. *Threads* five; as long as the bloffom; awl-fhaped; declining; unequal. *Tips* oblong; fixed fideways to the threads.
POINT. *Seed-buds* four; *Shaft* thread-fhaped; as long as the chives. *Summit* blunt; cloven.
S. VESS. None. The cup becoming more inflexible, contains the feeds.
SEEDS. Four: roundifh; obliquely tapering.

Wall
Italicum

VIPERGRASS. The ſtem upright; hairy; the ſpikes rough with hair. Bloſſoms nearly equal. Chives exceedingly long.— *Bloſſoms blue.*

Echium alterum ſive lycopſis anglica. *Ray's Syn.* 228.

Engliſh Viper Bugloſs. *Hudſon's Flor. Angl.* 70.

1. Bloſſoms nearly equal, ſcarcely longer than the empalement with ſoft hairs at the border.—*Blue; purple; or white.*

Lycopſis. *Bauh. pin.* 255. *Park.* 519. *Gerard.* 802. *Ray's Syn.* 227.

Wall Vipers Bugloſs.

In ſandy cornfields and road-ſides. P. July. Auguſt.

Common
Vulgare

VIPERGRASS. The ſtem rough with hairs and tubercles. The ſtem leaves ſpear-ſhaped, and rough with hair. Flowers in lateral ſpikes—*firſt red, afterwards blue; ſometimes purple; or white.*

Echium vulgare. *Bauh. pin.* 254. *Park.* 414. *Gerard.* 802. *Ray's Syn.* 227.

Viper Bugloſs.

Cows and Sheep are not fond of it. Horſes and Goats refuſe it.—The flowers are highly grateful to Bees.

61 LUNGWORT.   184 Pulmonaria.

EMPAL. *Cup* of one leaf; with five teeth and five angles; permanent.

BLOSS. One petal; funnel-ſhaped. *Tube* cylindrical; as long as the empalement. *Border* with five ſhallow clefts; blunt; not quite upright. *Mouth* open.

CHIVES. *Threads* five; very ſhort; in the mouth of the tube. *Tips* upright; approaching.

POINT. *Seed-buds* four; *Shaft* thread-ſhaped; ſhorter than the empalement. *Summit* blunt; notched at the end.

S. VESS. None. The *Empalement* unchanged contains the ſeeds in its baſe.

SEEDS. Four; roundiſh; blunt.

LUNGWORT. The cups as long as the tube of the blof Broadleaved
fom. Root-leaves betwixt egg and heart-fhaped; rough—*Blof*- Officinalis
foms purple when newly expanded, but afterwards blue. Stems rough.
Tube of the bloffom white; mouth hairy.
Pulmonaria foliis echii. *Gerard.* 808.
Symphytum maculofum, feu pulmonaria latifolia. *Bauh.*
pin. 259.
Pulmonaria latifolia. *Park.* 248.
Buglofs Cowflips. long leaved fage of Jerufalem. Spotted
Lungwort.
Hampfhire, in woods. P. May.
When burnt it affords a larger quantity of afhes than almoft
any other vegetable; often a feventh part of its own weight.—
Sheep and Goats eat it. Cows are not fond of it. Horfes
and Swine refufe it—It nourifhes an infect, called the *Chryfomela*
Nemorum.

LUNGWORT. The cup but half as long as the tube of the Sea
bloffom; leaves egg-fhaped; ftem branched; trailing—*Bloffoms* Maritima
purple.
Echium marinum. *Ray's Syn.* 228.
Bugloffum dulce ex infulis Lancaftriæ. *Park.* 765.
Sea Buglofs.
On fandy fhores. P. July.

62 GROMWELL. 181 Lithofpermum.

EMPAL. *Cup* with five divifions; oblong, ftraight, fharp;
permanent. *Segments* awl-fhaped, and keel-fhaped
on the back.
BLOSS. One petal funnel-fhaped; as long as the empalement. *Tube* cylindrical. *Border* with five fhallow
clefts; blunt, upright. *Mouth* open.
CHIVES. *Threads* five; very fhort. *Tips* oblong, ftanding in the mouth of the bloffom.
POINT. *Seedbuds* four. *Shaft* thread-fhaped; as long as
the tube of the bloffom. *Summit* blunt; cloven.
S.VESS. None. The feeds are contained at the bottom
of the open cup.
SEEDS. Four; egg-fhaped; tapering; hard; fmooth.

FIVE CHIVES.

Common
Officinale

GROMWELL. The leaves smooth; blossoms hardly longer than the cup. Leaves spear-shaped.—*The seeds are as hard as bone.* Blossoms *white.*

Lithospermum majus erectum. *Bauh. pin.* 258.
Lithospermum minus. *Gerard.* 609.
Lithospermum vulgare minus. *Park.* 432.
Lithospermum seu milium solis. *Ray's Syn.* 228.
Gromill. Graymill.
In dry gravelly soil. P. May. June.
Grew says the seeds have so much earth in their composition that they effervesce with acids.—Sheep and Goat eat it. Cows and Horses refuse it.

Lesser
Arvense

GROMWELL. The seeds wrinkled. Blossoms hardly longer than the cups—*White; on short fruit-stalks.* Roots *reddish.*

Lithospermum arvense radice rubra. *Bauh. pin.* 258. *Park.* 432. *Ray's Syn.* 227.
Anchusa degener facie milii solis. *Gerard.* 610.
Bastard Alkanet.
In cornfields, common. A. May. June.
The girls in the North of Europe paint their faces with the juice of the root upon days of festivity—The bark of the root tinges wax and oil of a beautiful red, similar to that which is obtained from the root of the foreign Alkanet that is kept in the shops.—Sheep and Goats eat it. Cows are not fond of it. Horses and Swine refuse it.

Alkanet
Purpureo-
cæruleum

GROMWELL. The seeds smooth: blossom greatly longer than the cup—*The barren stems creeping and sending forth roots. The flowering stems upright.* Blossoms *purple.*

Lithospermum minus repens latifolium. *Bauh. pin.* 258.
Lithospermum majus. *Gerard.* 609.
Lithospermum vulgare majus. *Park.* 258.
Lithospermum majus dodonæi, flore purpureo, semine anchusæ. *Ray's Syn.* 229.
Lesser creeping Gromwell.
On hills. P. June.

63 COMFREY.

63 COMFREY. 185 Symphytum

EMPAL. *Cup* with five divisions and five corners; upright; sharp; permanent.

BLOSS. One petal; bell shaped. *Tube* very short. *Border* tubular; distended; thicker than the tube; *Mouth* with five teeth; blunt; reflected. *Mouth of the Tube* furnished with five awl-shaped valves shorter than the border, approaching so as to form a cone.

CHIVES. *Threads* five; awl-shaped; standing alternately with the valves in the mouth of the tube. *Tips* upright; sharp, covered.

POINT. *Seed-buds* four. *Shaft* thread-shaped, as long as the blossom. *Summit* simple.

S.VESS. None. The *Cup* grows larger and wider.

SEEDS. Four; hunched, tapering; approaching at the points.

COMFREY. The leaves betwixt egg and spear-shaped; Common running along the stem—*Blossoms yellowish white, or purple.* Officinale
Symphytum majus vulgare. *Park.* 523.
Symphytum consolida major. *Bauh. pin.* 259. *Gerard.* 660.
Symphytum magnum. *Ray's Syn.* 230.
1. There is a variety with purple blossoms.
On the banks of rivers and wet ditches. P. May.
The particles of the dust appear in the microscope like two globules united together---the leaves give a greatful flavour to cakes and panadoes---The roots are glutinous and mucilaginous.
Cows and Sheep eat it. Horses, Goats and Swine refuse it.

64 BORRAGE.

FIVE CHIVES.

64 BORRAGE. 181 Borago.

EMPAL. *Cup* with five divifions; permanent.

BLOSS. One petal; wheel-fhaped; as long as the empalement. *Tube* fho:ter than the cup. *Border* with five
divifions; wheel-fhaped, flat. The *Mouth* crowned
with five projecting fubftances, which are blunt and
notched at the end.

CHIVES. *Threads* five; awl-fhaped; approaching. *Tips*
oblong; approaching; fixed by the middle to the
inner fide of the thread.

POINT. *Seedbuds* four. *Shaft* thread-fhaped: longer than
the chives. *Summit* fimple.

S. VESS. None. The cup grows larger and bladder-
fhaped.

SEEDS. Four; roundifh; wrinkled; keel-fhaped outwardly towards the point; globular at the bafe:
lying lengthways in a hollow of the receptacle.

Common
Officinalis

BORRAGE with all the leaves alternate; cups expanding;
fruit-ftalks terminating; fupporting feveral flowers—*This plant
originally came from Aleppo; but it is now found in many parts of
Europe.* Leaves *egg-fhaped; embracing the ftem.* Bloffoms *a
fine blue.* Tips *black.*

Borrago hortenfis. *Gerard.* 797. *Ray's Syn.* 228.
Buglofium latifolium, Borrago. *Bauh. pin.* 256.
The varieties are
1. Bloffoms white.
2. Bloffoms rofe-coloured.
3. Leaves variegated.
On walls and amongft rubbifh. P. June. Auguft.

By the experiments of Mr. Marggraff, *Mem. de Berlin.* 1747
pa. 72. it appears that the juice affords a true nitre---It is now
feldom ufed inwardly but as an ingredient in cool tankards for
fummer drinking, though the young and tender leaves are good
in fallads, or as a pot-herb.---It affords nourifhment to the
Lambda Moth, *Phalæna Gramma.*

65 BUGLOSS.

65 BUGLOSS. 190 Lycopfis.

EMPAL. *Cup* with five divifions: permanent. *Segments* oblong; fharp; open.

BLOSS. One petal; funnel-fhaped. *Tube* cylindrical; crooked. *Border* with five fhallow clefts; blunt. *Mouth* clofed by five prominent, convex, approaching valves.

CHIVES. *Threads* five, very fmall; fixed to the bend óf the Tube, *Tips* fmall, covered; the mouth of the bloffom being clofed.

POINT. *Seedbuds* four. *Shaft* thread-fhaped; as long as the chives. *Summit* blunt; cloven.

S. VESS. None. The *Cup* very large and bladder-fhaped inclofes the feed.

SEEDS. Four, rather long; covered by a dry hard wrinkled coat.

OBS. *The effential charaᏟer of this genus confifts in the curvature of the tube of the bloffom.*

BUGLOSS. The leaves rough with hair; fpear-fhaped. Wild Cups upright whilft the plant is in flower—*The whole plant is* Arvenfis *rough with ftrong hairs.* Bloffoms blue.
Bugloffum fylveftre minus. *Bauh. pin.* 256. *Park.* 765.
Bugloffa fylveftris minor. *Gerard.* 799. *Ray's Syn.* 227.
Small wild Buglofs.
Ploughed fields and road fides. A. June. September.
Cows, Horfes, Sheep and Goats eat it. Swine refufe it.

66 MADWORT. 189 Afperugo.

EMPAL. *Cup* of one leaf; permanent: with five upright, unequal fegments.

BLOSS One petal; funnel-fhaped. *Tube* cylindrical; very fhort. *Border* with five fhallow clefts; blunt; fmall. *Mouth* clofed by five convex, projecting, approaching valves.

CHIVES. *Threads* five; very fhort; fixed in the mouth of the Tube. *Tips* rather oblong; covered.

POINT. *Seedbuds* four; compreffed. *Shaft* thread-fhaped; fhort. *Summit* blunt.

S. VESS. None. The *Cup* very large; upright; compreffed; clofes upon and contains the feeds.

SEEDS. Four; oblong; compreffed; in diftant pairs.

MADWORT

Goofegrafs
Procumbens

MADWORT. The cup containing the ripe fruit, compreffed
—*Bloffoms purple.*

Bugloffum fylveftre, caulibus procumbentibus. *Bauh. pin.* 257.
Borago minor fylveftris. *Park.* 765.
Aparine major. *Gerard.* 1122.
Small wild Buglofs. Great Goofe-grafs. German Madwort.
In roads and amongft rubbifh. A. April. May.
Horfes, Goats, Sheep and Swine eat it. Cows are not fond
of it.

67 HOUNDSTONGUE. 183 Cynogloffum.

EMPAL. *Cup* with five divifions; oblong; fharp; permanent.

BLOSS. One petal; funnel-fhaped; as long as the
empalement. *Tube* cylindrical; fhorter than the
Border. *Border* with five fhallow clefts; blunt.
Mouth clofed by five convex, prominent, approaching valves.

CHIVES. *Threads* five; very fhort; fixed to the mouth
of the bloffom. *Tips* roundifh; naked.

POINT. *Seedbuds* four. *Shaft* awl-fhaped; as long as the
chives; permanent. *Summit* notched at the end.

S. VESS. None; but the feedcoats of the four feeds;
depreffed, roundifh, outwardly more blunt; rough;
not opening; flat upon the outer fide; fixed by
their points.

SEEDS. Four; fomewhat egg-fhaped; hunched, tapering; fmooth.

OBS. *The effential character of this genus confifts in the four feedcoats, each inclofing a fingle feed; and fixed to the fhaft.*

Stirking
Officinale

HOUNDSTONGUE. Chives fhorter than the bloffom.
Leaves broad; fpear-fhaped; fitting; downy—*Bloffoms purple
fometimes white.*

Cynogloffum majus vulgare. *Bauh. pin.* 257. *Gerard.* 804.
Park. 511. *Ray's Syn.* 226.
1. There is a variety with greener leaves. *Ray's Syn.* 226.
Great Houndftongue.
Road fides and amongft rubbifh. P. June.
Both the root and leaves have been fufpected to poffefs
narcotic properties, but others will not admit the fact. It is
difcarded from the prefent practice. The fmell of it is very dif-
agreeable.
Goats eat it. Cows, Horfes, Sheep and Swine refufe it.
It furnifhes food for the Scarlet Tyger Moth, *Phalæna
Domina.* 68 ALKANET

68 ALKANET. 182 Anchuſa.

EMPAL. *Cup* with five diviſions; oblong; cylindrical; ſharp; permanent.

BLOSS. One petal; funnel-ſhaped. *Tube* cylindrical; as long as the empalement. *Border* with five ſhallow clefts; blunt; not quite upright. *Mouth* cloſed by five convex, prominent, oblong, approaching valves.

CHIVES. *Threads* five; very ſhort; fixed to the mouth of the bloſſom. *Tips* oblong; covered; fixed ſideways to the threads.

POINT. *Seedbuds* four. *Shaft* thread-ſhaped; as long as the chives. *Summit* blunt; notched at the end.

S. VESS. None. The *Cup* grows larger and upright; and includes the feeds within it.

SEEDS. Four; rather long; blunt; hunched.

OBS. *When the bloſſom is fully expanded it is nearly ſalver-ſhaped.*

ALKANET with flowers growing in heads; upon fruit-ſtalks which are furniſhed with two leaves— *Evergreen Sempervirens*

The ſtems proceeding from the ſides of the crown of the root are upright and rough with hair. *Leaves* betwixt egg and ſtrap-ſhaped; on leaf-ſtalks; remote; ſpotted with white. *Fruit-ſtalks* riſing from the baſe of the leaves are furniſhed with two oppoſite, ſitting, floral leaves, betwixt ſpear and egg-ſhaped. Several flowers upon each fruit-ſtalk. The *Bloſſoms* blue and the tube ſhort.

Bugloſſum latifolium ſempervirens. *Bauh. Pin.* 256.
Borrago ſempervirens. *Gerard.* 797. *Ray's Syn.* 227.
Road ſides and amongſt rubbiſh. P. May. June.
Cows, Horſes, Sheep and Goats eat it. Swine are not fond of it.

69 SCORPIONGRASS. 180 Myofotis.

EMPAL. *Cup* with five fhallow clefts; oblong; upright: fharp: permanent.

BLOSS. One petal; falver-fhaped. *Tube* cylindrical; fhort. *Border* flat; with five fhallow cleits. *Segments* blunt; notched at the end. *Mouth* clofed with five convex, prominent, approaching valves.

CHIVES. *Threads* five: very fhort; fixed to the neck of the tube. *Tips* very fmall; covered.

POINT. *Seedbuds* four. *Shaft* thread-fhaped; as long as the tube of the bloffom. *Summit* blunt.

S. VESS. None. The *Cup* large and upright contains the feeds within it.

SEEDS. Four; egg-fhaped; tapering; fmooth.

OBS. *In fome fpecies the feeds are covered with hooked prickles.*

Moufe-Ear Scorpioides.   SCORPIONGRASS with naked feeds and the points of the leaves callous—*This varies confiderably in different fituations. In dry fituations the plant and flowers are fmaller; in moift ones both are larger and fometimes hairy. The* Bloffoms *vary from a full blue to a very pale one, or fometimes a yellow; and are in a long fpirally twifted fpike.*

Myofotis fcorpioides hirfuta. *Park.* 691. *Ray's Syn.* 229 *Gerard.* 337.

Echium fcorpioides arvenfe. *Bauh. pin.* 254.

In dry fields and on the margin of fprings and rills. P. April. Auguft.

When it grows in the water, and its tafte and fmell thereby rendered lefs obfervable Sheep will fometimes eat it, but it is generally fatal to them---Cows, Horfes, Swine and Goats refufe it.

70 PIMPERNEL 206 Anagallis.

EMPAL. *Cup* with five divisions; sharp; permanent. *Segments* keel-shaped.

BLOSS. One petal; wheel shaped. *Tube* none. *Border* with five divisions; flat. The *Segments* roundish egg-shaped; connected by the claws.

CHIVES. *Threads* five; upright hairy towards the bottom; shorter than the blossom. *Tips* simple.

POINT. *Seedbud* globular. *Shaft* thread-shaped; a little inclining. *Summit* knobbed.

S. VESS. *Capsule* globular: of one cell; cut round.

SEEDS. Several; angular. *Receptacle* very large; globular.

PIMPERNEL with undivided leaves and a trailing stem— Common *Leaves betwixt egg and spear-shaped.* Blossoms *red. They open* Arvensis *about eight in the morning; but close in the afternoon.*

Anagallis flore phæniceo. *Baub. pin.* 252. *Park.* 558. *Ray's Syn.* 282.

There are three varieties, viz.

1. Red flowered with larger leaves, four at a joint. *Ray's Syn.* 282.
2. Blue flowered, or Female Pimpernel. *Gerard.* 617. *Ray's Syn.* 282. *Baub. pin.* 252.
3. White flowered. *Ray's Syn.* 282.

Male Pimpernel. 2. Female Pimpernel.

In sandy cornfields. A. May. August.

Cows and Goats eat it. Sheep refuse it. Small Birds are highly delighted with the seeds.

71 LOOSESTRIFE. 205 Lysimachia.

EMPAL. *Cup* with five divisions; sharp; upright; permanent.

BLOSS. One petal; wheel-shaped. *Tube none. Border* with five divisions; flat. *Segments* oblong egg shaped.

CHIVES. *Threads* five; awl-shaped. *Tips* tapering.

POINT. *Seedbud* roundish. *Shaft* thread-shaped; as long as the chives. *Summit* blunt.

S. VESS. *Capf.* globular; with one cell and ten valves.

SEEDS. several; angular. *Receptacle* very large; globular, dotted.

LOOSE-

* Fruit-stalks with several flowers.

Yellow Vulgaris

LOOSESTRIFE. The flowers in panicled bunches, terminating—*Stem scored; hairy.* Leaves *spear-shaped, three or four at each joint.* Blossoms *yellow.*

Lysimachia lutea. *Gerard.* 474. *Ray's Syn.* 282.
Lysimachia lutea major. *Bauh. pin.* 245. vulgaris. *Park.* 544.
1. There is a variety with shorter and broader leaves. *Ray's Syn.* 282.
Yellow Willowherb.
Banks of rivers and shady marshes. P. June. July.—The leaves give a yellow dye to wool.
Cows and Goats eat it. Sheep are not fond of it. Horses and Swine refuse it.

Tufted Thyrsiflora

LOOSESTRIFE. The flowers on fruit-stalks in lateral bunches—*Blossoms cloven down to the base, with very short teeth intervening. The ends of the petals marked with tawny spots.* Blossoms *yellow.* Leaves, *two at each joint.*

Lysimachia lutea, flore globoso. *Gerard.* 475. *Park.* 544. *Ray's Syn.* 283.
Lysimachia bifolia flore globoso luteo. *Bauh. pin.* 245.
Marshes and banks of rivers, near King's Langley in Hertfordshire.
In the Isle of Anglesea. P. June.
Goats eat it. Cows and Sheep are not fond of it. Horses and Swine refuse it.

** Fruit-stalks with only flower.

Pimpernel Nemorum

LOOSESTRIFE with sharp egg-shaped leaves; solitary flowers and trailing stems—*Blossoms yellow.* Stem *cylindrical.*
Anagallis lutea. *Gerard.* 618. nemorum. *Bauh. pin.* 252.
Anagallis flore luteo. *Park.* 558.
Yellow Pimpernel of the woods.
In moist shady places. P. May. June.

Moneywort Nummularia

LOOSESTRIFE. The leaves somewhat heart-shaped; flowers solitary; stem creeping—*Four cornered.* Blossoms *yellow.*
Nummularia. *Gerard.* 630. vulgaris. *Park.* 555.
Nummularia major lutea. *Bauh. pin.* 309.
In moist meadows. P. June.
The plant is a little acrid and sub-astringent.
Cows and Sheep eat it. Goats are not fond of it. Horses refuse it.

LOOSESTRIFE. The leaves rather sharp and egg-shaped; Purple fruit-stalks longer than the leaves; stem creeping—*Chives very* Tenella *woolly.* Blossoms *red.*

Nummularia minor flore purpurascente. *Bauh. pin.* 310. *Park.* 565. *Gerard.* 630. *Ray's Syn.* 283.

Purple Moneywort.

In turfy bogs. P. July. August.

72 PRIMROSE. 197 Primula.

EMPAL. *Fence* small; of many leaves; including several flowers. *Cup* one leaf; tubular; sharp; upright; permanent, with five angles and five teeth.

BLOSS. One petal. *Tube* cylindrical; as long as the cup; terminated by a short hemispherical neck. *Border* expanding, with five shallow clefts. *Segments* inversely-heart-shaped; notched at the end; blunt. *Mouth* open.

CHIVES. *Threads* five; very short; within the neck of the blossom. *Tips* upright; approaching; tapering; within the tube.

POINT. *Seedbud* globular. *Shaft* thread-shaped; as long as the cup. *Summit* globular.

S. VESS. *Capsule* cylindrical; nearly as long as the cup, which covers it; of one cell: opening at the top with ten teeth.

SEEDS. Numerous; roundish. *Receptacle* oblong; egg-shaped; loose.

PRIMROSE. The leaves wrinkled and toothed—*A single* Common *flower only upon a fruit-stalk.* Blossoms *pale yellow.*   Vulgaris

Primula veris vulgaris. *Park.* 536. *Ray's Syn.* 284.

Primula veris minor. *Gerard.* 781.

Verbasculum sylvarum majus singulari flore. *Bauh. pin.* 241.

In woods and hedges, frequent. P. April.

Gerard reports that a dram and a half of the dried roots, taken up in autumn, operates as a strong but safe emetic.

Sheep and Goats eat it. Cows are not fond of it. Horses and Swine refuse it.

FIVE CHIVES.

Cowslip **1.**
Veris

PRIMROSE. The leaves wrinkled; toothed—*Stalk supporting several flowers, which are* yellow *and agreeably scented.*

Primula veris major. *Gerard.* 781.

Verbafculum pratenfe odoratum. *Bauh. pin.* 241.

Paralyfis vulgaris 'pratenfis, flore flavo fimplici odorato. *Park.* 535.

Pagils. Paigles. Cowflips.

In paftures.

The flowers are ufed for making cowflip wine. The Leaves may be eaten as a pot-herb, or in fallads. The root hath a fine fmell, like anife.

Oxlip **2.**
Elatior

PRIMROSE. Larger than the preceding; without fmell. Border of the bloffom flat: pale yellow.

Verbafculum pratenfe aut fylvaticum inodorum. *Bauh. pin.* 241.

Primula pratenfis inodora lutea. *Gerard.* 780. *Ray's Syn.* 284.

In high barren paftures. P. April. May.

Different as thefe varieties may feem, it would be wrong to rank them as diftinct fpecies, fays Linnæus: and the accurate Dr. Martyn in his *Catal. Hort. Botan. Cantab.* p. 30, fays, experience hath taught him to adopt the fame opinion; though at the fame time, with that candour which ever diftinguifhes real merit, he quotes a different opinion from another author.

The different kinds of Polyanthuffes, fo much admired by florifts, all originate from thefe.

Mealy
Farinofa

PRIMROSE. Leaves fmooth, fcolloped. Border of the bloffom flat—*The fruit-ftalks and cups are beautifully covered with the duft.* Bloffoms *red.*

Primula veris flore rubro. *Gerard.* 783.

Verbafculum umbellatum alpinum minus. *Bauh. pin.* 247. *Ray's Syn.* 285.

Paralyfis minor flore rubro. *Park.* 246.

Birds Eye.

In marfhes and bogs, upon mountains in the North. P May.

The flowers are beautiful, but they indicate a barren foil.

Horfes, Sheep and Goats eat it. Cows refufe it.

The great yellow under-winged Moth, *Phalæna Pronuba*, lives upon the different fpecies and varieties of the Primrofe.

73 FEATHERFOIL. 203 Hottonia.

EMPAL. *Cup* of one leaf, with five divisions. *Segments* strap-shaped; not quite upright.

BLOSS. One petal; salver-shaped. *Tube* as long as the empalement. *Border* with five clefts; flat. *Segments* oblong, egg-shaped; notched at the end.

CHIVES. *Threads* five; awl-shaped: short; upright; standing upon the tube and opposite to the segments of the blossom. *Tips* oblong.

POINT. *Seedbud* globular, tapering. *Shaft* thread shaped; short. *Summit* globular.

S. VESS. *Capsule* globular, tapering; with one cell; standing upon the empalement.

SEEDS. Many, roundish. *Receptacle* globular; large.

OBS. *Some flowers have six chives; and then the cup and the blossom is divided into six segments.*

FEATHERFOIL with numerous flowers; on fruit-stalks, Water growing in whorls—*The leaves like those of the Featherwort are* Palustris *concealed under water, only the spikes of flowers rising above the surface.* Leaves *winged.* Blossoms *white or tinged with pale purple.*
Hottonia. *Ray's Syn.* 285.
Millefolium aquaticum seu viola aquatica caule nudo. *Bauh. pin.* 141.
Millefolium aquaticum floridum seu viola aquatica. *Park.* 1256.
Viola palustris. *Gerard.* 826.
Water Violet.
In ponds and ditches. P. July. August.
Cows eat it. Swine refuse it.

74 BUCK-

74 BUCKBEAN. 202 Menyanthes.

EMPAL. *Cup* of one leaf with five upright divisions : permanent.

BLOSS. One petal ; funnel-shaped. *Tube* short. somewhat cylindrical at bottom but funnel-shaped upwards. *Border* cloven more than half way down into five *Segments.* The *Segments* blunt ; reflected ; expanding ; remarkably hairy.

CHIVES. *Threads* five ; awl shaped ; short. *Tips* sharp ; upright ; cloven at the base.

POINT. *Seedbud* conical. *Shaft* cylindrical ; nearly as long as the blossom. *Summit* cloven ; compressed.

S. VESS. *Capsule* egg-shaped ; of one cell : bound round by the cup.

SEEDS. Many ; egg-shaped ; small.

OBS. *The first species hath the segments of the petals fringed at the edges ; but the upper surface is not hairy.*

Lilly
Nymphoides

BUCKBEAN. The leaves heart-shaped, very entire ; the blossoms fringed at the edges.—*Flowers forming a simple rundle ; sitting upon the side of a leaf-stalk.* Blossoms *yellow.* Leaves *sometimes spotted.*

Nymphæa lutea minor, flore fimbriato. *Bauh. pin.* 194. *Rays Syn.* 368.

Fringed Water-lilly.

In large ditches and slow streams. P. June. July.

Trefoil
Trifoliata

BUCKBEAN with three leaves on each leaf-stalk.—*Flowers on fruit-stalk, opposite to the leaves. The segments of the blossoms stand wide and are fringed on the inner surface ; white or purplish.*

1. It varies in having broader or narrower leaves.

Menianthe palustre triphyllum, latifolium et angustifolium. *Ray's Syn.* 285.

Trifolium palustre. *Bauh. pin.* 327.

Trifo ium paludosum. *Gerard.* 1194. *Park.* 1212.

In ponds and pits, frequent. P. June. July.

An infusion of the leaves is extremely bitter, and is prescribed in Rheumatisms and Dropsies. A dram of them in powder purges and vomits : it is sometimes given to destroy worms. In a scarcity of hops this plant is used in the North of Europe to bitter the ale. The powdered roots are sometimes used in Lapland instead of bread, but they are unpalatable. Some people say that Sheep will eat it, and that it cures them of the rot ; but from the Upsal experiments it appears, that though Goats eat it, Sheep sometimes will and sometimes will not. Cows, Horses and Swine refuse it.

75 BIND-

75 BINDWEED. 215 Convolvulus.

EMPAL. *Cup* with five divisions, approaching; egg-shaped, blunt, small; permanent.

BLOSS. One petal; bell-shaped; expanding; large; plaited. The border slightly marked with five or ten notches.

CHIVES. *Threads* five; awl shaped; half the length of the blossom. *Tips* egg shaped, compressed.

POINT. *Seedbud* roundish. *Shaft* thread-shaped: as long as the chives. *Summits* two; oblong and somewhat broad.

S. VESS. *Capsule* inclosed by the cup: roundish; with one, two, or three valves.

SEEDS. Two. Roundish.

OBS. *The species of Bindweed furnish nutriment to the Unicorn, or Bindweed Hawk Moth,* Sphinx Convolvuli, *and the Elephant Moth,* Phalæna Elpenor.

BINDWEED. The leaves arrow-shaped; sharp at each angle. Generally one flower upon a fruit-stalk.—*Blossoms reddish or white; or striped; or purple.* Small Arvensis
Convolvulus minor vulgaris. *Park.* 171. *Ray's Syn.* 275.
Convolvulus minor arvensis. *Bauh. pin.* 294.
1. There is a variety with very small leaves, and another with very small flowers. *Ray's Syn.* 276.
In corn-fields, common. P. June. July.
Cows, Horses, Goats and Sheep eat it. Swine refuse it.

BINDWEED. The leaves arrow-shaped; the two angles at the base lopped. Fruit-stalk four cornered; supporting a single flower—*Edges of the leaves brown.* Blossoms *white; or purplish, or striped. Floral leaves two; close to the cup.* Great Sepium
Convolvulus major albus. *Bauh. pin.* 294. *Park.* 163. *Ray's Syn.* 275.
Smilax lævis major. *Gerard.* 861.
In moist hedges. P. July. August.
Though the root is a very acrid purgative to the human race, it is eaten by Hogs in large quantities without any detriment. Scammony is the inspissated juice of a species of Convolvulus so much resembling this, that they are with difficulty distinguished. Can it then be worth while to import Scammony from Aleppo at a considerable annual expence, when a medicine with the very same properties grows spontaneously in many of our hedges? Sheep, Goats and Horses eat it. Cows refuse it.

BINDWEED.

Sea
Soldanalle

BINDWEED. The leaves kidney-ſhaped, only one flower
upon a fruit-ſtalk—*Bloſſoms red.*

Convolvulus maritimus, ſoldanella dicta. *Ray's Syn.* 276.
Soldanella marina. *Gerard.* 861
Soldanella maritima minor. *Bauh. pin.* 293
Soldanella vulgaris. *Park.* 161.
Scottiſh Scurvy-graſs.
Common on the ſea ſhore. P. July.

Half an ounce of the juice, or a dram of the powder, is an
acrid purge. The leaves applied externally are ſaid to diminiſh
dropſical ſwellings of the feet.

76 THORNAPPLE.   246 Datura.

EMPAL. *Cup* of one leaf: oblong; tubular; diſtended; with
  five angles and five teeth; deciduous; but leaving
  a part of the baſe behind.

BLOSS. One petal; funnel-ſhaped. *Tube* cylindrical;
  generally longer than the cup. *Border* not quite
  upright, almoſt entire; with five angles, five tapering teeth, and five plaits.

CHIVES. *Threads* five; awl-ſhaped; as long as the cup.
  Tips oblong; blunt; compreſſed.

POINT. *Seedbud* egg-ſhaped. *Shaft* thread-ſhaped; ſtraight.
  Summit thick: blunt; compoſed of two flat ſubſtances.

S. VESS. *Capſule* nearly egg-ſhaped; with two cells and
  four valves: ſtanding upon the remains of the cup.
  Receptacle large; convex; dotted; fixed to the partition of the capſule.

SEEDS. Numerous, kidney-ſhaped.

OBS. *The ſeed-veſſel in moſt ſpecies is thick ſet with thorns.*

Whiteflowered
Stramonium

THORNAPPLE. The ſeed-veſſel upright; egg-ſhaped;
thorny. Leaves egg-ſhaped—*Indented.* Bloſſoms *white. At night
the leaves, particularly the upper ones, riſe up and incloſe the flowers.
The Bloſſoms have ſometimes a tinge of purple, or violet.*

Solanum pomo ſpinoſo oblongo, flore calathoide, ſtramonium
vulgo dictum. *Ray's Syn.* 266.
Solanum fœtidum, pomo ſpinoſo oblongo, flore albo. *Bauh. pin.*
164. Stramonium ſpinoſum. *Gerard.* 349.

Amongſt rubbiſh; common about London. A. July.

An ointment prepared from the leaves gives eaſe in external
inflammations and Hæmorrhoids. The Edinburgh College
direct an extract to be prepared by evaporating the expreſſed
juice of the leaves. The ſeeds or leaves given internally bring
on delirium, and in larger doſe would undoubtedly prove fatal.
Cows, Goats, Sheep and Horſes refuſe it.

77 HENBANE.

77 HENBANE. 247 Hyoſcyamus.

EMPAL. *Cup* of one leaf tubular; diſtended in the lower part. *Rim* with five clefts. ſharp; permanent.

BLOSS. One petal: funnel-ſhaped. *Tube* cylindrical; ſhort. *Border* not quite upright ; with five ſhallow clefts. *Segments* blunt; one broader than the reſt.

CHIVES. *Threads* five; awl-ſhaped; inclining. *Tips* roundiſh.

POINT. *Seedbud* roundiſh. *Shaft* thread ſhaped; as long as the chives. *Summit* ſomewhat globular.

S. VESS. *Capſule* egg-ſhaped, blunt; marked with a line upon each ſide. Cells two; cut round; covered with a lid that opens horizontally. Two capſules lying cloſely preſſed together. *Receptacle* in the ſhape of half an egg, fixed to the partition of the capſule.

SEEDS. Numerous; unequal.

HENBANE with indented leaves, embracing the ſtem. Common Flowers ſitting—*Bloſſoms purple and brown; clammy.* Niger

Hyoſcyamus vulgaris. *Ray's Syn.* 274. *Bauh. pin.* 169. *Park.* 362.

Hyoſcyamus niger. *Gerard.* 353.

Road ſides, and amongſt rubbiſh. B. June.

The ſeeds, the leaves and the roots taken internally, are all poiſonous: and many well atteſted inſtances of their bad effects are recorded. Madneſs, Convulſions, and death are the general conſequence. In a ſmaller doſe, they occaſion giddineſs and ſtupor. It is ſaid that the leaves ſcattered about a houſe will drive away mice. The Edinburgh College order the expreſſed juice of the plant to be evaporated to an extract; and perhaps in this ſtate it may be advantageouſly joined with opium, where the effects of that medicine are deſirable, and coſtiveneſs is to be avoided. There is no doubt of it being an uſeful medicine under proper management. The doſe is from half a ſcruple to half a dram.

Goats are not fond of it Horſes, Cows, Sheep and Swine refuſe it.

The Henbane Chryſomela, *Chryſomela Hyoſcyami*, and the Scarlet Bug, *Cimex Hyoſciami*, are found upon this plant.

78 MULLEIN. 245 Verbaſcum.

EMPAL. *Cup* of one leaf with five diviſions; ſmall; permanent. *Segments* upright; ſharp.

BLOSS. One petal; wheel-ſhaped. *Tube* cylindrical; very
ſhort. *Border* with five diviſions expanding. *Segments* egg-ſhaped, blunt.

CHIVES. *Threads* five; awl-ſhaped; declining: ſhorter
than the bloſſom. *Tips* roundiſh; compreſſed; upright.

POINT. *Seedbud* roundiſh. *Shaft* thread-ſhaped; inclining; as long as the chives. *Summit* rather thick and
blunt.

S. VESS. *Capſule* roundiſh with two cells, opening at the
top. *Receptacle* the ſhape of half an egg, fixed to the
partition of the ſeed-veſſel.

SEEDS. Numerous; angular.

OBS. *In moſt ſpecies the chives are inclining, and the bottom of the
threads cloathed with ſoft coloured hairs.*

Great
Thapſus

MULLEIN. The leaves running along the ſtem; downy
on both ſides; ſtem ſimple—*Bloſſoms, in a long terminating ſpike,
yellow.*
   Verbaſcum mas latifolium luteum. *Bauh. pin.* 239. *Ray's Syn.*
287.
   Verbaſcum album vulgare, ſive Thapſus barbatus communis.
Park. 60.
   Thapſus barbatus. *Geraid.* 773.
   Great White Mullein. Hightaper. Cows Lungwort.
   In dry ditch banks, common. B. July.
   Externally uſed it is emollient. It is ſaid to intoxicate fiſh ſo
that they may be taken with the hand.----In Norway they give
it to cows that are conſumptive.----The down ſerves for tinder.
Neither Cows, Goats, Sheep, Horſes or Swine will eat it. The
Water Betony Moth, *Phalæna Verbaſci,* and the Figwort Curculio,
Curculio Scrophularia, live upon this plant.

Hoary
Lychnitis

MULLEIN. The leaves oblong; Wedge-ſhaped.—*Bloſſoms
yellow or white, in lateral and terminating ſpikes.*
   Verbaſcum flore albo parvo. *Ray's Syn.* 287.
   Verbaſcum lichnitis flore albo parvo. *Bauh. pin.* 240.
   Verbaſcum mas, foliis longioribus. *Park.* 60.
   Verbaſcum lichnitis matthioli. *Gerard.* 775.
   1. There is a variety with a yellow flower. *Ray's Syn.* 287.
   White flowered Mullein.
   In ſandy places. B. July.
   Neither Cows, Goats, Sheep, Horſes or Swine will touch it.
MULLEIN.

MULLEIN. The leaves oblong; heart-fhaped, ftanding up. Black
on leaf-ftalks.—*The ftem is befet with hairs that are beautifully* Nigrum
branched. Bloſſoms *yellow;* Tips *purple.*

Verbafcum nigrum flore parvo, apicibus purpureis. *Ray's
Syn.* 288.

Verbafcum nigrum flore ex luteo purpurafcente. *Bauh. pin.*
240.

Verbafcum nigrum. *Gerard.* 775. vulgare. *Park.* 61.

Sage Leaved Black Mullein.

Hedges and road-fides. P. July.

This is a beautiful plant, and the flowers are grateful to Bees.
Swine will eat it. Sheep are not fond of it. Cows, Horſes and
Goats refuſe it.

MULLEIN. The leaves oblong; fmooth; embracing the
ftem. Fruit ftalks folitary—*Bloſſoms yellow.*   Moth
Blattaria

Blattaria lutea. *Ray's Syn.* 288.

Blattaria lutea, folio longo laciniato. *Bauh. pin.* 240.

Blattaria flore luteo. *Park.* 64.

Blattaria Plinii. *Gerard.* 776.

Yellow Moth Mullein.

In gravelly foils. A. June.

79 JACOBSLADDER. 217 Polemonium.

EMPAL. *Cup* beneath; of one glafs-fhaped leaf: perma-
    nent fharp with five fhallow clefts.

BLOSS. One petal, wheel fhaped. *Tube* fhorter than the
    cup; cloſed by five valves, placed at the top of it.
    Border with five divifions; large; flat. *Segments*
    roundifh, blunt.

CHIVES. *Threads* five; thread-fhaped; inclining; fhorter
    than the bloſſom: ftanding upon the valves of the
    tube. *Tips* roundifh; fixed to the thread fide-ways.

POINT. *Seedbud* egg-fhaped fharp. *Shaft* thread-fhaped;
    as long a the bloſſom. *Summit* with three clefts;
    rolled back.

S. VESS. *Capfule* covered. Egg-fhaped, but with three
    angles; three cells and three valves.

SEEDS. Several; irregular; rather fharp.

JACOBS.

Common
Cæruleum

JACOBSLADDER. With winged leaves; upright flowers; and the cup longer than the tube of the bloſſom — *Blloſſoms blue, or white. Little leaves betwixt egg and ſpear-ſhaped; eleven pair or more on each leaf.*

Valeriana cærulea. *Bauh. pin.* 164.
Valeriana græca. *Gerard.* 10;6. *Park.* 122.
Polemonium vulgare cæruleum et album. *Ray's Syn.* 288.
Greek Valerian. Ladder to Heaven.
About Malham Cove. P. June.
Its beauty hath obtained it a place in our gardens.
Cows, Goats and Sheep it. Horſes are not fond of it.

So ROSEBAY. 212 Azalea.

EMPAL. *Cup* with five diviſions; ſharp; upright: ſmall; coloured; permanent.

BLOSS. One petal; bell ſhaped; with five ſhallow clefts. *Segments* with the edges bent inwards.

CHIVES. *Threads* five; thread-ſhaped; growing on the receptacle; looſe. *Tips* ſimple.

POINT. *Seedbud* roundiſh. *Shaft* thread-ſhaped; as long as the bloſſom: permanent. *Summit* blunt.

S. VESS. *Capſule* roundiſh; with five cells and five valves.

SEEDS. Many; roundiſh.

Trailing
Procumbens

ROSEBAY. Branches ſpreading wide, and trailing—*Blloſſoms purpliſh fleſh colour.*

Chamæciſtus ſerpyllifolia, floribus carneis. *Bauh. pin.* 466.
On mountains in the North. S.
The Rev. Mr. Lightfoot found it near Arniſdale in the Highlands of Scotland. *Pennant's Tour.* 1772. p. 345.

81 PERRYWINKLE. 295 Vinca.

EMPAL. *Cup* with five divisions; upright; sharp; permanent.

BLOSS. One petal; salver-shaped. *Tube* longer than the cup; cylindrical in the lower part, wider above: marked with five lines, and five angles at the mouth. *Border* with five divisions; horizontal. The segments connected with the top of the tube: broadeit on the outward edge, and obliquely lopped.

CHIVES. *Threads* five: very short; first bent inwards and then backwards. *Tips* membranaceous; blunt; upright but bowed inwards; with the dust at the margins.

POINT. *Seedbuds* two; roundish; with two roundish bodies lying contiguous to them. *Shaft* common to both seedbuds; cylindrical; as long as the chives. *Summits* two, the lower one round and flat; the upper summit a concave knob.

S. VESS. Two *Bags*; cylindrical; long; tapering; upright: of one valve opening lengthways.

SEEDS. Numerous; oblong; cylindrical: furrowed.

PERRYWINKLE. The stems creeping; leaves betwixt Lesser spear and egg-shaped, flowers on fruit-stalks—*Cups shorter than* Minor *the tube of the blossom. Flowering stems upright. Leaf-stalks short.* Blossoms *blue.*

Vinca pervinca minor. *Gerard.* 894. *Ray's Syn.* 258.
Vinca pervinca vulgaris. *Park.* 380.
Clematis daphnoides minor. *Bauh. pin.* 301.
Near Hampstead. In woods and hedges but rare Common in our gardens. P. May.
Varieties.
1. Blossoms white.
2. Blossoms double, blue.
3. Blossoms double, white.
4. Blossoms double, purple.
5. Leaves with yellow stripes; blossoms blue, or white; single or double.
6. Leaves with white stripes; blossoms blue, or white; single or double.

PERRY-

Greater PERRYWINKLE. The ftems nearly upright. Leaves egg-
Major   fhaped; flowers on fruit-ftalks—Cup *as long as the tube of the*
bloffom. Leaf-ftalks long. Bloffoms *blue; with a blufh of purple.*
Clematis daphnoides major. *Bauh. pin.* 302. *Ray's Syn.* 268.
Clematis daphnoides latifolia, feu vinca pervinca major.
Park. 380.
Clematis daphnoides, feu pervinca major. *Gerard.* 894.
In woods and hedges. Not uncommon in gardens. P. May.
It is bitter and flightly aftringent.

52 NIGHTSHADE. 251 Solanum.

EMPAL. *Cup* of one leaf; permanent: with five fhallow
clefts; fharp; upright.

BLOSS. One petal; wheel fhaped. *Tube* very fhort.
Border large: plaited; with five fhallow clefts; flat,
but turned backwards.

CHIVES. *Threads* five; awl-fhaped; fmall. *Tips* oblong;
approaching fo as to touch; with two open pores
at the end.

POINT. *Seedbud* roundifh. *Shaft* fimple, longer than the
chives. *Summit* blunt.

S. VESS. *Berry* roundifh, gloffy; with a hollow dot at
the end; and two cells. *Receptacle* convex on both
fides: flefhy.

SEEDS. Several, roundifh, difperfed among the pulp.

Woody   NIGHTSHADE. The fruit bearing ftems without prickles;
Dulcamara zigzag. The upper leaves halberd-fhaped; the flowers in tufted
bunches—*Its blue bloffoms are fometimes changed to flefh colour, or
white.* Berry *red.*
Solanum lignofum feu dulcamara. *Park.* 350. *Ray's Syn.* 265.
Solanum fcandens, feu dulcamara. *Bauh. pin.* 167.
Amara dulcis. *Gerard.* 350.
Bitter-fweet.
In moift brakes and hedges. P. June. July.
Varieties.
1. Bloffoms white.
2. Leaves variagated with white.
3. Leaves ftriped with yellow.
And Mr. Ray fays it hath a peculiar habit when it grows near
the fea.
Boerhaave fays it is a medicine far fuperior to China and Sar-
faparilla as a fweetner and reftorative. Linnæus fays an infufion
of the young twigs is an admirable medicine in acute Rheu-
matifms, Inflammations, Fevers and Suppreffion of the Lochia.
Dr. Hill fays he has found it very efficacious in the Afthma.—
Sheep and Goats eat it. Horfes, Cows and Swine refufe it.

NIGHT-

NIGHTSHADE. The stem herbaceous; without prickles; leaves egg shaped, toothed and angular. Flowers in bunches; nodding.—*Blossoms white. Berries black.* Garden Nigrum

Solanum vulgare. *Park.* 346. *Ray's Syn.* 264.
Solanum officinarum. *Bauh. pin.* 166.
Solanum hortense. *Gerard.* 339.
Common Nightshade.
Amongst rubbish. A. June. July.

From one to three grains of the leaves infused in boiling water, and taken at bed-time, occasions a copious perspiration; increases the secretion by the kidneys, and generally purges more or less the following day. These properties judiciously applied, render it capable of doing essential service in several diseases, as may be seen in Mr. Gattaker's Treatise on the Solanum. But its effects on the Nervous System are so uncertain, and sometimes so considerable, that it must be ever administered with the greatest caution. The leaves externally applied abate inflammation, and assuage pain. The flowers smell like musk.—Horses, Cows, Goats, Sheep and Swine refuse it.

83 DWALE. 249 Atropa.

EMPAL. *Cup* one leaf: permanent: with five divisions; hunched. *Segments* sharp.

BLOSS. One petal; bell-shaped. *Tube* very short. *Border* distended: egg-shaped; longer than the cup. *Mouth* small; with five clefts; open. *Segments* nearly equal.

CHIVES. *Threads* five; awl-shaped; springing from the base of the blossom, and as long as the blossom; approaching at the base but bowed outwards and diverging towards the top. *Tips* rather thick; rising.

POINT. *Seedbud* half egg-shaped. *Shaft* thread-shaped; inclining; as long as the chives. *Summit* a knob; transversely oblong: rising.

S. VESS. *Berry* of two cells; globular; sitting upon the cup, which enlarges. *Receptacle* fleshy; kidney-shaped; convex on both sides.

SEEDS. Numerous, kidney-shaped.

DWALE.

Deadly
Belladonna

DWALE. The ſtem herbaceous; the leaves egg-ſhaped; entire—*Hairy, ſoft*. Bloſſoms *dark purple*. Fruit *firſt red, afterwards black*.

Belladonna. *Ray' Syn.* 265.
Solanum lethale. *Park.* 346. *Gerard.* 340.
Solanum melanoceraſus. *Bauh. pin.* 166.
Deadly Nightſhade.

In woods, hedges, amongſt lime-ſtone and rubbiſh. P. June,
July.

The whole plant is poiſonous; and children allured by the beau-
ful appearance of the berries, have too often experienced their fatal effects. Buchanan, the Scotch Hiſtorian, gives an account of the
deſtruction of the army of Sweno, when he invaded Scotland,
by the juice of theſe berries being mixed with the drink which
the Scots, by their truce, engaged to ſupply them with. The
Danes became ſo intoxicated, that the Scots fell upon them
in their ſleep, and killed the greateſt part of them; ſo that there
were ſcarcely men enough left to carry off their king. Whatever credit is due to this ſtory, there is no doubt but thoſe who
eat the berries are attacked with Stupor or Delirium, and become variouſly convulſed, and that death is the certain conſequence, if not prevented by timely and plentiful vomiting, ſo
as to evacuate the poiſon. Mr. Ray ſays, that tumours of the
breaſts, even of the cancerous kind, are reſolved by a topical
application of the freſh leaves. There is no doubt but their
external application may be productive of good effects in ſeveral
caſes, but the following ſtory related by the ſame author, ſhews
us that their application is dangerous when the ſkin is broken.
A lady of quality who had a ſmall ulcer a little below one
of her eyes, which was ſuppoſed to be of a cancerous nature,
put a ſmall bit of the green leaf of this plant upon it. In the
morning the Uvea in that eye was ſo affected that the pupil
would not contract, even in the brighteſt light; whilſt the
other eye retained its uſual powers. The leaf being removed,
the eye was gradually reſtored to its former ſtate. This could
not be an accidental effect, for it was repeated three ſeparate
times, and the ſame circumſtances attended each application.
Ray's Hiſt. Plant. 680. In the Philoſ. Tranſ. vol. 50. p. 77,
there is a caſe of a woman cured of a Cancer in her breaſt by
taking a tea-cup full of an infuſion of the dried leaves every
morning. The complaint became worſe at firſt, but afterwards
the ſymptoms abated, and in ſix months ſhe was perfectly well.
The infuſion was made by pouring ten tea-cups of water upon
twenty grains of the dried leaves, letting it ſtand all night in a
warm place. The relation is very well atteſted. It hath ſince
been tried in our hoſpitals, with the effect of mitigating the
ſymptoms, but hardly ever perfecting a cure. Mr. Gattaker
gives the preference to the Garden Nightſhade in theſe caſes, as he
found

found that to increafe the fecretions, and produce fimilar good
effects, without affecting the nervous fyftem fo difagreeably as
the deadly Dwale generally does.—*See his Treatife on the Nightfhade.*

84 MARSHWORT. 222 Samolus.

EMPAL. *Cup* with five divifions; fuperior; blunt at the
bafe: permanent. *Segments* upright.

BLOSS. One petal; falver-fhaped. *Tube*, open; very
fhort; but as long as the cup. *Border* flat; with
five blunt divifions. *Valves* very fhort; approaching;
fixed to the bottom of the clefts in the border.

CHIVES. *Threads* five: fhort; one betwixt each fegment
of the bloffom. *Tips* approaching; covered.

POINT. *Seedbud* beneath. *Shaft* thread-fhaped; as long
as the chives. *Summit* a knob.

S VESS. *Capfule* egg-fhaped; of one cell, and five valves
opening half way down; bound round by the cup.

SEEDS. Many; egg-fhaped; fmall. *Receptacle* large; globular.

MARSHWORT. As there is only one fpecies known Linnæus Pimpernel
gives no defcription of it—*The* Leaves *oblong egg-fhaped; very* Valerandi
entire. Bloffoms *white.*

Samolus valerandi. *Ray's Syn.* 283.
Anagallis aquatica, rotundo folio non crenato. *Bauh. pin.* 252.
Anagallis aquatica rotundifolia. *Gerard.* 620.
Anagallis aquatica 3. lob. folio fubrotundo non crenato.
Park. 1237.
Round leaved Water Pimpernel.
In marfhes and moift meadows. P. June.
Cows, Goats and Sheep eat it. Horfes refufe it.

85 RAMPION. 220 Phyteuma.

EMPAL. *Cup* one leaf, with five divifions; fharp, not quite
upright; fuperior.

BLOSS. One petal; ftarry; expanding; with five divifions. *Segments* ftrap-fhaped; fharp; bent back.

CHIVES. *Threads* five; fhorter than the bloffom. *Tips*
oblong.

POINT. *Seedbud* beneath. *Shaft* thread-fhaped; as long
as the bloffom, bent backwards. *Summit* with three
divifions; oblong; rolled back.

S VESS. *Capfule* roundifh: cells three.

SEEDS. Several; fmall roundifh.

RAMPION.

Roundheaded
Orbicularis

RAMPION with flowers in roundish heads and serrated leaves.
Root leaves heart-shaped—*Blossoms purple; or bluish.*

Rapunculus corniculatus montanus. *Gerard.* 455. *Ray's Syn.* 278.

Rapunculus alopecuroides orbiculatus. *Park.* 648.

Rapunculus folio oblongo, spica orbiculari. *Bauh: pin.* 92.

Horned Rampions, with a round head or spike of flowers.

In dry pastures. On the Downs in Sussex. P. July. August.

86 BELLFLOWER. 218 Campanula.

EMPAL. *Cup* with five divisions; sharp; not quite upright; superior.

BLOSS. One petal; bell-shaped; close at the base: with five clefts; shrivelling. *Segments* broad; sharp, open. *Honey-cup* in the bottom of the blossom, composed of five sharp valves; approaching and covering the receptacle.

CHIVES. *Threads* five; hair-like; very short; growing upon the ends of the honey-cup valves. *Tips* compressed; longer than the threads.

POINT. *Seedbud* beneath; angular. *Shaft* thread-shaped; longer than the chives. *Summit* thick; oblong; with three divisions which are rolled backwards.

S. VESS. *Capsule* somewhat round; angular; of three or five cells, and letting out the seed at as many lateral holes.

SEEDS. Numerous; small. *Receptacle* resembling a pillar.

OBS. *The figure of the seed-vessel is different in different species.*

* *Leaves somewhat smooth and narrow.*

Mountain
Uniflora

BELLFLOWER. The stem, solitary; upright, supporting only one flower. Cup as large as the blossom—
On mountains in the North, and in Wales. P. July. August.

Round-leaved
Rotundifolia

BELLFLOWER. The root leaves kidney-shaped; stem leaves strap-shaped—*Blossom blue; or purple; sometimes white.*

Campanula rotundifolia. *Gerard.* 452. *Ray's Syn.* 277.

Campanula minor sylvestris rotundifolia. *Park.* 651.

Campanula minor rotundifolia vulgaris. *Bauh. pin.* 93.

Lesser round leaved Bellflower.

On heaths and barren pastures. P. August. October.

The juice of the petals stains blue, but with the addition of alum, green.

Cows, Horses, Goats and Sheep eat it. Swine refuse it.

BELL-

BELLFLOWER. The leaves ſtiff and ſtraight. Root-leaves Field
betwixt ſpear and egg-ſhaped; panicle expanding—*At the baſe* Patula
of each ſegment of the cup there is a little livid tooth. Bloſſom
blue.
In woods, hedges and cornfields. P. July. Auguſt.

BELLFLOWER. The leaves waved at the edges; root- Rampion
leaves betwixt ſpear-ſhaped and oval. Panicle compact—*Stem* Rapunculus
angular, rough. Fruit-ſtalks *generally growing by threes, and the*
middle one the longeſt. Bloſſoms *blue, or whitiſh.*
Rapunculus eſculentus. *Bauh. pin.* 92. *Ray's Syn.* 276.
Rapunculus eſculentus vulgaris. *Park.* 648.
Rapuntium parvum. *Gerard.* 453.
In cornfields, near Croyden in Surry; not common. P.
Auguſt.
The roots are eaten raw in ſallads, or boiled like ſparagus.
In gardens they are blanched.

* * *Leaves rough, and rather broad.*

BELLFLOWER. The leaves betwixt egg and ſpear-ſhaped; Broad-leaved
ſtem undivided; cylindrical. Flowers ſolitary; fruit-ſtalks Latifolia
nodding—*Cups ſmooth.* Leaves *ſerrated.* Bloſſoms *blue; ſome-*
times pale red; or white; or aſh-coloured.
Campanula maxima foliis latiſſimis. *Bauh. pin.* 94. *Ray's*
Syn. 276.
Trachelium majus Belgarum. *Park.* 643. ſive giganteum.
Gerard. 448.
Giant Throatwort.
In high grounds. P. July. Auguſt.
The beauty of its flowers frequently procures it a place in our
gardens. The whole plant abounds with a milky liquor.—
Horſes, Sheep and Goats eat it.

BELLFLOWER. The ſtem angular; lower leaves on leaf- Canterbury
ſtalks: cups fringed; fruit-ſtalks divided into three parts. Trachelium
—*Leaves betwixt egg and heart-ſhaped; toothed, and rough.*
Bloſſoms *blue; ſometimes pale red, or white.*
Campanula vulgatior foliis urticæ, vel major et aſperior.
Ray's Syn. 276. *Bauh. pin.* 94.
Trachelium majus. *Gerard.* 448. flore purpureo. *Park.* 644.
Great Throatwort or Canterbury Bells.
In woods and hedges rare. In gardens frequent. P. July.
Auguſt.
In gardens the bloſſoms frequently grow double, or triple.
The whole plant contains a dirty yellow juice.—Cows eat it.
Horſes and Goats refuſe it.

Throatwort
Glomerata

BELLFLOWER. The stem simple, angular. Flowers sitting, and forming a head at the end of the stem—*Leaves oblong egg shaped; sitting; rather blunt; the lower leaves on leaf-stalks. Blossoms blue; reddish; or white.*

Campanula pratensis flore conglomerato.   *Bauh. pin.* 94. *Ray's Syn.* 277.

Trachelium minus.   *Park.* 644. *Gerard.* 449.

Lesser Throatwort, or Canterbury Bells.

On chalk hills.   P. July.

*** *Capsules covered by the reflected segments of the cup.*

Venus
Hybrida

BELLFLOWER. The stem with straight and stiff branches at the base. Leaves scolloped and oblong. Cups incorporated and longer than the blossoms. Capsule prism-shaped—*Blossoms purple; deeply divided.*

Campanula arvensis erecta, vel speculum veneris minus. *Park.* 1331. *Gerard.* 439. *Ray's Syn.* 278.

Lesser Venus looking Glass.   Codded Corn Violet.

In cornfields, but not common.   A. June. July.

Ivy-leaved
Hederacea

BELLFLOWER. The leaves on leaf-stalks; smooth; heart-shaped; with five lobes. Stem flexible—*Blossoms blue. Is not this plant derived from the seedbud of one of the Bellflowers fertilized by the dust of the Ivy-leaved Speedwell?*

Campanula cymbalariæ foliis.   *Gerard.* 452. *Park.* 652. *Ray's Syn.* 277.

Campanula cymbalariæ foliis vel folio hederaceo. *Bauh. pin.* 93.

In moist shady places.   P. May. August.

The Sword-grass Moth, *Phalæna exsoleta*, feeds upon the different species of Bellflower.

87 HONEYSUCKLE. 233 Lonicera.

EMPAL. *Cup* superior; with five divisions, small.

BLOSS. One petal; tubular. *Tube* oblong; hunched. Border with five divisions. *Segments* rolled backwards, one segment more deeply separated than the others.

CHIVES. *Threads* five; awl-shaped; nearly as long as the blossom. *Tips* oblong.

POINT. *Seedbud* beneath: roundish. *Shaft* thread-shaped; as long as the blossom. *Summit* a blunt knob.

S.VESS. Berry with two cells, and crowned with the cup at the top.

SEEDS. Nearly round, compressed.

OBS. *The* Woodbine Honeysuckle *hath the segments of the blossom cut nearly to an equal depth and the berries distinct. The* upright Honeysuckle *hath the segments of the blossom cut nearly to an equal depth, and two berries sitting upon the same base.*

HONEYSUCKLE with flowers in egg-shaped, terminating, tiled knobs. Leaves distinct—*Blossoms red on the outside; yellowish within.* Berries *red.* *(Woodbine Periclymenum)*

Caprifolium Germanicum. *Ray's Syn.* 458.
Periclymenum. *Gerard.* 891.
Peryclymenum, five caprifolium vulgare. *Park.* 1460.
Periclymenum non perfoliatum Germanicum. *Bauh. pin.* 302.

In hedges. S. May—July.

The varieties are
1. Blossoms white.
2. Leaves with yellow stripes.
3. Leaves indented like those of oak.
4. Leaves indented and variegated.

The beauty and fragrance of its flowers make it a welcome guest in our gardens, hedges and arbours. Cows, Goats and Sheep eat it Horses refuse it.

HONEY.

Upright
Xylosteum

HONEYSUCKLE with fruit-stalks supporting two flowers; berries distinct, leaves very entire; downy—*Blossoms white. Berries red.*

Chamæcerasus dumetorum, fructu gemino rubro. *Bauh. pin.* 451.

Upright Alpine Honeysuckle.

Under the Roman Wall, on the west side of Shewing Sheels in Northumberland. *Wallis.* p. 149. S. July.

In very dry soils it makes good hedges—The clear parts betwixt the joints of the shoots are used in Sweden as tubes for tobacco pipes—The wood is very hard, and makes excellent teeth for rakes. Goats and Sheep eat it. Cows and Horses refuse it.

The insects that have been observed to feed upon the Honeysuckles are the Privet Hawk Moth, *Sphinx Ligustri,* the brown feathered Moth, *Phalæna didactyla,* the Small Bee Moth, *Sphinx tipuliformis,* and the many feathered Moth, *Phalæna Hexadactyla.*

88 BUCKTHORN. 265 Rhamnus.

EMPAL. *Cup* none; except you call the blossom the cup.

BLOSS. One petal; funnel-shaped; closed at the base. Rough outwardly, but coloured within. *Tube* cylindrical; turban-shaped. *Border* expanding; divided; sharp. Five small *Scales,* one at the base of each division of the blossom, approaching inwards.

CHIVES. *Threads* as many as there are segments in the blossom; awl shaped; growing upon the blossom under the scale. *Tips* small.

POINT. *Seedbud* roundish. *Shaft* thread-shaped; as long as the chives. *Summit* blunt; divided into fewer segments than the blossom.

S.VESS. *Berry* roundish, naked, divided into fewer cells than the blossom hath segments.

SEEDS. Solitary; roundish; hunched on one side, compressed on the other.

OBS. *In the* Alder Buckthorn, the Summit *is notched; the* Berry *hath four seeds, and the blossom five clefts. In the* Purging Buckthorn *the* Summit *hath four clefts, the* Berry *hath four seeds, the* Blossom *four clefts, and the chives and pointal are upon different plants.*

BUCKTHORN. The thorns terminating. The leaves egg- Purging
shaped; the stem upright. Blossoms with four segments; the Catharticus
chives four. The chives and pointal upon distinct plants.—
Blossoms pale green. Berries *black.*

Rhamnus catharticus. *Bauh. pin.* 478. *Ray's Syn.* 466.
Rhamnus solutivus. *Gerard.* 1337. Sive spina infectoria
vulgaris. *Park.* 243.
In woods and hedges. S. April. May.

A purgative syrup is prepared from the berries and kept in the
shops. about an ounce of it is a moderate dose; but it generally
occasions so much sickness and griping that it is falling into dif-
ufe---the flesh of birds that feed upon the berries is said to be
purgative—the juice of the unripe berries is the colour of saffron,
and is used for staining maps or paper. These are sold under
the name of French Berries—The juice of the ripe berries mixed
with alum, is the sap green of the painters; but if they are gathered late in the autumn the juice is purple—the bark affords a
beautiful yellow dye. Goats, Sheep and Horses eat it. Cows
refuse it.

BUCKTHORN without thorns. Leaves very entire. Alder
Chives and pointals in the same flower---*The inner bark is yellow; the* Frangula
outer sea-green, and the middle bark red as blood.

Frangula, seu alnus nigra baccifera. *Park.* 240. *Ray's Syn.*
465.
Alnus nigra bacifera. *Bauh. pin.* 428.
Alnus nigra five frangula. *Gerard.* 1469.
Black berry bearing alder.
Woods and wet hedges. S. April. May.

From a quarter to half an ounce of the inner bark, boiled in
small beer, is a sharp purge. In dropsies, or constipations of
the bowels of cattle, it is a very certain purgative—The berries
gathered before they are ripe, dye wool green. The bark dyes
yellow. Charcoal prepared from the wood is preferred by the
makers of gunpowder—The flowers are particularly grateful to
Bees. Goats devour the leaves voraciously, and Sheep will eat
them.

The Brimstone Butter Fly, *Papilio Rhamni,* and the Blue
Argus, *Papilio Argus,* live upon both the species of Buckthorn.

FIVE CHIVES.

89 SPINDLE. 271 Evonymus.

EMPAL. *Cup* one leaf, with five divisions; flat. *Segments* roundish; concave.

BLOSS. *Petals* five; egg shaped; flat; expanding; longer than the cup.

CHIVES. *Threads* five; awl-shaped; upright; shorter than the blossom; standing upon the seedbud. *Tips* double.

POINT. *Seedbud* tapering to a point. *Shaft* short; simple. *Summit* blunt.

S.VESS. *Capsule* succulent; coloured; with five sides; five angles; five cells and five valves,

SEED. Solitary; egg-shaped; inclosed in a seedcoat, not much unlike a berry.

OBS. *In some species and in some individual flowers there are only four petals; four chives, &c.*

Gatteridge
Europæus

SPINDLE. Blossoms for the most part with only four petals —*greenish white.* Fruit *angular, purplish; sometimes white.*
Evonymus vulgaris. *Park.* 241. *Ray's Syn.* 468.
Evonymus vulgaris granis rubentibus. *Bauh. pin.* 428.
Evonymus Theophrasti. *Gerard.* 1468.
Spindle-tree. Prickwood. Gatteridge-tree. Louse Berry.
In woods and hedges. S. April — May.
The berries vomit and purge violently. They are fatal to Sheep. In powder and sprinkled upon the hair they destroy lice---If the wood is cut when the plant is in blossom, it is tough and not easily broken. and in that state is used by watch-makers for cleaning watches, and to make skewers and toothpickers. Cows, Goats and Sheep eat it. Horses refuse it.
The small Ermine Moth, *Phalæna Evonymella*. lives upon this shrub.

90 C U R R A N T. 281 Ribes.

EMPAL. *Cup* of one permanent leaf, with five shallow clefts; distended. *Segments* oblong; concave; coloured;. reflected.

BLOSS. *Petals* five; small; blunt; upright; growing to the rim of the cup.

CHIVES. *Threads* five; awl-shaped; upright; standing on the cup. *Tips* fixed side-ways to the threads; compressed; opening at the edges.

POINT. *Seedbud* beneath: roundish. *Shaft* cloven. *Summits* blunt.

S.VESS. *Berry*, globular of one cell; dimpled. *Receptacles* two; longitudinal, lateral, opposite.

SEEDS. Several; roundish; somewhat compressed.

CURRANT without prickles; flowers rather flat; in smooth Red pendant bunches—*Blossoms greenish white.* Berries red, or white. Rubrum
Ribes vulgaris fructu rubro. *Gerard.* 1593. *Ray's Syn.* 456.
Ribes fructu rubro. *Park.* 1561.
Grossularia sylvestris rubra. *Bauh. pin.* 455.
Currants.
In woods in the northern counties. Very common in gardens.
S. May.
Variety.
1. Ribes vulgaris fructu dulci. *Ray's Syn.* 456.
Grossularia vulgaris fructu dulci. *Bauh. pin.* 455.
Sweet Currants.
In woods in Yorkshire and Leicestershire.
2. Ribes fructu parvo. *Ray's Syn.* 456.
Small Currants.
In Wimbleton Park, Surry; and many places in Lancashire.
3. Gooseberry leaved Currant.
4. Variegated leaves.
    The fruit is universally acceptable, either as nature presents it, or made into jelly. The juice is a most agreeable acid in punch. The white fruit is sweeter than the red. If an equal weight of picked Currants and pure sugar is put over the fire, the liquor that separates spontaneously is a most agreeable jelly. Cows, Goats and Sheep eat the leaves. Horses are not fond of them.
    This plant is very apt to be infested by the Currant Louse, *Aphis Ribes*, and then the green leaves become red, pitted and hunched.

Mountain
Alpinum

CURRANT without prickles; bunches upright; floral leaves longer than the flowers—

Ribes alpinus dulcis. *Ray's Syn.* 456.

Sweet Mountain Currants.

In dry hedges in Yorkſhire. S. May.

The fruit hath a flat ſweetiſh taſte, and is only agreeable to children. The wood being hard and tough makes good teeth for rakes. Cows, Goats, Sheep and Horſes eat it.

Black
Nigrum

CURRANT without prickles; flowers oblong, in hairy bunches—*Buds glandular*; Floral Leaves *woolly, and as long as the fruit-ſtalks.* Leaf-ſtalks *a little woolly and glandular.* Leaves *glandular.* Glands *very ſmall; yellow.*

Ribes nigrum vulgo dictum folio olente. *Ray's Syn.* 456.

Ribes fructu nigro. *Park.* 1562. *Gerard.* 1593.

Groſſularia non ſpinoſa fructu nigro. *Bauh. pin.* 455.

Squinancy Berries.

Wet hedges and banks of rivers. S. May.

The berries have a very peculiar flavour that many people diſlike; but their juice is frequently boiled down into an extract, with the addition of a ſmall proportion of ſugar: in this ſtate it is called Rob; and is much uſed in ſore throats, but chiefly in thoſe of the inflammatory kind. Some people put them into brandy, for the ſame purpoſe that other people uſe black cherries. The tender leaves tinge common ſpirits, ſo as to reſemble brandy. An infuſion of the young roots is uſeful in fevers of the eruptive kind; and in the dyſenteric fevers of cattle. Goats and Horſes eat the leaves.

All the ſpecies are eaten by the Magpye or Currant Moth, *Phalæna groſſularia.*

91 IVY. 283 Hedera.

EMPAL. *Fence* of the simple rundle very small; with many teeth. *Cup* very small, with five teeth; binding round the seedbud.

BLOSS. *Petals* five; oblong; expanding; bent inwards at the points.

CHIVES. *Threads* five; awl-shaped; upright; as long as the blossom. *Tips* fixed side-ways to the threads; forked at the base.

POINT. *Seedbud* turban-shaped; bound round by the *Cup*. *Shaft* simple; very short. *Summit* simple.

S. VESS. *Berry* globular; with one cell.

SEEDS. Five; large; hunched on one side; angular on the other.

IVY with some leaves egg-shaped; and others divided into lobes—*Glossy*. Blossoms *greenish white*. Berries *black*. Common Helix

Hedera arborea. *Bauh. pin.* 305.

Hedera arborea five scandens et corymbosa communis. *Park* 678.

Hedera helix. *Gerard.* 858.

In woods and hedges. S. October.

1. The leaves are sometimes variegated with white or yellow.

The roots are used by Leather-cutters to whet their knives upon.—Its ever-green leaves adorn our walls and cover the naked trunks of trees.—Apricots and Peaches covered with Ivy during the month of February, have been observed to bear fruit plentifully. *Philof. Tranf.* No. 475.—The leaves have a nauseous taste. Haller says they are given in Germany as a specific in the Atrophy of children. Common people apply them to Issues. The berries have a little acidity. In warm climates a resinous juice exsudes from the stalks.—Horses and Sheep eat it. Goats and Cows refuse it.

92 KNOTGRASS. 290 Illecebrum.

EMPAL. *Cup* of five leaves and five angles.. The *Leaves* coloured; tapering; diflant at the points: permanent.

BLOSS. None.

CHIVES. *Threads* five; hair-like; within the cup. *Tips* fimple.

POINT. *Seedbud* egg-fhaped; fharp; ending in a fhort cloven *Shaft. Summit* fimple; blunt.

S. VESS. *Capfule* roundifh; tapering at each end; with five valves, and one cell: covered by the cup.

SEED. Single; very large; fomewhat round but fharp at each end.

Whorled Verticillatum
KNOTGRASS with flowers in naked whorls, and trailing ftems—*Bloffoms white.*
Polygonum ferpyllifolium verticillatum. *Ray's Syn.* 160.
Polygala repens. *Park.* 1333.
Polygala repens nivea. *Bauh. pin.* 215.
Verticillate Knotgrafs.
In wet paftures in Cornwall. P. July.

93 SALTWORT. 291. Glaux.

EMPAL. None; unlefs you call the bloffom the cup.

BLOSS. Petal fingle; upright; bell-fhaped; permanent: with five blunt *Segments,* rolled back.

CHIVES. *Threads* five; awl-fhaped; upright; as long as the bloffom. *Tips* roundifh.

POINT. *Seedbud* egg-fhaped. *Shaft* thread-fhaped: as long as the chives. *Summit* a knob.

S. VESS. *Capfule* globular; tapering; of one cell and five valves.

SEEDS. Five; roundifh. *Receptacle* large; globular; with hollows where the feeds lie.

SALTWORT.

SALTWORT. As there is only one known species Linnæus Sea
gives no defcription of it.—*The* Stems *trailing; jointed.* Leaves Maritima
fitting. Bofloms *at the bafe of the leaves; purple; fometimes
greenifh white; or white; or ftriped.*

Glaux maritima. *Bauh pin.* 215 *Ray's Syn.* 285.
Glaux maritima minor. *Park.* 1283.
Glaux exigua maritima. *Gerard.* 562.
Sea Milkwort. Black Saltwort.
On the fea coaft. P. July.
Cows eat it.

94 FLUELLIN. 292 Thefium.

EMPAL. *Cup* one leaf; permanent; turban-fhaped; with
five fhallow clefts. *Segments* half fpear-fhaped; upright; blunt.

BLOSS. None; unlefs you call the *Cup* a bloffom, from
its being coloured on the infide.

CHIVES. *Threads* five; awl-fhaped; inferted at the bafe
of the fegments of the cup; fhorter than the cup.
Tips roundifh.

POINT. *Seedbud* beneath; at the bottom of the cup.
Shaft thread-fhaped: as long as the chives. *Summit*
rather thick and blunt.

S. VESS. None. The *Cup* contains the feed in its bottom without opening.

SEED. Single; fomewhat round; covered by the clofing
cup.

FLUELLIN. The panicle leafy; leaves ftrap-fhaped.—*Blof-* Flax-leaved
foms white. Linophyllon

Linaria adulterina. *Gerard.* 555. *Ray's Syn.* 202.
Linaria montana flofculis albicantibus. *Bauh pin.* 213.
Pfeudo linaria montana alba. *Park.* 459.
Baftard Fluellin.
In high paftures. P. June.—July.

95 FELWORT.

Order II. Two Pointals.

95 FELWORT. 321 Swertia.

EMPAL. *Cup* with five divisions; flat; permanent. *Segments* spear-shaped.

BLOSS. *Petal* single. *Tube* none. *Border* flat; with five divisions. *Segments* spear shaped, larger than the cup; connected by the claws. *Honey-cups* five; each of them like two hollow dots in the inner side of the base of each segment; encompassed with small upright bristles.

CHIVES. *Threads* five; awl-shaped, not quite upright; shorter than the blossom. *Tips* fixed side-ways to the threads.

POINT. *Seedbud* oblong; egg-shaped. *Shaft* none. *Summits* two; simple.

S. VESS. *Capsule* cylindrical; tapering at each end; with one cell, and two valves.

SEEDS. Numerous; small.

Marsh   FELWORT. The blossoms with five segments: the root
Perennis   leaves oval.—
    Gentiana palustris latifolia. *Bauh. pin.* 188.
    Gentiana pennei minor. *Gerard.* 433.
    Marsh Gentian.
    In Wales. P. August.

96 GENTIAN.

96 GENTIAN. 322 Gentiana.

EMPAL. *Cup* with five divisions; sharp; permanent. *Segments* oblong.

BLOSS. Petal single; tubular in the lower part; tube close. The upper part with five clefts; flat, shrivelling: variously shaped.

CHIVES. *Threads* five; awl-shaped; shorter than the blossom. *Tips* simple.

POINT. *Seedbud* oblong; cylindrical; as long as the chives. *Shafts* none. *Summits* two; egg-shaped.

S. VESS. *Capsule* oblong; cylindrical; tapering; slightly cloven at the end: of one cell and two valves.

SEEDS. Numerous; small. *Receptacles* two, each growing length-ways to a valve.

OBS. *The figure of the fruit is constant; but the flowers vary in different species both as to the number and shape of the parts. In one species the neck of the blossom is open, in another it is closed with soft hairs. In some the segments of the blossom are fringed; in others the border is bell-shaped, upright and plaited. Some have a starry appearance, with small segments betwixt the larger; others are funnel-shaped, &c.*

GENTIAN. The blossoms with five clefts; bell-shaped; Calathian opposite; on fruit-stalks. Leaves strap shaped—*Blossoms blue.* Pneumonanthe
Gentiana palustris angustifolia. *Bauh. pin.* 188. *Ray's Syn.* 274.
Gentiana autumnalis pneumonanthe dicta. *Park.* 406.
Pneumonanthe. *Gerard.* 438.
Calathian Violet.
In moist pastures. P. August.

GENTIAN. The blossoms with five clefts; funnel-shaped. Centaury
Stem forked; pointal simple—*Blossoms pale red; sometimes white.* Centaurium
Centaurium minus. *Bauh. pin.* 278. *Ray's Syn.* 286.
Centaurium minus vulgare. *Park.* 272.
Centaurium parvum. *Gerard.* 547.
Lesser Centory.
In dry barren pastures. A. June—August.
This plant is extremely bitter. It is the basis of the famous Portland Powder, which cures the Gout, when taken in a large quantity, and a long time together; but it brings on Schirrosity of the Liver, Palsy and Apoplexy. A tincture of the leaves and the upper part of the root is a good medicine in Weak Stomachs and Cachectic Habits. A decoction of the whole plant destroys Lice, and cures the Itch.—Cows are not fond of it.

GENTIAN.

Autumnal
Amarella

GENTIAN. The bloffoms with five clefts; falver-fhaped; bearded at the mouth.—*Blossoms blue.*
Gentianella pratenfis flore lanuginofo. *Bauh. pin.* 188. *Ray's Syn.* 275.
Gentianella fugax autumnalis elatior, centaurii minoris foliis. *Ray's Syn.* 275.
Gentianella fugax minor. *Gerard.* 437.
Fellwort.
Dry paftures. A. July. Auguft.
Sheep eat it. Horfes refufe it.

Dwarf
Campeftris

GENTIAN. The bloffom with four clefts; mouth bearded; —*Stem upright*; Bloffom *blue*; *terminating*; *woolly within. Root leaves fpear-fhaped*; *lying on the ground.*; Chives *four.*
Gentianella fugax verna feu præcox. *Ray's Syn.* 275.
Gentianella alpina verna minor. *Bauh. pin.* 188.
Vernal Dwarf Gentian.
Mountainous paftures. A. July.
Poor people fometimes ufe it inftead of Hops.

Marfh
Filiformis

GENTIAN. The bloffoms with four clefts without any beard. Stem thread-fhaped; forked.—*Blossoms yellow, funnel-shaped; on long fruit-stalks.* Chives *four.*
Centaurium paluftre luteum minimum. *Ray's Syn.* 286.
Marfh Centory.
In marfhes in Cornwall. A. July.

97 KELPWORT. 311 Salfola.

EMPAL. *Cup* of five leaves; egg-fhaped; concave; permanent.
BLOSS. None. Unlefs you call the cup the bloffom.
CHIVES. Threads five; very fhort; ftanding upon the leaves of the cup.
POINT. *Sedbud* globular. *Shaft* fhort; with two or three divifions. *Summits* bent back.
S. VESS. *Capfule* egg-fhaped; of one cell; lapped up in the cup.
SEEDS. Single; very large; fpiral like a fnail fhell.

KELPWORT. Herbaceous; drooping. Leaves rough; awl- Prickly
fhaped; ending in a thorn. Cups at the bafe of the leaves; Kali
with leafy borders—*Bloffoms greenifh.*

Kali fpinofum cochleatum. *Bauh. pin.* 289. *Ray's Syn.* 159.
Tragus feu tragum matthioli. *Park.* 1034.
Tragon Matthioli, feu potius tragus improbus Matthioli.
Gerard 959.
Prickly Glafs-wort.
Cows, Horfes, Goats, Sheep and Swine refufe it.

KELPWORT. An upright fhrub; with thread-fhaped blunt Upright
leaves.— Fruticofa
Blitum fruticofum maritimum, vermicularis frutex dictum.
Ray's Syn. 156.
Sedum minus fruticofum. *Bauh. pin.* 284.
Vermicularis frutex minor. *Gerard.* 523.
Vermicularis fruticofa altera. *Park.* 731.
Chenopodium fruticofum. *Hudfon.* 93.
Shrub Stone-crop or Glafs-wort.
On the fea-fhore. S. Auguft.

98 BLITE. 309 Chenopodium.

EMPAL. *Cup* five leaves; concave; permanent. Leaves
egg-fhaped; concave; membranaceous at the edges.
BLOSS. None.
CHIVES. *Threads* five : awl-fhaped; as long as the leaves
of the cup and ftanding oppofite to them. *Tips*
roundifh; double.
POINT. *Seedbud* round and flat. *Shaft* fhort; divided.
Summits blunt.
S. VESS. None. The *Cup* clofes upon the feed; hath
five fides; five compreffed angles, and falls off when
the feeds are ripe.
SEED. Single; round; depreffed.

OBS. *In fome fpecies the* Shaft *hath three clefts.*

* Leaves angular.

Mercury
Bonus Henricus

BLITE. The leaves triangular;' nearly arrow-shaped; very entire. Flowers in compound spikes, not leafy; growing at the base of the leaves.—*Little spikes alternate; sitting. Flowers congregated; sitting.* Leaves *waved at the edge; the ribs of the leaves sprinkled with a soapy mealiness.* Blossoms *greenish white.*

Blitum perenne, bonus henricus dictum. *Ray's Syn.* 156.
Bonus henricus. *Gerard.* 329.
Lapathum unctuosum. *Park.* 1225.
Lapathum Sylvestre latifolium; seu unctuosum folio triangulo. *Bauh. pin.* 115.
Common English Mercury. All-good.
Among rubbish : on road-sides. P. August.

The young shoots peeled and boiled, may be eaten as sparagus, which they resemble in flavour. They are gently laxative. The leaves are often boiled in broth. The roots are given to Sheep that have a Cough.—Goats and Sheep are not fond of it. Cows, Horses and Swine refuse it.

Upright
Urbicum

BLITE. The leaves triangular; a little toothed : flowers crowded in very stiff, straight, and long bunches lying close to the stem—*Blossoms pale green.*
Chenopodium erectum, foliis triangularibus dentatis, spicis e foliorum alis plurimis, longis, erectis, tenuibus. *Ray's Syn.* 155. Among rubbish. A. August. Sept.
Goats and Sheep eat it. Horses and Cows refuse it.

Goosefoot
Rubrum

BLITE. The leaves triangularly heart-shaped; toothed; rather blunt. Bunches of flowers upright; compound; somewhat leafy; shorter than the stem. *Spikes congregated, sitting; with little strap-shaped leaves intervening : red when ripe.* Leaves *thick; shining.* Stems *drooping and pressed down to the ground.*
Blitum pes anserinus, acutiore folio. *Ray's Syn.* 154.
Atriplex sylvestris latifolia acutiore folio. *Bauh. pin.* 119.
Atriplex sylvestris latifolia altera. *Gerard.* 328.
Sharp-leaved Goose-foot.
On dung-hills and among rubbish. A. August.

Sowbane
Murale

BLITE. The leaves egg-shaped; toothed; shining; sharp. Bunches of flowers branching; naked—*Stem upright;* Fruit *in a sort of panicle; green or reddish.*
Blitum pes anserinus dictum. *Ray's Syn.* 154.
Atriplex sylvestris latifolia. *Bauh. pin.* 119.
Pes anserinus. *Gerard.* 528. *Park.* 749.
1. There is a variety with trailing stems and indented leaves.
Common Goose-foot, or Sow Bane.
In gardens and amongst rubbish. A. August.
Cows eat it; but it is said to be poisonous to Swine.

BLITE.

BLITE. The leaves triangularly spear-shaped; indented and toothed; wrinkled; smooth; uniform. Bunches of flowers terminating—*Stem five or six feet high ; very much branched ; flowering late in the year.* Leaves *pale green ; resembling those of the following species but broader.* Autumnal Serotinum
Blitum ficus folio. *Ray's Syn.* 155.
Late Flowered Blite.
On dung-hills and amongst rubbish. A. August—September.

BLITE. The leaves betwixt triangular and diamond-shaped; gnawed at the edges ; entire at the base. Upper leaves oblong; bunches of flowers upright—*Little spikes alternate ; sitting ; crowded. The whole plant is white when fully grown.* Seed *black.* Orach Album
Blitum atriplex sylvestre dictum. *Ray's Syn.* 154.
Atriplex folio sinuato candicante. *Bauh. pin.* 119.
Atriplex sylvestris vulgatior sinuata major. *Park.* 748.
Atriplex vulgaris. *Gerard.* 326.
1. There is a variety with a roundish leaf. *Ray's Syn.* 155.
Common Orache.
Gardens, dung-hills. Amongst rubbish. A. August.
Cows, Goats, and Sheep eat it. Horses refuse it. Swine are extremely fond of it.

BLITE. The leaves diamond-shaped ; indented and toothed. Flowers in branching and somewhat leafy bunches.— *Stem upright ; green ; but rather purplish at the angles.* Branch Leaves *spear-shaped ; very entire ; with one or two teeth. The* Cup *of the fruit with five elegantly sharp angles.* Green Viride
Chenopodium foliis integris racemosum. *Ray's Syn.* 155.
1. There is a variety with a thicker and blunter leaf. *Ray's Syn.* 156.
In gardens and cultivated places very common. A. July—August.
Goats, Sheep and Swine eat it.

BLITE. The leaves heart-shaped, but tapering at the angles. Flowers in branching naked bunches.—*Did not this plant originate from the seedbuds of the* Green Blite, *impregnated by the dust of the* White-flowered Thorn-apple ? Maple leaved Hybridum
Chenopodium stramonii folio. *Ray's Syn.* 154.
Common on dung-hills and amongst rubbish. A. August.
Cows and sheep eat it. Horses, Goats and Swine refuse it.

Oak-leaved
Glaueum

BLITE. The leaves oblong; egg-shaped serpentine at the edges. Flowers in naked, simple, congregated bunches.—
Chenopodium angustifolium laciniatum minus. *Ray's Syn.* 155.
Amongst rubbish. A. Auguft.
Cows and Horses eat it.

* * *Leaves entire.*

Stinking
Vulvaria

BLITE. The leaves very entire; betwixt diamond and egg-shaped. Flowers congregated; at the base of the leaves.
Blitum fœtidum vulvaria dictum. *Ray's Syn.* 156.
Atriplex olida. *Gerard.* 327.
Atriplex olida, five sylveftris fœtida. *Park.* 749.
Atriplex fœtida. *Bauh. pin.* 119.
Stinking Orache.
Road-fides, old walls. A. Auguft.
From its rank fœtid smell it hath got the reputation of being an Antihyfteric.—Cows Horses, Goats and Sheep eat it. Swine refuse it. The Sword-grass Moth, *Phalæna Exfoleta*, is found upon this species.

All-feed
Polyfpermum

BLITE. The leaves very entire; egg-shaped; Stem trailing. Flowers in forked naked tufts, at the base of the leaves.
Chenopodium betœ folio. *Ray's Syn.* 157.
Blytum Polyfpermum. *Bauh. pin.* 118. *Park.* 753.
Round Leaved Blite.
Gardens. Dunghills. A.
Cows and Sheep eat it. Goats and Horses refuse it. It is a most grateful food to fish.

Sea
Maritimum

BLITE. The leaves awl-shaped, semi-cylindrical.—*Seeds gloffy.*
Blitum Kali minus album dictum. *Ray's Syn.* 156.
Kali minus. *Gerard.* 535. album. *Park.* 749.
Kali minus album femine fplendente. *Bauh. pin.* 289.
White Glafs-wort.
Common on the fea-fhore. A. Auguft.
This is an excellent pot-herb.—The Spotted Buff-moth, *Phalæna Lubricipeda*, feeds upon moft of the fpecies of Blite.

99 BEET. 310 Beta.

EMPAL. *Cup* of five leaves; concave; permanent; *Leaves* oblong egg-shaped; blunt.

BLOSS. None.

CHIVES. *Threads* five; awl-shaped; as long as the leaves of the cup and standing opposite to them. *Tips* nearly round.

POINT. *Seedbud* in a manner below the receptacle. *Shafts* two; very short; upright. *Summits* acute.

S. VESS. *Capsule* in the bottom of the cup; of one cell; soon falling off.

SEED. Single; kidney-shaped; compressed; lying in the substance of the base of the cup.

BEET. The flowers collected into little balls. Leaves of the cup toothed at the base—*Stem upright; leaves horizontal.* Root red. Common Vulgaris

Beta rubra vulgaris. *Bauh. pin.* 118.

Beta alba; *Gerard.* 318.

Beta communis, feu viridis. *Ray's Syn.* 157.

In gardens, and on the sea-coast. About Nottingham plentifully. B. August.

The Gardiners produce several varieties by cultivation, viz. the yellow, the green, the red, and the turnip-rooted red.—The root of the red Beet is used as a pickle, and is sometimes eaten boiled. It is also employed to improve the colour of Claret. The white Beet is a pot-herb, but it is now almost banished from our tables, being too insipid, and is never used at all but in conjunction with Onions and other savory vegetables. The juice or powder of the root taken up the nostrils, excites sneezing, and occasions a considerable discharge of mucus.—Mr. Margraff found that a good sugar may be obtained from the juice of the fresh roots by the methods practiced abroad for preparing it from the Sugar-cane.

BEET. With flowers in pairs—*Stems drooping. This differs from the common Beet in flowering the first year; in the leaves growing oblique or vertical; in the leaves of the cup being equal, and not toothed.* Sea Maritima

Beta sylvestris maritima. *Bauh. pin.* 157. *Park.* 750.

Beta rubra. *Gerard.* 318.

On the sea-coast, common. B. August.

100 RUPTUREWORT. 308 Herniaria.

EMPAL. *Cup* of one leaf; with five divisions; sharp; expanding; coloured within; permanent.

BLOSS. None.

CHIVES. *Threads* five; awl shaped; small; within the segments of the cup. *Tips* simple. There are five other threads without tips, alternating with the segments of the cup.

POINT. *Seedbud* egg shaped. *Shaft* hardly any. *Summits* two; tapering; as long as the shaft.

S. VESS. *Capsule* small; at the bottom of the cup; covered; scarcely opening.

SEED. Solitary; egg-shaped but tapering; shining.

OBS. *In one of the foreign species there are only four chives and four segments in the cup.*

Smooth
Glabra
RUPTUREWORT. Smooth—*Blossoms yellowish.*
Herniaria. *Gerard.* 569. *Ray's Syn.* 160.
Polygonum minus seu millegrana major. *Bauh. pin.* 281.
Millegrana major, seu Herniaria vulgaris. *Park.* 446.
In gravelly soil. A. July.
It is a little saltish and astringent. It increases the secretions by the kidneys. The juice takes away Specks in the Eye.—Cows, Sheep and Horses eat it. Goats and Swine refuse it.

Rough
Hirsuta
RUPTUREWORT. Rough with hair—*Blossoms greenish yellow.*
Herniaria hirsuta. *Ray's Syn.* 161.
In gravelly soil; not common. A. July—August.

Sea
Lenticulata
RUPTUREWORT. With some small roundish leaves; and flowers in terminating bunches—*Greenish white.*
Polygonum maritimum longius radicatum nostras serpyllifolio circinato crasso nitente *Ray's Syn.* 161.
Polygonum minus lentifolium. *Bauh. pin.* 282.
Polygonum minus monspeliense. *Park.* 446.
On the sea-shore. P. August.

101 ELM. 316 Ulmus.

EMPAL. *Cup* one leaf; turban-fhaped; wrinkled; permanent. *Border* with five clefts; upright; coloured on the infide.

BLOSS. None.

CHIVES. *Threads* five; awl-fhaped; twice as long as the cup. *Tips* with four furrows; upright; fhort.

POINT. *Seedbud* round and flat; upright. *Shafts* two; reflected; fhorter than the chives; *Summits* downy.

S. VESS. *Berry* oval; large; not pulpy; compreffed and membranaceous.

SEED. Single; fomewhat globular, but a little compreffed.

ELM. With doubly ferrated leaves; unequal at the bafe— Common *The flowering buds are beneath the leafy buds.* Flowers *in very fhort* Campeftris *broad-topped fpikes.* Bark *of the trunk cracked and wrinkled.*

Ulmus vulgatiffima folio lato fcabro. *Gerard.* 1480. *Ray's Syn.* 468.

Ulmus campeftris et Theophrafti. *Bauh. pin.* 426.

1. Leaf broad and fmooth. Bark fmooth.

Ulmus glabra. *Hudfon.* 95. *Gerard.* 1481. *Park.* 1404. *Ray's Syn.* 469.

Broad-leaved Elm. 1. Wych Hazel.

In hedges. S. April.

1. The leaves of both are fubject to be broader or narrower and variegated with white or yellow ftripes.

A decoction of the inner bark drank freely hath been known to carry off the water in Dropfies.—It cures the Lepra icthyofis Sauvagefii. *Lettfom's Medical Memoirs.* Sect. 3.—The bark dried and ground to powder, hath been mixed with meal in Norway to make bread, in times of fcarcity.—The flowers have a violet fmell.—The wood being hard and tough, is ufed to make axle-trees, mill-wheels, keels of boats, chairs and coffins.—The tree is beautiful, and well adapted to make fhady walks, and it does not deftroy the grafs, and its leaves, are acceptable to Cows, Horfes, Goats, Sheep and Swine; for this purpofe it fhould be grafted upon the Wych Hazel, and then the roots will not fend out fuckers, which the common Elm is very apt to do, and give a great deal of trouble to keep the ground clear of them.—It loves an open fituation, and a black or clayey foil. It bears to be tranfplanted. The great Tortoife-fhell Butterfly, *Papilio polychloros*—Comma Butterfly, *Papilio c. album*—Spotted Buff Moth, *Phalæna lubricipeda.*—Emperor Moth, *Phalæna Pavonia.* Spotted Elm Moth, *Phalæna betularia*—Creamfpot Tyger Moth,

L 3 *Phalæna*

Phalæna vellica—Elm Bug, *Cimex ulmi*—Fine ſtreaked Bugkin, *Cimex ſtriatus*—Elm Frog-hopper, *Cicada Ulmi*, and the Elm Louſe, *Aphis Ulmi*, are found feeding upon the Elm-tree: the latter generally curl up the leaves ſo as to make them a ſecure ſhelter againſt the weather.

102 ERYNGO. 324 Eryngium.

EMPAL., *Common Receptacle* conical; florets ſitting; ſeparated by chaff. *Fence* of the receptacle flat; of many leaves, longer than the florets. *Cup* of five leaves; upright; ſharp; longer than the bloſſom; ſitting on the ſeedbud.

BLOSS. *General*, uniform; roundiſh. *Florets* all fertile. *Individuals* of five oblong petals; with the points bent inwards towards the baſe; and contracted by a line running length-ways.

CHIVES. *Threads* five; hair-like; ſtraight; longer than the florets. *Tips* oblong.

POINT. *Seedbud* beneath; rough with hair. *Shafts* two; thread-ſhaped; ſtraight; as long as the chives. *Summits* ſimple.

S. VESS. *Fruit* egg-ſhaped; diviſible into two parts.

SEEDS. Oblong; cylindrical.

OBS. *In ſome ſpecies the ſeeds fall out of the ſeed-veſſel; in others they continue incloſed within it.*

Sea
Maritimum

ERYNGO. The root leaves nearly circular; plaited; thorny; Flowering heads on fruit-ſtalks. Chaff with three points—*Petals whitiſh, or blue.*

Eryngium maritimum. *Bauh. pin.* 386.
Eryngium marinum. *Gerard.* 1162. *Park.* 986. *Ray's Syn.* 222.
Sea Holly.
On the ſea-ſhore. P. July—Auguſt.

The leaves are ſweetiſh, with a light aromatic warmth and pungency. The roots are ſuppoſed to have the ſame aphrodiſiac virtues as the bulbous roots in the twentieth Claſs. They are kept in the ſhops, candied. The young flowering ſhoots eaten like ſparagus are very grateful and nouriſhing.

ERYNGO.

ERYNGO. The root-leaves embracing the ſtem; ſpear- Common
ſhaped; but a little winged. Fence awl-ſhaped; longer than Campeſtre
the flowering head—*Petals blue; ſometimes white, or yellowiſh.*
Eryngium vulgare. *Ray's Syn.* 222.
Eryngium mediterraneum. *Gerard.* 1162.
Eryngium mediterraneum, ſeu campeſtre. *Park.* 986.
On the ſea-coaſt. By the ſide of Watling-ſtreet near a village
called Brook-hall not far from Daventry in Northamptonſhire.
P. July. Auguſt.

103 PENNYWORT. 325 Hydrocotyle.

Rundle Simple.

EMPAL. *Fence* frequently of four leaves; ſmall. *Cup*
hardly perceptible.
BLOSS. *General;* uniform in figure, but not in ſituation.
Florets all fertile. *Individuals;* of five petals, egg-
ſhaped; ſharp; entire; expanding.
CHIVES. *Threads* five; awl-ſhaped; ſhorter than the
bloſſom. *Tips* very ſmall.
POINT. *Seedbud* beneath; upright; compreſſed; round ·
the fruit-ſtalks fixed to its centre; *Shafts* two; awl-
ſhaped; very ſhort. *Summit* ſimple.
S. VESS. None. *Fruit* compreſſed; round; diviſible
croſs-ways into two parts.
SEEDS. Two; compreſſed; in the ſhape of a half moon.

PENNYWORT. The leaves with central leaf-ſtalks. About Marſh
five flowers in a rundle—*Petals reddiſh white.* Vulgaris
Hydrocotyle vulgaris. *Ray's Syn.* 222.
Cotyledon paluſtris. *Gerard.* 529. *Park.* 1214.
Ranunculus aquaticus cotyledonis folio. *Bauh. pin.* 180.
White-rot.
In marſhy grounds. P. May.
The Farmers ſuppoſe it occaſions the Rot in Sheep. (*See*
BUTTERWORT.)

104 S A N I C L E.　326 Sanicula.

Rundle with very few fpokes ; (generally four.) *Rundlets* with many fpokes crowded into heads.

EMPAL. *General Fence* going half way round on the outer fide.　*Partial Fence* going quite round ; fhorter than florets.　*Cup* fcarcely perceptible.

BLOSS. *General*, uniform. The *Florets* in the centre barren.　*Individuals* ; petals five, compreffed ; bent inwards fo as to clofe the flower.

CHIVES. *Threads* five ; fimple ; upright ; twice as long as the petals.　*Tips* roundifh.

POINT. *Seedbud* beneath ; rough with ftiff hairs ; *Shafts* two ; awl-fhaped ; reflected.　*Summits* fharp.

S. VESS. None. *Fruit* egg-fhaped ; fharp ; rough ; dividing into two.

SEEDS. Two. Convex and prickly on one fide ; flat on the other.

Eurppean
Europæa
　　　　SANICLE. The root-leaves fimple.　Florets all fitting—*Petals whitifh*.　Leaves *dark-green, and fhining*.
　　　Sanicula officinarum. *Bauh. pin.* 319.
　　　Sanicula feu diapenfia. *Gerard.* 948. *Ray's Syn.* 221.
　　　Sanicula vulgaris, five diapenfia. *Park.* 532.
　　　In woods and hedges.　P. May—June.
　　　A French proverb fays, " He who is poffeffed of Bugle and " Sanicle, may difmifs his Surgeons ;" but modern practice gives no countenance to fuch an affertion.　The leaves are flightly bitter and aftringent.
　　　Sheep eat it.　Goats are not fond of it. Horfes refufe it.

105 MADNEP.

105 MADNEP. 345 Heracleum.

Rundle very large; confifting of numerous flat *Rundlets.*

EMPAL. *General Fence* of many leaves; fhedding. *Partial Fence* going half way round on the outer fide. Leaves from three to feven; betwixt ftrap and fpear-fhaped; the outer leaves longer than the reft, *Cup* but faintly marked.

BLOSS. *General*; unequal, irregular in its fhape. *Florets* nearly all fertile. *Individuals*: in the center; of five equal petals, bent and hooked inwards; notched at the end. In the circumference; of five unequal petals. The outer ones largeft, with the deepeft notches; hooked; oblong.

CHIVES. *Threads* five; longer than the petals. *Tips* fmall.

POINT *Seedbud* beneath; fomewhat egg-fhaped. *Shafts* two; fhort; near together. *Summits* fimple.

S.VESS. None. *Fruit* oval, compreffed; notched; fcored in the middle each way.

SEEDS. Two; egg-fhaped, compreffed; with a leafy edge.

OBS. In fome fpecies the florets in the circumference have only pointals without chives, and produce feeds; the central florets have chives without pointals and are barren. In the Cows Madnep the florets have all chives and pointals. The general fence is fometimes altogether wanting. In the two Britifh fpecies the florets in the center are compofed of unequal petals as well as thofe in the circumference.

MADNEP. The little leaves with winged clefts; fmooth. Florets uniform—*White or tinged with red; all fertile.* Cows
Sphondylium. *Gerard.* 1009. *Ray's Syn.* 205. vulgare. Sphondylium
Park. 953.
Sphondylium vulgare hirfutum. *Bauh. pin.* 157.
1. There is a variety with larger and more jagged leaves. *Park.* 953.
Cow Parfnep.
In hedges, very common. B. July.
In Poland and Lithuania, the poor people prepare a liquor from the leaves and feeds; which undergoes a fermentation, and is drank inftead of ale—The ftalks when peeled, are eaten by the Camfchatkians—The Ruffians take the leaf-ftalks of the root-leaves, peel them, and hang them in the fun to dry a little; then they tye them in little bundles, and hang them up again 'till they become yellow: in this ftate they put them into bags, and a meally fubftance like fugar forms upon the furface of them.
This

This they fhake off, and treat their guefts with it as a great de
cacy. They likewife diftill an ardent fpirit from it. *Gmelin.*
Sibir. 1. p. 214. the peelings of the ftalks are acrid—The leaves
are a favourite food of Rabbits and Hogs. Cows, Goats and
Sheep eat them, but horfes are not fond of them.

Jagged
Anguftifolium
    MADNEP. The little leaves ftrap-fhaped and flender, four
at each joint of the leaf-ftalk. Florets uniform—*White, or
greenifh.*
    Sphondylium hirfutum, foliis anguftioribus. *Baub. pin.* 157.
    In hedges. P. July.

106 DROPWORT.   352 Oenanthe

Rundle with few fpokes. *Rundlets* with many very
fhort fpokes.

EMPAL. *General Fence* of many leaves; fimple; fhorter
than the rundle. *Partial Fence* of many leaves;
fmall. *Cup* with five awl-fhaped teeth; permanent.
BLOSS. *General*; irregular; unequal. *Florets* in the circumference barren. *Individuals*; in the center fertile; petals five; nearly equal; heart-fhaped but
bent inwards. In the circumference with five petals;
large; unequal; bent inwards; cloven.
CHIVES. *Threads* five; fimple. *Tips* roundifh.
POINT. *Seedbud* beneath. *Shafts* two; awl-fhaped; permanent. *Summits* blunt.
S.VESS. None. *Fruit* nearly egg-fhaped; crowned with
the cup; divifible into two parts.
SEEDS. Two; fomewhat egg-fhaped; convex on one
fide; fcored; flat on the other; toothed at the
point.

    OBS. *In this genus the cup is more evident than in the other plants
whofe flowers grow in rundles; and the general fence is often wanting in fome of the fpecies.*

Water
Fiftulofa
    DROPWORT. Sending forth fuckers. Stem leaves winged;
thread-fhaped; hollow—*The firft rundle is cloven into three parts;
the others into many; fo that the plant changes its appearance confiderably in the courfe of the fummer.* Petals white.
    Oenanthe aquatica. *Baub. pin.* 162. *Ray's Syn.* 210.
    Oenanthe paluftris, feu aquatica. *Park.* 895.
    Filipendula aquatica. *Gerard.* 1060.
    In ponds and ditches, frequent. P. July.
    Cows and Horfes refufe it: though from experiments made
on purpofe, it does not appear to be in the leaft degree
noxious to the former.   DROP-

DROPWORT. All the leaves with numerous blunt and nearly equal clefts *Stem yellowish red. Some leaves winged; but more doubly winged. The little leaves wedge-shaped; smooth; streaked; jagged at the edges.* Fruit-stalks *angular; scored.* General Fence *not always present.* Petals *white; sharp; bent inwards.* Tips *purple or brown. General blossom not very unequal.* — Hemlock Crocata

Oenanthe cicutæ facie lobelii. *Park.* 894. *Ray's Syn.* 210.
Oenanthe chærephylli foliis. *Bauh. pin.* 162.
Filipendula cicutæ facie. *Gerard.* 1059.
Deadtongue.
Banks of rivers. P. June.

The whole of this plant is poifonous, and Dr. Poultney re-marks, that the root is the moft virulent of all the vegetable poifons that Great Britain produces. many inftances of its fatal effects are recorded; for which, and for an elegant engraving of it fee *Martyn's Philof. Tranf.* v. 10. p. 772. Alfo vol. 50. p. 856. See alfo *Gent. Mag.* for July 1747 for March 1755 and for September 1758.—An infufion of the leaves, or three teafpoonfuls of the juice of the root taken every morning, effected a cure in a very obftinate cutaneous difeafe; but not without occafioning very great difturbances in the conftitution. *Philof. Tranf.* v. 62. p. 469.—Sheep eat it. Cows and Horfes refufe it.

DROPWORT. The root leaves with wedge-fhaped clefts. Stem leaves entire; ftrap-fhaped; channelled: very long—*Stem about fix inches high; angular.* Rundle *irregular. General and partial fence like awl-fhaped briftles.* Petals *white; not quite equal.* — Pimpernel Pimpinelloides

Oenanthe apii folio. *Bauh. pin.* 162.
In flow ftreams, ponds and ditches. P. June.

107 PRICKLENEP. 329 Echinophora.

Rundle of many fpokes: The intermediate ones fhorteft.
  Rundlets of many florets; thofe in the center fitting.
EMPAL. *General Fence* of feveral fharp leaves. *Partial Fence* turban-fhaped; of one leaf, with fix clefts, fharp; unequal. *Cup* very fmall; with five teeth; permanent.
BLOSS. *General*; irregular in its fhape and unequal. *Florets* which have only chives, barren. Central florets fertile. *Individuals* of five unequal petals, ftanding open.
CHIVES. *Threads* five; fimple. *Tips* roundifh.
POINT: *Seedbud* beneath; oblong; involved in the fence. *Shafts* two; fimple. *Summits* fimple.
S.VESS. None; but inftead thereof the fence grows hard and fharp pointed and includes the feed.
SEED. Single; oblong egg-fhaped.

PRICKLENEP

Marine
Spinofa

PRICKLENEP. The little leaves very entire: awl-fhaped
and terminating in a thorn—*Petals white*; '*or reddifh.*
  Echinophora maritima fpinofa. *Ray's Syn.* 220.
  Crithmum maritimum fpinofum. *Bauh. pin.* 288.
  Crithmum maritimum fpinofum, feu paftinaca marina.
Park. 1286. E
  Crithmum fpinofum. *Gerard.* 533.
  Prickley Sampire. Sea Parfnep.
  Sea coaft. P. July.

108 HENSFOOT. 331 Caucalis.

Rundle, unequal; of very few fpokes. *Rundlets* unequal,
with more fpokes; the five outermoft of which are
the longeft.

EMPAL. *General Fence* with the leaves undivided; fhort;
membranaceous at the edges; betwixt egg and fpear-
fhaped; equal in number to the rays of the rundle.
Partial Fence with leaves fimilar to the foregoing;
longer than the fpokes; generally five in number.
Cup with five teeth, ftanding out.

BLOSS. *General*; irregular in its fhape, and unequal.
Florets in the center barren. *Individuals* in the center, without pointals; fmall; petals five; equal;
heart fhaped, but bent inwards. In the circumference, with both chives and pointals. Petals
five; heart-fhaped; bent inwards; the outermoft
the largeft, and cloven.

CHIVES. *Threads* hair-like; five in every floret. *Tips*
fmall.

POINT. *Seedbud* beneath: in the florets of the circumference oblong and rough. *Shafts* two; awl-fhaped.
Summits two; blunt; expanding.

S. VESS. *Fruit* oblong egg-fhaped; fcored lengthways;
rough with briftly hairs.

SEEDS. Two; oblong; flat on one fide, convex on the
other; armed with awl-fhaped prickles placed along
the fcores.

HENSFOOT. The general fence for the most part wanting. Fine-leaved
Rundle cloven. Partial fence of five leaves.—— Leptophylla

Caucalis arvensis echinata, parvo flore et fructu. *Bauh. pin.*
152.

Caucalis tenuifolia purpurea. *Park.* 920.

Fine-leaved Bastard Parsley.

In cornfields. A. July.

HENSFOOT. The general rundle cloven into three parts. Broad-leaved
Rundlets with five seeds. Leaves winged; serrated——*Blossoms* Latifolia
pale red.

Tordylium latifolium. *Hudson.* 98. *Berkenhout.* 77.

Caucalis arvensis echinata latifolia. *Bauh. pin.* 152. *Ray's Syn.* 219.

Caucalis apii foliis, flore rubro. *Gerard.* 1021.

Echinophora arvensis latifolia purpurea. *Park.* 920.

Purple flowered great bastard Parsley.

In cornfields. A. June—July.

109 CARROT. 333 Daucus.

Rundle of many spokes: flat during the continuance
of the blossoms, but afterwards concave and approaching. *Rundlets* similar to the foregoing.

EMPAL. *General Fence* of many leaves; as long as the
rundle; the leaves strap-shaped with winged clefts.
Partial Fence more simple; as long as the rundlet.
Cup hardly perceptible.

BLOSS. *General*, irregular in its shape and unequal. *Florets*
in the center barren. *Individuals* with five petals;
heart-shaped but bent inwards; the outermost petal
the largest.

CHIVES. *Threads* five; hair-like. *Tips* simple.

POINT. *Seedbud* beneath; small. *Shafts* two; reflected.
Summits blunt.

S.VESS. None. *Fruit* egg-shaped; divisible into two
generally beset with inflexible hairs.

SEEDS. Two; somewhat egg-shaped; convex and rough
with hairs on one side; flat on the other.

CARROT

Wild   CARROT. The seeds rough with strong hairs. Leaf-stalks
Carota   stringy on the underside—*Petals white, or purplish. The central
              floret is often red, and brings forth a seed to all appearance perfect;
              though Linnæus says the central florets are barren.*
                   Daucus vulgaris. *Ray's Syn.* 218.
                   Pastinaca tenuifolia sylvestris. *Bauh. pin.* 155. *Gerard.* 1028.
              Park. 902.
                   Birds Nest.
                   Varieties are.
                   1. Yellow and white rooted Carrot. *Ray's Syn.* 218.
                   2. Red rooted Carrot. *Ray's Syn.* 218.
                   3. Marine, shining, fine, leaved Carrot. *Ray's Syn.* 218.
                   In fields, common. Cultivated in gardens. B. June. July.
                   This in its cultivated state is the well known Garden Carrot,
              whose roots are eaten either boiled or raw. When raw, they are
              given to children troubled with worms. They seem to pass
              through most people but little changed—they are a grateful food
              to all kind of cattle, and well worthy of a more general cultiva-
              tion for the purpose of the farmer. Crickets are very fond of
              Carrots, and are easily destroyed by making a paste of powdered
              arsenic, wheat meal and scraped Carrots, which must be placed
              near their habitations. A poultice made of the roots, hath been
              found to mitigate the pain, and abate the stench of foul and
              cancerous ulcers. The seeds have been sometimes used as diure-
              tics and carminatives.

110 HARTWORT. 330 Tordylium.

Rundle unequal; of many spokes. *Rundlets* unequal, of many parts; very short; flat.

EMPAL. *General Fence*: the little leaves slender; undivided; frequently as long as the rundle. *Partial Fence* going half way round; outwardly longer than the rundlet. *Cup* with five teeth.

BLOSS. *General*, irregular in its shape and unequal. *Florets* all fertile. *Individuals* in the center, with five equal petals; heart-shaped but bent inwards. In the circumference; like the others; but the outermost petal which is the largest deeply divided.

CHIVES. *Threads* hair-like; five in every floret. *Tips* simple.

POINT. *Seedbud* beneath: in all the florets; roundish. *Shafts* two; small. *Summits* blunt.

S.VESS. *Fruit* roundish; almost flat; a little scolloped at the edge; divisible into two parts.

SEEDS. Two; roundish; almost flat; but raised and scolloped at the edge.

OBS. *This genus differs from the* Hensfoot *principally in the florets being all fertile.*

HARTWORT The rundles crowded; little leaves betwixt Hedge egg and spear-shaped, with winged clefts—*General fence of many* Anthriscus *leaves; simple and regular. Sometimes of only one leaf.* Stem *set with stiff sharp hairs, which are bent downwards and contiguous to the stem.* Petals *not very unequal; white: reddish beneath.*

Caucalis minor, flosculis rubentibus. *Gerard.* 1022. *Ray's Syn.* 219.

Caucalis minor, flore rubente. *Park.* 921.

Caucalis femine aspero, flosculis rubentibus. *Bauh. pin.* 153.

Caucalis anthriscus. *Hudson.* 99.

1. General fence of only one leaf.

Caucalis segetum minor, anthrisco hispido similis. *Ray's Syn.* 220.

Caucalis arvensis. *Hudson.* 98.

Hedge Parsley, 1. Small Corn Parsley.

In hedges, common. B. August.

Horses are extremely fond of it.

Knotted
Nodofum

HARTWORT. Rundles fimple; fitting. Seeds rough on the outer fides.—

Caucalis nodofo echinato femine. *Bauh. pin.* 153. *Gerard:* 1022. *Park.* 921. *Ray's Syn.* 220.

Knotted Parfley.

Cornfields. A. May.

111 HARESTRONG. 339 Peucedanum.

Rundle of many very long, flender fpokes. *Rundlets* expanding.

EMPAL. *General Fence* of many leaves: ftrap-fhaped, fmall, reflected. *Partial Fence*, ftill fmaller. *Cup* with five teeth; very fmall.

BLOSS. *General*, uniform. *Florets* in the center barren. *Individuals*; petals five; equal; oblong; entire; bent inwards.

CHIVES. *Threads* five; hair-like. *Tips* fimple.

POINT. *Seedbud* beneath: oblong. *Shafts* two; fmall. *Summits* blunt.

S.VESS. None. *Fruit* egg-fhaped; divifible into two; fcored on each fide, encompaffed round by a membranaceous border.

SEEDS. Two; oblong egg-fhaped; compreffed; convex on one fide and marked by three rifing ridges: bound round by a broad flat membrane; notched at the end.

Fennel-leaved
Officinale

HARESTRONG with leaves five times divided into threes; thread and ftrap-fhaped—*Petals yellowifh.*

Peucedanum. *Gerard.* 1054. *Ray's Syn.* 206. Germanicum. *Bauh. pin.* 146.

Peucedanum vulgare. *Park.* 880.

Hogsfennel. Sulphurwort.

Salt marfhes. P. June.

The roots have a ftrong fœtid fmell, and an acrid, bitterifh, unctuous tafte. Wounded in the fpring they yield a confiderable quantity of yellow juice, which dries into a gummy refin and retains the ftrong fmell of the root. Its virtues have not yet been afcertained with any precifion.

HART-

HARESTRONG. The little leaves with winged clefts; Meadow
fegments oppofite. General fence of two leaves—*Petals yellow*; Silaus
white on the outfide.
  Sèfeli pratenfe noftras. *Park.* 905. *Ray's Syn.* 216.
  Saxifraga anglica facie fefeli pratenfis. *Gerard.* 1047.
  Sefeli caruifolia. *Hudfon.* 106.
  Meadow Saxifrage.
  Meadowes and moift paftures. P. Auguft.

112 HEMLOCK. 336 Conium

Rundle of many fpokes; expanding. *Rundlets* the fame.

EMPAL. *General Fence* of many leaves; very fhort; unequal. *Partial Fence* the fame. *Cup* hardly perceptible.
BLOSS. *General,* uniform. *Individuals*; Petals five; unequal; heart-fhaped but bent inwards.
CHIVES. *Threads* five; fimple. *Tips* roundifh.
POINT. *Seedbud* beneath. *Shafts* two; reflected. *Summits* blunt.
S. VESS'. None. *Fruit* nearly globular; with five fcolloped ridges; divifible into two parts.
SEEDS. Two; convex on one fide; almoft hemifpherical;
  fcored; flat on the other fide.

  OBS. *The Partial Fence confifts of one leaf, divided into three
  parts.*

  HEMLOCK. The feeds fcored—*Stems and branches fpotted* Spotted
with brown or black. Petals *white.*   Maculatum
  Cicuta major. *Bauh. pin.* 160.
  Cicuta. *Gerard.* 1061. *Ray's Syn.* 215.
  Cicuta vulgaris major. *Park.* 933.
1. There is a variety with finer divifions in the leaves.
  Kex.
  In hedges, orchards, and amongft rubbifh. P. June—July.
  The whole plant is poifonous, and many inftances are recorded of its deleterious effects; but modern experience hath proved
it to be lefs virulent than was formerly imagined. Dr. Storck
of Vienna was the firft who ventured to give it internally in
confiderable quantities, and from his account of its good effects
in a variety of cafes, particularly in cancers and fchirrus tumours,
it hath been very generally employed in this kingdom. He directs an extract to be prepared from it by evaporating the expreffed juice over the fire, and then adding a fufficient quantity of
the powdered leaves to form it into pills. He generally gave
                          M   from

from two to twelve grains for a dofe, but fome have taken it in much larger quantities. By the accounts from Vienna, the expectations of the world were raifed to a very high pitch; and, as is then generally the cafe, thefe expectations have been difappointed; fo that after a very extenfive application of it for feveral years, it is now likely to be entirely difregarded, as a medicine of little or no ufe. Perhaps, however, the truth will be found to lie in the medium betwixt the two opinions. Many reafons may be affigned to fhew that it hath yet undergone but an imperfect trial; fome of which I fhall beg leave to mention.

1ft. The *Wild Cicely*, or *Cow-weed*, hath often been gathered inftead of the intended plant; and it is not improbable but other plants may have been likewife miftaken for it, for many of the *Rundle-bearing* or *Umbelliferous Plants* are fo much alike, that it requires more fkill in Botany to diftinguifh them, than we can fuppofe the common collectors of medicinal plants to poffefs. They know plants that they have been taught to collect and to diftinguifh, but this was one that had never before been in requeft. After fome time, however, the fpots upon the ftalks were pointed out as a criterion, and then there was lefs probability of miftakes.

2dly. The feafon of its higheft perfection hath never yet been fufficiently afcertained. Some fuppofe it ought to be gathered juft as its bloffoms open, others fay not till the feeds are forming; but even thefe fuppofitions have been too little attended to.

3dly. Perhaps the plants of this natural order have their properties more changed by foil and fituation, than thofe of any other natural order; yet this circumftance hath been entirely difregarded.

4thly. No vegetable that is ufed in medicine is more liable to ferment than this. I have feen it, when collected in a bag and carried only two miles on horfeback, heat and ferment to fuch a degree, that the yield of extract was much lefs than ufual and the properties of it greatly impaired.

5thly. At its firft introduction it was very common to take only the clearer part of the expreffed juice, and to throw away the feculencies. Whenever this is done, the medicine is fpoiled.

6thly. Too little attention hath been given to the degree of heat applied during the evaporation. Many things lofe their peculiar properties when expofed to more than the heat of boiling water, and others are greatly changed even by that. Many reafons may be alledged to prove that Hemlock is one of thefe.

When further experience hath pointed out the beft feafon for gathering the plant, and in what foils and fituations it poffeffes the greateft virtues, I believe a medicine much more efficacious than that commonly ufed, may be obtained by attending to the following directions.

Let several people be employed to gather the plant; and as fast as it is cut, let others carry it in hand baskets to the press; but it must lie light and loosely in the baskets. Let the juice be immediately squeezed out; and as fast as it runs from the press, it must be put over the fire, and boiled 'till three parts out of four of the whole liquor is wasted. Then it must be put into a water bath, and evaporated to the consistence of honey. If it is now taken and spread thin upon a board or marble slab, and exposed to the sun and the air, it will soon be of a proper consistence to be formed into pills. From five to ten grains of this extract is a proper dose; few constitutions will bear more without experiencing disagreeable effects.

Such a medicine as this I believe will be found an useful addition to our materia medica: not that I have seen it cure cancers either in an ulcerated state or otherwise, but I have never given it without a mitigation of pain and an amendment of the discharge. Dr. Fothergill in the 3d vol. of the Medical Observations, hath given us a variety of cases to which I can with pleasure refer the reader, as they perfectly correspond with my own experience in a pretty extensive hospital practice for several years past. Dr. Butter uses it in the chincough; but I have had no opportunity of trying it in that disease.

After all it may be said, that it acts merely as a narcotic, and only effects what small doses of opium will do, in a less disagreeable manner. But 'till further experience gives a sanction to this opinion it must not be too hastily adopted; and if it should prove at last that it is only a narcotic, surely there are many cases in which a narcotic that does not occasion costiveness is preferable to one that does. Sheep eat the leaves; Horses, Cows and Goats refuse them. Thrush feed upon the seeds.—There is a good engraving of this plant in the *Gent. Mag.* 1762. p. 273.

113 PIGNUT. 335 Bunium

Rundle with fewer than twenty spokes. *Rundlets* very
short; crowded.

EMPAL. *General Fence* of many strap-shaped short leaves.
Partial Fence bristly; as long as the rundlet. *Cup*
hardly discernible.

BLOSS. *General* uniform. *Florets* all fertile. *Individuals*;
petals five; equal; heart shaped, bent inwards.

CHIVES. *Threads* five; shorter than the petals; *Tips*
simple.

POINT. *Seedbud* beneath: oblong. *Shafts* two; reflected.
Summits blunt.

S.VESS. None. *Fruit* egg-shaped; divisible into two
parts.

SEEDS. Two; egg shaped; convex on one side; flat on
the other.

Bulbous-rooted   PIGNUT. As there is only one species known Linnæus
Bulbocastanum   gives no description of it—*The Leaves winged.* Roots *bulbous.*
Petals *white.*
    Bulbocastanum. *Ray's Syn.* 200.
    Bulbocastanum majus, folio apii *Bauh. pin.* 162.
    Bulbocastanum majus et minus. *Gerard.* 1065.
    Nucula terrestris major. *Park.* 893.
    Earth Nut. Kipper Nut. Hawknut. Jurnut.
    In orchards and pasture. P. May—June.
    The roots eaten either raw or boiled are very little inferior to
Chesnuts, and would be an agreeable addition to our winter
deserts.

114 SPIGNEL. 338. Athamanta

Rundle of many fpokes, expanding. *Rundlets* with fewer fpokes.

EMPAL. *General Fence* of many ftrap-fhaped leaves; a little fhorter than the fpokes. *Partial Fence* ftrap-fhaped; as long as the fpokes. *Cup* not difcernible.

BLOSS. *General,* uniform. *Florets* all fertile. *Individuals*; petals five; heart-fhaped; bent inwards; nearly equal.

CHIVES. *Threads* five; hair-like; as long as the petals. *Tips* roundifh.

POINT. *Seedbud* beneath. *Shafts* two; diftant. *Summits* blunt.

S.VESS. None. *Fruit* oblong egg-fhaped; fcored; divifible into two parts.

SEEDS. Two; egg-fhaped; convex and fcored on one fide, flat on the other.

SPIGNEL. The leaves doubly winged, flat. Rundles in Mountain form of half a globe. Seeds rough with hair—*Leaf-ftalks a little* Libanotis *compreffed, and channelled betwixt the little Leaves.* Shafts purple. Stem *about two feet high, with ftrongly marked angles; unequally furrowed; generally fimple and rather fmooth; bending a little towards the top.* Leaves *fmooth on the upper furface, pale and full of veins beneath; fcarce fenfibly hairy at the edges.* Petals white.
Apium petræum, feu montanum album. *Ray's Syn.* 219.
Daucus montanus apii folio albicans. *Bauh. pin.* 157.
Mountain Stone Parfley.
Dry paftures. P. Auguft—September.
Sheep and Swine eat it. Cows refufe it.

SPIGNEL. The little leaves hair-like: feeds fcored; fmooth Common —*Petals white.* Meum
Meum foliis anethi. *Bauh. pin.* 148.
Meum. *Gerard.* 1052. *Ray's Syn.* 207. vulgatius. *Park.* 888.
Meu. Bald or Bawd-money.
On hilly paftures. P. May.
The roots and feeds are aromatic and acrid. They have been ufed as ftomachics and carminatives. Sometimes they are given to cure the tertian ague; and there is no doubt but they will often anfwer as well as pepper and other acrid aromatics.

115 THOROUGHWAX. 328 Bupleurum.

Rundle with fewer than ten ſpokes. *Rundlets* with abou
ten upright expanding ſpokes.

EMPAL. *General Fence* of many leaves. *Partial Fence* larger
of five leaves; the leaves expanding; egg-ſhaped
ſharp. *Cup* not diſcernible.

BLOSS. *General*, uniform. *Florets* all fertile. *Individuals*
petals five; very ſhort; entire; rolled inwards.

CHIVES. *Threads* five; ſimple. *Tips* roundiſh.

POINT. *Seedbud* beneath. *Shafts* two; reflected; ſmall
Summits very ſmall.

S.VESS. None. *Fruit* roundiſh; compreſſed; ſcored;
diviſible into two.

SEEDS. Two; oblong egg-ſhaped; convex and ſcored or
one ſide; flat on the other.

Round-leaved
Rotundifolium

THOROUGHWAX without any general fence. Leave:
perforated—*Smooth*; *bluiſh green*. Petals *yellowiſh*.
Bupleurum perfoliatum rotundifolium annum. *Ray's Syn.* 221.
Perfoliata vulgaris. *Gerard.* 536. *Park.* 580.
Perfoliata vulgatiſſima, ſeu arvenſis. *Bauh. pin.* 272.
In cornfields. A. July.

Fine-leaved
Tenuiſſimum

THOROUGHWAX. The rundles ſimple; alternate; with
five leaves, and about three florets—*Stem very much branched.*
Branches alternate. General Fence *of three ſhort leaves.* Partial
Fence *with ſhort briſtle-ſhaped leaves. Rundles at the baſe of the
leaves.* Leaves *ſtrap-ſhaped; ſharp;* Petals *yellowiſh.*
Bupleurum minimum. *Park.* 587. *Ray's Syn.* 221.
Bupleurum anguſtiſſimo folio. *Bauh. pin.* 278.
Leaſt Hares Ear.
In paſtures, not common. A. July—Auguſt.

116 KEX. 348 Sium.

Rundle various in various species. *Rundlets* flat and expanding.

EMPAL. *General Fence* of many reflected leaves; shorter than the rundle. The leaves spear-shaped. *Partial Fence* of many leaves; strap-shaped; small. *Cup* hardly perceptible.

BLOSS. *General,* uniform. *Florets* all fertile. *Individuals*; petals five; equal; heart-shaped; bent inwards.

CHIVES. *Threads* five; simple. *Tips* simple.

POINT. *Seedbud* beneath: very small. *Shafts* two; reflected. *Summits* blunt.

S. VESS. None. *Fruit* roundish, egg-shaped; scored; small; divisible into two.

SEEDS. Two; nearly egg shaped; convex and scored on one side; flat on the other.

OBS. *The general fence in the* Creeping Kex *is often wanting.*

KEX with winged leaves and terminating rundles—*Little* Greater
leaves serrated. Petals *white.*   Latifolium
   Sium latifolium, foliis variis. *Ray's Syn.* 211.
   Sium latifolium. *Bauh. pin.* 154.
   Sium majus latifolium. *Gerard.* 256.
   Sium dioscoridis, seu pastinaca aquatica major. *Park.* 1240.
   Great Water Parsnep.
   In rivers and fens. P. July—August.
   Horses and Swine eat it. Sheep are not fond of it. The roots are noxious to cattle.

KEX with winged leaves, and rundles on fruit-stalks at the Lesser
base of the leaves. General fence with winged clefts—*Leaves* Angustifo-
halberd-shaped at the base. Rundles opposite to the leaves.   lium
   Sium majus angustifolium. *Gerard.* 256.
   Sium minus alterum. *Park.* 1241.
   Sium five apium palustre foliis oblongis. *Bauh. pin.* 154.
Ray's Syn. 211.
   Sium erectum. *Hudson.* 103.
   Upright Water Parsnep.
   Ditches and rivulets. P. June.
   It certainly possesses active properties that ought to be enquired into.

Creeping
Nodiflorum

KEX. With winged leaves and rundles fitting, at the bafe, of the leaves—*The general fence is feldom prefent.* Petals *white.*
Sium umbellatum repens. *Gerard.* 256. *Ray's Syn.* 211.
Creeping Water Parfnep.
In rivers and ditches, common. P. July. Auguft.

117 SAMPHIRE. 340 Crithmum.

Rundle of many fpokes; hemifpherical. *Rundlets* the fame.

EMPAL. *General Fence* of many leaves. The *Leaves* fpear-fhaped; blunt; refleded. *Partial Fence* betwixt fpear and ftrap-fhaped; as long as the rundlet. *Cup* hardly perceptible.

BLOSS. *General*; uniform. *Florets* all fertile. *Individuals*; petals five; egg-fhaped; bent inwards; nearly equal.

CHIVES. *Threads* five; fimple; longer than the petals. *Tips* roundifh.

POINT. *Seedbud* beneath. *Shafts* two; refleded. *Summits* blunt.

S.VESS. None. *Fruit* oval; compreffed; divifible into two.

SEEDS. Two; oval; flat, but fomewhat compreffed; fcored on one fide.

Sea
Maritimum

SAMPHIRE. The little leaves flefhy; fpear-fhaped—*Petals yellow.*
Crithmum marinum. *Gerard.* 553. *Ray's Syn.* 217.
Crithmum feu fæniculum marinum minus. *Bauh pin.* 288.
Crithmum marinum vulgare. *Park.* 1286.
On the fea coaft. P. Auguft.
Shakefpear, fpeaking of Dover cliffs, remarks the growth of famphire.

" Come on fir, here's the place—ftand ftill. How fearful
" And dizzy 'tis to caft one's eyes fo low !
" The crows and chows, that wing the midway air,
" Shew fcarce fo grofs as beetles. Half way down
" Hangs one that gathers famphire ; dreadful trade ! "
KING LEAR.

Poor people on the fea coaft eat it as a pot-herb. It is very generally ufed as a pickle.

118 LOVAGE. 346 Ligufticum.

Rundle of many fpokes. *Rundlets* the fame.

EMPAL. *General Fence* of feven unequal membranaceous leaves. *Partial Fence* of about four membranaceous leaves. *Cup* of five teeth; but hardly perceptible.

BLOSS. *General,* uniform. *Florets* all fertile. *Individuals*; petals five; equal; flat; entire; rolled inwards, and keel-fhaped on the infide.

CHIVES. *Threads* five; hair-like; fhorter than the petals. *Tips* fimple.

POINT. *Seedbud* beneath. *Shafts* two; ftanding clofe together. *Summits* fimple.

S.VESS. None. *Fruit* oblong; angular; with five furrows: divifible into two.

SEEDS. Two; oblong; glofly; marked on one fide with five ridges; flat on the other.

LOVAGE. With doubly threefold leaves—*Petals white.* Parfley Ligufticum fcoticum apii folio. *Ray's Syn.* 214. Scoticum
Scottifh Sea Parfley.
On clifts on the fea fhore; not common. B. July.
This plant is much valued in the ifle of Sky. The root is reckoned a good carminative, and an infufion of the leaves is thought a good purge for calves. It is befides ufed as food; either raw, as a fallad; or boiled as greens. *Pennant's Tour.* 1772. P. 310.
Horfes, Sheep and Goats eat it, Cows refufe it.

LOVAGE. The leaves doubly compound; jagged. Root- Saxifrage leaves fpear-fhaped; very entire; growing by threes.— Cornubienfe
Smyrnium tenuifolium noftras. *Ray's Syn.* 209. tab. 8.
Cornwall Saxifrage.
In Cornwall. P. July.

119 ANGELICA. 347 Angelica.

Rundle of many fpokes; nearly globular. *Rundlets* exactly
   globular.
EMPAL. *General Fence* fmall; of three or five leaves. *Partial Fence* fmall; of eight leaves. *Cup* with five teeth;
   hardly difcernible.
BLOSS. *General*, uniform. *Florets* all fertile. *Individuals*;
   petals five; fpear-fhaped, rather flat; but a little
   bent inwards; fhedding.
CHIVES. *Threads* five; fimple; longer than the petals.
   Tips fimple.
POINT. *Seedbud* beneath. *Shafts* two; reflected. *Summits*
   blunt.
S.VESS. None. *Fruit* roundifh; angular; folid; divifible into two.
SEEDS. Two; egg-fhaped; flat on one fide and encompaffed with a border; convex on the other fide,
   with three furrowed lines.

OBS. *In the* Wild Angelica *the general fence is often wanting.*

Wild
Sylveftris

ANGELICA. Little leaves all alike; betwixt egg and fpear-
fhaped: ferrated—*Petals greenifh white.*
   Angelica fylveftris. *Gerard.* 999. *Park.* 940. *Ray's Syn.*
208.
   Angelica fylveftris major. *Bauh pin.* 145.
   In woods and wet hedges. P. June. July.
   It is warm, acrid, bitter and aromatic; but the fpecies culti-
vated in our gardens poffeffing thefe properties in a higher
degree, this hath been long neglected.
   Cows, Goats and Swine eat it. Horfes refufe it.
   The Royal William or Swallow-tailed Butterfly. *Papilio
Machaon* is found upon this plant.

120 HONEWORT. 349 Sifon.

Rundle unequal; with fewer than fix fpokes. *Rundlets* unequal, with fewer than ten fpokes.

EMPAL. *General Fence* of four leaves; unequal. *Partial Fence* the fame. *Cup* hardly perceptible.

BLOSS. *General*, uniform. *Florets* all fertile. *Individuals*; equal; of five petals; fpear-fhaped; flat, but a little bent inwards.

CHIVES. *Threads* five; hair-like; as long as the petals. *Tips* fimple.

POINT. *Seedbud* beneath; nearly egg-fhaped. *Shafts* two; reflected. *Summits* blunt.

S. VESS. None. *Fruit* egg-fhaped; fcored; divifible into two.

SEEDS. Two; egg-fhaped; convex and fcored on one fide, and flat on the other.

HONEWORT. With winged leaves and upright rundles— Hedge *Terminating*. Petals *white*. Amomum
  Sifon quod amomum officinis noftris. *Bauh pin.* 154.
  Sifon vulgare, vel amomum germanicum. *Park.* 914.
  Sium aromaticum fue fifon officinalis. *Ray's Syn.* 211.
  Petrofelinum macedonicum fuchfii. *Gerard.* 1016.
  Baftard Stone Parfley.
  In woods and moift hedges. P. Auguft. September.
  The feeds are a mild warm aromatic. They give out their virtues to rectified fpirit, and tincture it green.

HONEWORT. With winged leaves and nodding rundles— Corn *Petals white*. Segetum
  Sium arvenfe, five fegetum. *Ray's Syn,* 211,
  Selinum fii foliis. *Gerard.* 1018.
  Selinum fegetale. *Park.* 932.
  Corn Parfley.
  In cornfields and hedges in clay lands. B. July.

HONEWORT. With creeping ftems and cloven rundles— Marfh *The leaves above the water winged; but thofe beneath the furface* Inundatum *more finely divided.*
  Sium pucillum. *Ray's Syn,* 212.
  Leaft Water Parfnep.

121 CICELY

121 CICELY. 355 Æthusa.

Rundle expanding; the inner spokes gradually shorter, and
　　those in the center the shortest of all. *Rundlets*
　　small; expanding.
EMPAL. *Partial Fence* with three or five leaves; going half
　　way round upon the outerside, strap-shaped; very
　　long; pendant. *Cup* hardly perceptible.
BLOSS. *General,* nearly uniform. *Florets* all fertile. *Indi-*
　　viduals; petals five; unequal, heart-shaped, but
　　bent inwards.
CHIVES. *Threads* five; simple. *Tips* roundish.
POINT. *Seedbud* beneath. *Shafts* two; reflected. *Summits*
　　blunt.
S.VESS. None. *Fruit* roundish, egg-shaped; scored; di-
　　visible into two.
SEEDS. Two; roundish, scored: on one side, which is
　　about a third part flat.

Hogs
Cynapium

　　CICELY. As there is only one species known Linnæus
gives no description of it—*The* Stem *furrowed, branched.*
Leaves *winged, smooth, glossy.* Petals *whitish.* Seeds *very large.*
　　Cicutaria tenuifolia. *Gerard.* 1063. *Ray's Syn.* 215.
　　Cicuta minor seu fatua. *Park.* 933.
　　Cicuta minor petroselino similis. *Bauh pin.* 160!
　　Fools Parsley. Lesser Hemlock.
　　In cornfields and gardens. A. August—September.
　　This plant, from its resemblance to common Parsley, hath
sometimes been mistaken for it, and when eaten it occasions sick-
ness. If the curled-leaved Parsley only, was cultivated in our
gardens, no such mistakes would happen in future.
　　Cows, Horses, Sheep, Goats and Swine eat it. It is noxious
to Geese.

122 CORIANDER

122 CORIANDER. Coriandrum. 356.

Rundle of few fpokes. *Rundlets* of many. *General Fence* for the moſt part abſent; but ſometimes there is a ſingle leaf.

EMPAL. *Partial Fence* of three ſtrap-ſhaped leaves, going half way round. *Cup* with five teeth; ſtanding out.

BLOSS. *General*; irregular in its ſhape and unequal. *Florets* in the center barren. *Individuals*; of the center with both chives and pointals; petals five; equal; heart-ſhaped, but bent inwards. *Individuals* of the circumference with both chives and pointals. Petals five; heart-ſhaped but bent inwards; the outermoſt the largeſt and deeply divided; thoſe on each ſide of it have large ſegments.

CHIVES. *Threads* five; ſimple. *Tips* roundiſh.

POINT. *Seedbud* beneath. *Shafts* two; diſtant. *Summits* in the florets of the circumference roundiſh.

S.VESS. None. *Fruit* globular; diviſible into two.

SEEDS. Two; hemiſpherical: concave.

CORIANDER. With globular fruit.—*Petals white.*   Cultivated
Coriandrum. *Gerard.* 10 2. *Ray's Syn.* 221.   Sativum
Coriandrum majus. *Bauhpin.* 158. vulgare. *Park.* 918.
Road ſides and dunghills. A. June.
　The leaves have a ſtrong diſagreeable ſmell. The ſeeds are tolerably grateful when dry. The Edinburgh College uſe them as correctors in the bitter infuſion and the preparations of ſenna, for nothing ſo effectually covers the diſagreeable taſte of that medicine. Confectioners incruſt the ſeeds with ſugar. They have been conſidered as ſuſpicious if not deleterious, but I have known ſix drams of them taken at once, without any remarkable effect.

123 SHEPHERDS-

123 SHEPHERDSNEEDLE. 357 Scandix.

Rundle long; with few fpokes: *Rundlets* with more. *General Fence* wanting.

EMPAL. *Partial Fence* of five leaves; as long as the rundlets. *Cup* not diftinguifhable.

BLOSS. *General*; irregular in its fhape and unequal. *Florets* in the center barren. *Individuals*; petals five; heart-fhaped, bent inwards; the inner ones fmall; the outer ones larger.

CHIVES. *Threads* five; hair-like. *Tips* roundifh.

POINT. *Seedbud* beneath: oblong. *Shafts* two; awl-fhaped; diftant; permanent; as long as the fmalleft petal. *Summits* in the unequal florets, blunt.

S. VESS. None. *Fruit* awl-fhaped; very long; divifible into two.

SEEDS. Two; awl-fhaped; convex and furrowed on one fide; flat on the other.

OBS. *In the firft fpecies the feeds are thread-fhaped, hiding a kernel at the bafe.*

Common Pecten Veneris

SHEPHERDSNEEDLE. The feeds with an exceeding long bill—*Petals white.*

Scandix femine roftrato vulgaris. *Bauh. pin.* 152. *Ray's Syn.* 207.

Scandix vulgaris, feu pecten veneris. *Park.* 916
Pecten veneris, feu fcandix. *Gerard.* 1040.
Venus comb. Crake needle.
Corn fields. A. June. July.

Chervil Cerefolium

SHEPHERDSNEEDLE. The feeds fhining; egg-fhaped below, but awl-fhaped upwards. Rundles lateral; fitting.— *Petals white.* Seeds *black. About four fpokes in each rundle.* Florets *all fertile.*

Chærophyllum fativum. *Bauh pin.* 152.
Cerefolium vulgare fativum. *Gerard.* 1038.
Common chervil.
Meadows and rich ground. A. May.
It is cultivated in our gardens as a pot-herb, and for fallads. It is flightly aromatic and aperient.
Cows are extremely fond of it. Sheep and Goats eat it. Horfes refufe it.

SHEPHERDSNEEDLE. The feeds egg-fhaped ; covered *Hemlockleaved* with ftrong hairs. Florets uniform. Stem fmooth—*Petals white.* *Anthrifcus*
Myrrhis fylveftris feminibus afperis. *Bauh. pin.* 160. *Ray's Syn.* 220.
Myrrhis fylveftris neapolitana atque etiam anglicana. *Park.* 935.
Cerefolium, five myrrhis nova equicolorum columnæ. *Gerard.* 1028.
Small Hemlock Chervil with rough feeds.
Cornfields, road-fides ; very common. A. May—June.
Cows, Goats and Sheep eat it.

124 C O W - W E E D.   358 Chærophyllum.

Rundle expanding. *Rundlets* with nearly the fame number of Spokes.

EMPAL. *General Fence*, none. *Partial Fence* of about five
leaves ; fpear-fhaped ; concave ; reflected ; nearly
as long as the rundlets. *Cup* not difcernible.

BLOSS. *General*, pretty uniform. *Florets* in the centre generally barren. *Individuals*; petals five ; heart-fhaped ;
bent inwards, flattifh ; with a fharp point bending
inwards ; the *outermoft* petals rather the largeft.

CHIVES. *Threads* five ; fimple as long ; as the rundlet.
Tips roundifh.

POINT. *Seedbud* beneath ; *Shafts* two ; reflected. *Summits*
blunt.

S. VESS. None. *Fruit* oblong egg-fhaped ; taper; fmooth ;
divifible into two.

SEEDS. Two ; oblong ; growing fmaller upwards. Convex on one fide, flat on the other.

OBS. *The feeds in the center are often barren.*

COW WEED. The ftem fmooth ; fcored ; with crooked *Wild* fweiled joints—*Petals entire ; flat; white.* Fruit-ftalks *cylindrical.* *Sylveftre* *Outer leaf of the partial fence much larger than the reft.*
Cicutaria vulgaris. *Ray's Syn.* 207.
Cicutaria alba Lugdunenfis. *Gerard.* 1038.
Myrrhis fylveftris. *Park.* 935. Seminibus lævibus. *Bauh. pix.* 160.
Wild Cicely.
In hedges. A. May. June.
The roots eaten as Parfneps have been found poifonous. The rundles afford an indifferent yellow dye ; the leaves and fialks a beautiful green. Its prefence indicates a fertile and gratetul foil. Neither Horfes, Sheep or Goats are fond of it. Cows and Swine refufe it.

COW-

Chervil
Temulum

COW-WEED. Stem rough; with crooked swelled joints—
Rundles before flowering drooping. Stem *but little furrowed; mark-
ed with purple spots, and set with white hairs which stand out.*
Rundlets *in the center barren.* Petals *very white.*

Cherophyllum sylvestre. *Bauh. pin.* 152.
Cherefolium sylvestre. *Gerard.* 1038. *Park.* 915. *Ray's Syn,*
207.
Wild Chervil.
Common in hedges. A. July—August.
Cows and Sheep refuse it.

125 WATERWORT.   353 Phellandrium.

Rundle with many spokes. *Rundlets* the same. *General*
Fence none.

EMPAL. *Partial Fence* of seven leaves; sharp; as long as
the rundlet. *Cup* of five teeth; permanent.

BLOSS. *General*; nearly uniform. *Florets* all fertile. *Individuals*; unequal. Petals five; tapering; heart-
shaped; bent inwards.

CHIVES. *Threads* five; hair-like; longer than the petals. *Tips* roundish.

POINT. *Seedbud* beneath: *Shafts* two; awl-shaped; upright; permanent; *Summits* blunt.

S. VESS. None. *Fruit* egg-shaped; smooth; crowned with
the cup and the pointals; divisible into two parts.

SEEDS. Two; egg-shaped; smooth.

OBS. *The florets in the center are smaller than the others.*

Skeleton
Aquaticum

WATERWORT, with the branchings of the leaves strad-
dling—*Stem very thick, hollow, scored,* Petals *white.*
Phellandrium vel cicutaria aquatica quorundam. *Ray's Syn.*
215.
Cicutaria palustris. *Gerard.* 1063.
Cicutaria palustris tenuifolia. *Bauh. pin.* 161.
Water Hemlock.
Rivers, ditches, pools. B. June.
The seeds are recommended in Intermmittent Fevers.—The
leaves are sometimes added to discutient Cataplasms—It is generally esteemed a fatal poison to Horses, occasioning them to become paralytic; but this effect is owing to an insect, (*Curculio
paraplecticus*) which generally inhabits within the stems. The
usual antidote. is pig dung.——In the winter the roots and
stem, dissected by the influence of the weather, afford a very
curious skeleton or net-work.—Horses, Sheep and Goats eat it.
Swine are not fond of it. Cows refuse it. The black Chrysomela, *Chrysomela Phellandria,* and the Gilt Leptura are found
upon the roots, and the *Curculio paraplecticus* within the stems.
126 COW-

126 COWBANE. 354 Cicuta.

Rundle roundifh, with many equal fpokes. *Rundlets* roundifh with many briftle-fhaped fpokes.

EMPAL. *General Fence* none. *Partial fence* of many leaves; little leaves briftly; fhort. *Cup* fcarcely evident.

BLOSS. *General*; uniform. *Florets* all fertile. *Individuals*; petals five; egg-fhaped; nearly equal : bent inwards.

CHIVES. *Threads* five ; hair-like ; longer than the petals. *Tips* fimple.

POINT. *Seedbud* beneath;. *Shafts* two ; thread fhaped ; longer than the petals; permanent. *Summits.*roundifh.

S. VESS. None. *Fruit* nearly egg-fhaped ; flightly furrowed ; divifible into two.

SEEDS. Two ; fomewhat egg-fhaped; convex and fcored on one fide, flat on the other.

COWBANE, with rundles oppofite the leaves. Leaf-ftalks with Water blunt borders—*Leaves with about feven pair of little leaves, which* Virofa *are varioufly divided and indented.* Petals *yellowifh pale green.*

Sium alterum olufatri facie. *Gerard.* 256. *Ray's Syn.* 212.

Sium majus alterum anguftifolium. *Park.* 1241.

Sium erucæ folio. *Bauh. pin.* 154.

Long-leaved Water Hemlock.

In fhallow waters. P. July.

This is one of the rankeft of our vegetable poifons. Numerous inftances are recorded of its fatality to the human fpecies in *Wepfer's* Treatife; and in the *Enum : Stirp. Helv.* p. 436. (See alfo an account of its deleterious effects, and an elegant engraving of it in *Martyn's Philof. Tranf.* vol. 10.) Early in the fpring when it grows in the water, Cows often eat it and are killed by it ; but as the fummer advances and its fmell becomes ftronger, they carefully avoid it. Though a certain and fatal poifon to Cows, Goats devour it greedily and with impunity. Horfes and Sheep eat it with fafety.

127 ALEXANDERS.   363 Smyrnium.

Rundle unequal; daily increasing. *Rundlets* upright.
Fences none.

Fᴍᴘᴀʟ. *Cup* hardly perceptible.

Bʟᴏss. *General,* uniform. *Florets* in the centre barren.
Individuals; petals five; fpear-fhaped; keeled on
the under-fide; flightly bent inwards.

Cʜɪᴠᴇs. *Threads* five; fimple; as long as the petals.
Tips fimple.

Pᴏɪɴᴛ. *Seedbud* beneath. *Shafts* two; fimple. *Summits*
two; fimple.

S. Vᴇss. None. *Fruit* nearly globular; fcored; divifible
into two.

Sᴇᴇᴅs. Two; crefcent-fhaped; convex on one fide and
marked with three angles: flat on the other.

Common   ALEXANDERS. The ftem leaves growing by threes; on
Olufatrum   leaf-ftalks; ferrated—*Sheaths of the leaves ragged and fringed.*
Partial fence very fhort. Petals *greenifh white.*

Smyrnium. *Ray's Syn.* 208.

Hippofelinum. *Gerard.* 1019.

Hippofelinum feu Smyrnium vulgare. *Park.* 930.

Hippofelinum Theophrafti, feu Smyrnium diofcoridis. *Bauh.
pin.* 154.

Upon rocks on the fea-coaft, and about Nottingham. B.
May—June.

It was formerly cultivated in our gardens, but its place is
now better fupplied by Celery.

128 CARAWAY.

128 CARAWAY. 365 Carum.

Rundle with ten spokes; long, and oftenunequal. *Rundlets* crowned. *Fences* none; or only one leaf for the general fence.

EMPAL. *Cup* hardly perceptible.

BLOSS. *General*; uniform. *Florets* in the center barren. *Individuals*; unequal. *Petals* five; unequal; blunt; heart-shaped; bent inwards at the end.

CHIVES. *Threads* five; hair-like; as long as the petals; shedding. *Tips* very small; roundish.

POINT. *Seedbud* beneath. *Shafts* two; very small. *Summits* simple.

S. VESS. None. *Fruit* oblong egg-shaped; scored; divisible into two.

SEEDS. Two; oblong; egg-shaped; convex on one side and scored. Flat on the other.

OBS. *Some of the florets in the center have neither Chives nor Pointals.*

CARAWAY. As there is only one species known, Linnæus Aromatic gives no description of it.—*The* Leaves *doubly winged.* Rundles Carui *terminating.* Petals *white.*

Carum, seu careum. *Gerard.* 1034. *Ray's Syn.* 213.
Cuminum pratense carui officinarum. *Bauh. pin.* 158.
In meadows and gardens. B. May—June.
Parkinson says, the young roots are better eating than Parsnips—The tender leaves may be boiled with pot herbs—The seeds are used in cakes; incrusted with sugar as sweet meats, and distilled with spirituous liquors for the sake of the flavour they afford.—The seeds were formerly recommended by Dioscorides to pale faced girls, and in more modern days their use in that case is not forgotten. They are no despicable remedy in Tertian Agues. They abound with an essential oil, which is antispasmodic and carminative.—Sheep, Goats and Swine eat it. Cows and Horses are not fond of it.

129 PARSNEP.   362 Paſtinaca.

Rundle of many ſpokes; flat. *Rundlets* of many ſpokes. *Fences* none.

EMPAL. *Cup* hardly perceptible.

BLOSS. *General*, uniform. *Florets* all fertile. *Individuals*; petals five; ſpear-ſhaped; entire; rolled inwards.

CHIVES. *Threads* five; hair-like. *Tips* roundiſh.

POINT. *Seedbud* beneath. *Shafts* two; reflected. *Summits* blunt.

S. VESS. None. *Fruit* oval; compreſſed and flat; diviſible into two.

SEEDS. Two; oval; nearly flat on each ſide; bound round with a border.

Cultivated
Sativa

PARSNEP, with ſimply winged leaves.—*Little leaves ſerrated.* Petals *yellow.*

Paſtinaca ſylveſtris latifolia.   *Bauh. pin.* 155.   *Ray's Syn.* 206.
Pnſtinaca latifolia ſylveſtris.   *Gerard.* 1025.   *Park.* 944.

1. There is a variety occaſioned by cultivation in which the leaves are broader.

Paſtinaca latifolia ſativa.

Wild Parſnep.   1. Garden Parſnep.

Ploughed fields and gardens. B. July.

The roots when cultivated, are ſweeter than Carrots and are much uſed by thoſe who abſtain from animal food in Lent: They are highly nutritious. In the North of Ireland they are brewed, inſtead of Malt, with Hops, and fermented with Yeaſt. The liquor thus obtained is agreeable. The ſeeds contain an eſſential oil; and will often cure Intermittent Fevers——Hogs are fond of the roots and quickly grow fat with them.

130 FENNEL.   364 Anethum.

Rundle of many ſpokes. *Rundlets* the ſame. *Fences* none.

EMPAL: *Cup* hardly perceptible.

BLOSS. *General*; uniform. *Florets* all fertile. *Individuals*; petals five, rolled inwards; entire; very ſhort.

CHIVES. *Threads* five; hair-like. *Tips* roundiſh.

POINT. *Seedbud* beneath; *Shafts* two; placed cloſe together, but not very diſcernible. *Summits* blunt.

S. VESS. None. *Fruit* nearly egg-ſhaped; compreſſed; ſcored; diviſible into two.

SEEDS. Two; ſomewhat egg ſhaped; convex and ſcored on one ſide; flat on the other.

FENNEL. The fruit egg-shaped—*Leavs divided into numer-* Common
ous slender thread-shaped segments. Petals *yellow.* Fæniculum
    Fæniculum vulgare. *Park.* 884. *Gerard.* 1032. *Ray's Syn.* 217.
    Fæniculum vulgare germanicum. *Bauh. pin.* 147.
1. There is a variety cultivated in gardens. *Sweet Fennel*
    Fæniculum dulce. *Bauh. pin.* 147.
    Finckle. 1. Fennel.
    On chalk-hills. 1. In Gardens. B. August.
    The tender buds are used in sallads—The leaves are commonly presented at table as sauce to fish—The stalks are blanched in Italy for winter use—The seeds abound with an essential oil, which is carminative and diuretic, but not heating.—The Royal William or Swallow-tail Butterfly, *Papilio machaon* feeds upon it.

131 GOUTWEED. 368 Ægopodium.

Rundle of many spokes : convex : *Rundlets* the same, but flat. *Fences* none.

EMPAL. *Cup* hardly discernible.
BLOSS. *General,.* uniform. *Florets* all fertile. *Individuals*; petals five; inversely egg-shaped; equal; concave : bent inwards at the point.
CHIVES. *Threads* five; simple; twice as long as the petals. *Tips* roundish.
POINT. *Seedbud* beneath; *Shafts* two; simple; upright; as long as the petals. *Summits* roundish.
S. VESS. None. *Fruit* oblong egg-shaped; scored; divisible into two.
SEEDS. Two; oblong egg-shaped; convex and scored on one side, flat on the other.

    GOUTWEED. The upper stem leaves growing by threes— Gerards
Petals white. Podagraria
    Podagraria vulgaris. *Park.* 943.
    Angelica sylvestris minor, sive erratica. *Bauh pin.* 155. *Ray's Syn.* 20
    Herba Gerardi. *Gerard.* 1001.
    Herb Gerard. Goutweed. Ashweed.
    In orchards, gardens, and pastures. P. May.
    The leaves may be eaten early in the spring with other pot-herbs.—Cows, Sheep and Goats eat it. Horses are not fond of it.

132 SMALLAGE. 367 Apium.

Rundle *with few spokes.* Rundlets *with many.*

EMPAL. *General Fence* none; or elfe of one or more leaves. *Partial Fence* the fame. *Cup* hardly perceptible.

BLOSS. *General,* uniform. *Florets* almoft all fertile. *Individuals;* petals circular; equal; bent inwards.

CHIVES. *Threads* five; fimple. *Tips* roundish.

POINT. *Seedbud* beneath. *Shafts* two; reflected. *Summits* blunt.

S. VESS. None. *Fruit* egg-fhaped; fcored; divifible; into two;

SEEDS. Two; egg-fhaped; fcored on one fide, flat on the other.

OBS. *In the* Celery Smallage *the general fence is often wanting.*

Celery
Gravcolens

SMALLAGE. The ftem leaves wedge-fhaped—*Root leaves winged. Little leaves divided into three lobes, and ferrated.* Petals *white. The fpokes of the* General Rundle *from five to eleven. Thofe of the* Rundlets *from eleven to fixteen.* Rundles *fitting.*

Apium paluftre et officinarum. *Bauh. pin.* 154. *Ray's Syn.* 214.

Apium vulgare, feu paluftre. *Park.* 926.

Elofelinum feu paludapium. *Gerard.* 1014.

Ditches and marfhes. B. Auguft.

1. The Garden Celery is only a variety of the above, with no other perceptible difference than the leaves being of a darker green.

The root of the variety (1) and the lower part of the ftalks bleached by covering them with earth, are very generally ufed raw in fallads; or boiled in foup; or ftewed. They are faid to be hurtful to people fubject to nervous complaints. They are certainly good antifcorbutics—The root in its wild ftate is fœtid, acrid and noxious.—The feeds yield an effential oil.—Sheep and Goats eat it. Cows are not fond of it. Horfes refufe it.

133 BREAKSTONE. 366 Pimpinella.

Rundle of many fpokes. *Rundlets* of ftill more. *Fences* none.

EMPAL. *Cup* not very difcernible.

BLOSS. *General*; nearly uniform; *Florets* all fertile. *Individuals*; petals five; nearly equal; heart-fhaped but beat inwards.

CHIVES. *Threads* five; fimple; longer than the petals. *Tips* roundifh.

POINT. *Seedbud* beneath. *Shafts* two; very fhort. *Summits* nearly globular.

S. VESS. None. *Fruit* oblong egg-fhaped; divifible into two.

SEEDS. Two; oblong; narrow towards the point; convex and fcored upon one fide; flat upon the other.

BREAKSTONE. The leaves winged; the fegments of the Burnet root leaves nearly circular; thofe of the ftem leaves ftrap-Saxifraga fhaped—*Petals white.*
  Pimpinella faxifraga. *Gerard,* 1044.
  Pimpinella faxifraga minor, foliis fanguiforbæ. *Ray's Syn.* 213.
  Pimpinella faxifraga major altera. *Bauh. pin.* 159.
  Pimpinella faxifraga major noftras. *Park.* 946.
  Burnet faxifrage.
1. There is a variety that is fmaller.
  Pimpinella faxifraga minor. *Bauh. pin.* 160.
  Pimpinella faxifraga hircina minor. *Park.* 947.
  Dry gravelly foils. P. Auguft.
Horfes, Cows, Goats, Sheep and Swine eat it.

BREAKSTONE. The leaves winged; little leaves heart-Greater fhaped; the odd one at the end divided into three lobes. *Hud-*Major *fon's Flor. Angl.* 110.
  Pimpinella faxifraga. *Ray's Syn.* 213.
  Pimpinella faxifraga hircina major. *Park.* 947.
  Pimpinella faxifraga major umbella candida. *Bauh. pin.* 159.
  Bipinella, five faxifraga minor. *Gerard.* 1044.
  Great Burnet Saxifrage.
1. There is a variety with purplifh bloffoms.
  Pimpinella faxifraga major, umbellarubente. *Baul. pin.*159.
  In woods and hedges. P. Auguft.
All the above fpecies and varieties of BREAKSTONE partake nearly of the fame qualities. The root is very acrid, burning the mouth like pepper. It affords a blue oil. Its acrimony hath

N 4

occa-

occasioned it to be used to cure the Tooth-ache and to cleanse the skin from Freckles. It is chewed to promote the secretion of saliva, and is used in gargles for dissolving viscid mucus in the throat. In Germany it is prescribed in the Asthma and Dropsy—The Royal William Butterfly, *Papilio Machaon*, is found upon both species.

Order III. Three Pointals.

134 MEALTREE. 370 Viburnum.

EMPAL. *Cup* with five teeth; superior: very small; permanent.

BLOSS. One petal; bell-shaped; with five shallow clefts. *Segments* blunt; reflected.

CHIVES. *Threads* five; awl-shaped; as long or twice as long as the blossom. *Tips* oblong; double.

POINT. *Seedbud* beneath: roundish. *Shaft* none, but instead thereof a turban-shaped gland. *Summits* three.

S. VESS. *Berry* roundish; one cell.

SEED. Single, roundish; hard as bone.

Pliant
Lantana

MEALTREE, with heart-shaped, serrated leaves; full of veins; and downy on the under side—*Blossoms white*; Berries *black*.

Viburnum. *Park.* 1448. *Bauh pin.* 429. *Ray's Syn.* 460.
Lantana sive viburnum. *Gerard.* 1490.
Pliant Mealy-tree. Way-faring tree.
In hedges. S. May.
Varieties.
1. Leaves oval.
2. Leaves variegated.
The bark of the root is used to make bird-lime.

Marsh
Opulus

MEALTREE, with gashed leaves, and leaf-stalks beset with glands—*Blossoms white*. Berries *red*.

Opulus ruellii. *Ray's Syn.* 460.
Sambucus aquatica, flore simplici. *Bauh pin.* 456.
Sambucus palustris seu aquatica. *Park.* 208.
Sambucus aquatilis seu palustris. *Gerard.* 1424.
Water Elder.
In woods and wet hedges. S. May—June.
1. The *Guelder Rose* is a variety of this, and there is another variety with variegated leaves.
Cows, Goats and Sheep eat it. Horses are not fond of it.

135 ELDER.

135 E L D E R.   372 Sambucus.

EMPAL. *Cup* superior; of one leaf; very small; with five divisions: permanent.

BLOSS. One petal; wheel-shaped, but somewhat concave; with five clefts; blunt. *Segments* reflected.

CHIVES. *Threads* five; awl shaped; as long as the blossom. *Tips* roundish.

POINT. *Seedbud* beneath; egg-shaped; blunt. *Shaft* none; but instead thereof a distended gland. *Summits* three; blunt.

S. VESS. *Berry* roundish; of one cell.

SEEDS. Three. Convex on one side; angular on the other.

ELDER. The tufts of flowers with three divisions; props Dwarf leafy; stem herbaceous—*Blossoms white above; purple beneath·* Ebulus Chives *white*; Tips *purple*; *one fixed on each side of every thread, so that there is properly speaking ten tips.*

Sambucus humilis, seu ebulus. *Bauh. pin.* 456. *Ray's Syn.* 461.

Ebulus, five sambucus humilis. *Gerard.* 1426. *Park.* 209.

Wall-wort.   Dane-wort.

Hedges and road-sides.   S. July.

Varieties.

1. Leaves jagged.
2. Leaves variegated.

This hath the same medical properties with the next species, but in some respects more violent, and therefore less manageable. A dram and a half of the root is a strong purge—The berries give out a violet colour—The green leaves drive away mice from granaries, and the Silesians strew them where their pigs lie, under a persuasion that they prevent some of the diseases to which they are liable.—Neither Cows, Goats, Sheep, Horses or Swine will eat it.

Common
Nigra

ELDER. The tufts of flowers with five divisions; stem woody—*Blossoms white.* Berries *black.* Tips *yellow; arrow-shaped; one on each thread.*

Sambucus. *Gerard.* 1422. vulgaris. *Park.* 207. *Ray's Syn.* 461.
Sambucus fructu in umbella nigra. *Bauh. pin.* 456.

1. There is one variety with white berries, (viz.) the Sambucus fructu albo. *Gerard.* 1422. *Park.* 208.
Sambucus fructu in umbella viridi. *Bauh. pin.* 456.

2. And another variety with jagged leaves.
Sambucus laciniato folio. *Bauh. pin.* 456. *Park.* 208. *Gerard.* 1234.

In woods and damp hedges. (1.) in several parts of Stafford-shire. S. April—May.

Cultivation produces the following variations;

3. Berries green.
4. Leaves cut and jagged.
5. Leaves striped with white.
6. Leaves spotted with white.
7. Leaves striped with yellow.

The whole plant hath a narcotic smell; it is not well to sleep under its shade—The *Wood* is hard, tough, and yellow. It is commonly made into skewers for butchers; tops for angling rods, and needles for weaving nets. It is not a bad wood to turn in the lathe—The inner green *Bark* is purgative, and may be used with advantage where acrid purgatives are requisite. In smaller doses it is diuretic, and hath done eminent service in obstinate glandular obstructions, and in dropsies. If Sheep that have the rot are placed in a situation where they can get at the bark and the young shoots, they will soon cure themselves—The *Leaves* are purgative like the bark, but more nauseous. They are an ingredient in several cooling ointments. If turnips, cabbages, fruit-trees, or corn (which are subject to blights, from a variety of insects), are whipped with the green leaves and branches of Elder, the insects will not attack them. *Philos. Transf.* v. 62. p. 348. A decoction of the *Flowers,* taken internally, is said to promote expectoration in pleurisies. If the flowers are fresh gathered they loosen the belly. Externally they are used in fomentations to ease pain and abate inflammation. Many people use the flowers to give a flavour to vinegar—A rob prepared from the *Berries* is a gentle opener, and promotes perspiration. The juice of the berries is employed to give a red colour to raisin or sugar wines. The berries are poisonous to poultry—The *Pith* being exceedingly light, is cut into balls, which are much used in many nice electrical experiments—Sheep eat it. Horses, Cows and Goats refuse it. The flowers kill Turkeys. The Elder Louse, *Aphis Sambuci,* and the Swallow-Tail Butter Fly, *Phalæna Sambucaria,* are found upon this species.

136 BLADDERNUT.

136 BLADDERNUT. 374 Staphylæa.

EMPAL. *Cup* with five divisions; concave; roundish; coloured; nearly as large as the blossom.

BLOSS. *Petals* five; oblong; upright; resembling the cup. *Honey-cup* concave, distended; situated at the bottom of the flower, upon the receptacle.

CHIVES. *Threads* five; oblong; upright; as long as the cup, *Tips* simple.

POINT. *Seedbud* rather thick, with three divisions. *Shafts* three; simple; somewhat longer than the chives. *Summits* blunt, contiguous.

S.VESS. *Capsules* three; bladder-shaped; limber; joined by a seam lengthways; tapering at the points; opening inwardly.

SEEDS. Two; hard as bone; somewhat globular; obliquely tapering; with a circular hole at the side, near the point.

OBS. *In the British species there are but two divisions in the pointal and in the capsule.*

BLADDERNUT with winged leaves—*Blossoms in whorls;* Tree *white.* Pinnata
  Staphylodendron. *Ray's Syn.* 468.
  Nux veficaria· *Park.* 1417· *Gerard.* 1437·
  Piftacea fylveftris. *Bauh. pin.* 401.
  In hedges near Pontefract, Yorkshire. S. June.

137 CHICKWEED. 380 Alfine.

EMPAL. *Cup* with five leaves; concave; oblong; tapering.

BLOSS. *Petals* five; equal; longer than the cup.

CHIVES. *Threads* five; hair-like. *Tips* roundish.

POINT. *Seedbud* nearly egg-shaped. *Shafts* three; thread-shaped. *Summits* blunt.

S.VESS. *Capsule* egg-shaped; of one cell; covered by the cup.

SEEDS. Numerous, roundish.

OBS. *In the* Common Chickweed, *the chives soon fall off, so that it is not unusual to find flowers with fewer than five chives.*

CHICKWEED.

Common
Media

CHICKWEED with divided petals; and leaves betwixt egg and heart-shaped—*Flowers upright, and open from nine in the morning to noon; but if it rains that day they do not open: After rain they become pendant, but in the course of a few days rise again. The Capsule opens with six valves, letting fall the seeds; which are round; compressed; yellow; and rough with little tubercles. Blossoms white.*

This species is a notable instance of what is called the *Sleep of Plants*; for every night the leaves approach in pairs, so as to include within their upper surfaces, the tender rudiments of the new shoots; and the uppermost pair but one at the end of the stalk, are furnished with longer leaf-stalks than the others, so that they can close upon the terminating pair and protect the end of the branch.

Alsine media. *Bauh. pin.* 250. seu minor. *Gerard.* 611.

Alsine vulgaris seu morsus gallinæ. *Ray's Syn.* 347.

In rich cultivated ground; very common. A. April—October.

The young shoots and leaves when boiled, can hardly be distinguished from Spring Spinach, and are equally wholesome.— Swine are extremely fond of it; Cows and Horses eat it; Sheep are indifferent to it, and Goats refuse it. Small Birds eat the seeds.

The Cream Spot Tyger Moth, *Phalæna Villica,* is found upon it.

Order IV. Four Pointals.

138 PARNASSUS.   384 Parnassia.

EMPAL. *Cup* with five divisions; permanent. *Segments* oblong; sharp; expanding.

BLOSS. *Petals* five; nearly circular; scored; concave, expanding: margins imperfect. *Honey-cups* five; each being a concave heart-shaped substance, furnished with thirteen little shafts or pillars set along the edge; and each pillar terminated by a little globe.

CHIVES. *Threads* five: awl-shaped; as long as the petals. *Tips* depressed; fixed side-ways to the threads.

POINT. *Seedbud* egg-shaped; large. *Shafts* none; but instead thereof an open hole. *Summits* four; blunt; permanent: growing larger as the seed ripens.

S.VESS. *Capsule* egg-shaped; with four angles; one cell and four valves. *Receptacle* in four parts; connected with the valves.

SEEDS. Numerous; oblong.

OBS. *The essential character of this genus is strongly marked by the honey-cups.*

PARNASSUS.

PARNASSUS. As there is only one species known Linnæus Meadow gives no description of it—*The root leaves heart-shaped; on long* Paluſtris *leaf-stalks. Stem leaves sitting. Whilſt it is in flower, the seedbud having neither ſhaft nor summit, is open at the top. The chives lay their tips alternately upon the hole, and having discharged their duſt, return back to the petals. The Petals are white, with remarkable yellow ſtreaks; and the peculiar appearance of the honey-cups is as beautiful as it is uncommon.*

Parnaſſia vulgaris et paluſtris. *Ray's Syn.* 355.
Gramen Parnaſſi minus. *Gerard.* 840.
Gramen Parnaſſi vulgare. *Park.* 429.
Gramen Parnaſſi flore albo ſimplici. *Bauh. pin.* 309.
Graſs of Parnaſſus.
In moiſt meadows. P. July—Auguſt.

Horſes and Goats eat it. Sheep are not fond of it. Cows and Swine refuſe it.

Order V. Five Pointals.

139 F L A X   389 Linum.

EMPAL. *Cup* of five leaves; ſmall; ſpear-ſhaped; upright permanent.

BLOSS. Funnel-ſhaped, *Petals* five; oblong; large; blunt; gradually expanding and growing broader upwards.

CHIVES *Threads* five; awl-ſhaped; upright; as long as the cup; alternating with theſe are the rudiments of five more. *Tips* ſimple; arrow ſhaped.

POINT. *Seedbud* egg-ſhaped. *Shafts* five;—thread-ſhaped upright; as long as the chives. *Summits* ſimple; reflected.

S VESS. *Capſule* globular; with five imperfect angles ten cells and five valves.—

SEEDS. Solitary: flattiſh egg ſhaped, tapering; gloſſy.

OBS. *In the* Little Flax *one fifth in the numbe of parts is wanting.*

FLAX. The cups and capſules furniſhed with ſharp points; Common petals ſcolloped; leaves alternate; ſpear-ſhaped. Stem ſolitary Uſitatiſſimum —*The inner edges of the cup a little fringed.* Bloſſoms *pale blue.*

Linum arvenſe. *Bauh. pin.* 214.
Linum ſylveſtre vulgatius. *Park.* 1334.
Linum ſativum. *Gerard.* 556. *Ray's Syn.* 362.
In cornfields. A. July.

This

This valuable plant originally came from those parts of Ægypt that are exposed to the inundations of the Nile. It would require a volume to recite its uses; the limits of this work will only permit me to mention a few of them. The *Seeds* yield by expression only, a large proportion of oil, which is an excellent pectoral as is likewise the mucilaginous infusion. They make an easy and useful poultice in cases of external inflammation; and they are the food of several small Birds. After the oil is expressed the remaining farinaceous part called Oil Cake, is given to Oxen who soon grow fat upon it. The *Oil* itself differs in several respects from other expressed oils; it does not congeal in winter, nor does it form a solid soap with fixed alkaline salts; and it acts more powerfully as a menstruum upon sulphureous bodies. When heat is applied during the expression it gets a yellowish colour, and a peculiar smell. In this state it is used by the painters and the varnishers—The fibres of the *Stem* are manufactured into linen; and this linen when worn to rags, is made into paper.

Perennial
Perenne

FLAX. The cups and capsules blunt; leaves alternate; spear shaped; very entire—*Blossoms blue.*

Linum sylvestre cæruleum perenne erectius, flore et capitulo majore. *Ray's Syn.* 362.

1. There is a variety which is smaller, viz. the lesser trailing perennial Flax.

Linum sylvestre cæruleum perenne procumbens, flore et capitulo minore. *Ray's Syn.* 362.

Perennial blue Flax.

In barren soil. P. June,

Wild
Tenuifolium

FLAX. The cups furnished with a sharp point. Leaves scattered; betwixt strap and bristle-shaped; rough with points turned backwards—*Blossoms blue, or reddish; with purple streaks.*

Linum sylvestre angustifolium, floribus dilute purpurascentibus vel carneis. *Bauh. pin.* 214. *Ray's Syn.* 362.

Narrow-leaved wild Flax.

On the sea coast. P. June—July.

Purging
Catharticum

FLAX. The leaves opposite; betwixt egg and spear-shaped; stem forked, blossoms sharp—*Before the flowers open they hang down.* Blossoms *white.*

Linum sylvestre catharticum. *Gerard.* 560. *Ray's Syn.* 362.

Linum pratense flosculis exiguis. *Bauh. pin.* 214.

Chamælinum clusii flore albo, seu linum sylvestre catharticum. *Park.* 1336.

In dry pastures. A. May—June.

An infusion of two drams of the dried plant is an excellent purge. —Horses, Sheep and Goats eat it.

FLAX.

FIVE POINTALS. 191

FLAX. The leaves oppofite: ftem forked. Bloffoms with Little
four chives and four pointals—*White.*   Radiola
  Radiola vulgaris ferpyllifolia. *Ray's Syn.* 345. Tab. 15. fig. 3.
  Millegrana minima. *Gerard.* 569. five herniaria minor.
Park. 447.
  Polygonum minimum feu millegrana minima. *Bauh. pin.* 282.
Leaft Rupture Wort, or Allfeed.
Wet gravelly foil. A. Auguft.

140 SUNDEW. 391 Drofera

EMPAL. *Cup* one leaf; with five clefts; fharp; upright;
  permanent.
BLOSS. Funnel-fhaped. *Petals* five; nearly egg-fhaped;
  blunt, fomewhat larger than the cup.
CHIVES. *Threads* five; awl-fhaped; as long as the cup.
  Tips fmall.
POINT. *Seedbud* roundifh. *Shafts* five; fimple; as long
  as the chives. *Summits* fimple.
S.VESS. *Capfule* nearly egg-fhaped; of one cell, opening
  with five valves at the top.
SEEDS. Numerous; very fmall; nearly egg-fhaped.

  SUNDEW. The ftalk fpringing direЄtly from the root; leaves Round-leaved
circular.—*Leaf-ftalks fringed at the bafe.* Bloffoms *white.* Rotundifolia
  Ros folis folio rotundo. *Bauh. pin.* 357. *Ray's Syn.* 356.
  Ros folis major. *Gerard.* 1557. *Park.* 1052.
  Rofa folis. Red Rot.
  In moffy bogs. P. July—Auguft.
  The whole plant is acrid, and fufficiently cauftic to erode the
fkin; but fome ladies know how to mix the juice with
milk, fo as to make it an innocent and fafe application to remove
freckles and funburn. The juice that exfudes from it unmixed,
will deftroy warts and corns—The plant hath the fame effeЄt
upon milk as the common *Butterwort* hath; and like that too is
fuppofed to occafion the rot in Sheep. Query, is not the four
coagulated milk of the Syrians, called *Leban,* or *Leven,* at firft
prepared with fome plant of this kind? See *Ruffel's* Nat. Hift.
of Aleppo.

  SUNDEW. The ftalk fpringing direЄtly from the root. Long-leaved
Leaves oblong—*Bloffoms white. This fpecies differs from the pre-* Longifolia
ceding only in the fhape of the leaves.
  Ros folis folio oblongo. *Bauh. pin.* 357. *Ray's Syn.* 356.
  Ros folis fylveftris longifoliis. *Park.* 1052.
  Ros folis minor. *Gerard.* 1557.
  In moffy bogs. P. July—Auguft.

It.

Its properties are the same as in the preceding species. The Generic name, SUNDEW, seems to be derived from a very striking circumstance in the appearance of these plants; the leaves are fringed with hairs, supporting small drops or globules of a pellucid liquor like dew, which continue even in the hottest part of the day, and in the fullest exposure to the sun. Perhaps the acrimony of the plant resides in this secreted liquor.

141 SIBBALD. 393 Sibbaldia.

EMPAL. *Cup* one leaf; with ten shallow clefts; upright at the base: permanent. *Segments* alternately narrower; half spear-shaped; equal; expanding.

BLOSS. Petals five; egg-shaped; standing on the cup.

CHIVES. *Threads* five; hair-like; shorter than the petals; standing on the cup. *Tips* small; blunt.

POINT. *Seedbuds* five; egg-shaped; very short. *Shafts* as long as the chives and standing upon the sides of the seedbuds. *Summits* somewhat globular.

S.VESS. None. The *Cup* closes upon the seeds.

SEEDS. Five; rather long.

Cinquefoil Procumbens   **SIBBALD.** The little leaves with three teeth—*Blossoms yellow.*
Pentaphylloides pumila foliis ternis, ad extremitates trifidis. *Ray's Syn.* 256.
Bastard Cinquefoil.
On Benlomond, a mountain on the borders of Lochlomond in Scotland. P. July—August.

142 THRIFT. 388 Statice.

EMPAL. *Common Cup* different in different species. *Cup* one leaf; funnel-shaped. *Tube* narrow. *Border* entire; plaited; skinny.

BLOSS. Funnel-shaped. *Petals* five; narrow at the base; broader upwards; blunt; expanding.

CHIVES. *Threads* five; awl-shaped; shorter than the blossom and fixed to the claws of the petals. *Tips* fixed side-ways to the threads.

POINT. *Seedbud* extremely small. *Shafts* five; thread-shaped; distant. *Summits* sharp.

S.VESS. None. The tube of the cup closes and contains the seed; but the border continues expanded.

SEED. Single: very small; roundish; crowned with the cup.

THRIFT.

THRIFT. The flowers forming a head upon a fimple ftalk. Sea
Leaves ftrap-fhaped—*Bloſſoms pink.*   Armeria
   Statice montana minor. *Ray's Syn.* 203.
   Caryophyllus montanus minor. *Bauh. pin.* 211.
   Caryophyllus marinus minimus lobelii. *Gerard.* 602.
   Gramen marinum minus. *Park.* 1276.
   Sea Gilliflower. Sea Cuſhion.
   On the fea coaſt. P. June.
   It is now very generally introduced into gardens, as an edg-
ing for borders.—Horſes and Goats eat it. Sheep are not fond
of it.

THRIFT. The ſtalk cylindrical, fupporting a panicle. Lavender
Leaves fmooth; without ſtrings; but with ſharp points on the Limonium
under fide—*Bloſſoms blue.*
   Limonium. *Gerard.* 411. *Ray's Syn.* 201.
   Limonium majus vulgatius. *Park* 1234.
   Limonium maritimum majus. *Bauh. pin* 192.
   Sea Lavender.
   Sea coaſt. P. July—Auguſt.
   Sheep and Goats eat it.

THRIFT. The ſtalk bearing the panicle, trailing. The Matted
barren branches naked and bent backwards. Leaves wedge- Reticulata
fhaped; without any ſharp points.—
   Matted Sea Lavender.
   Sea coaſt. September.

Order VI. Many Pointals.

143 MOUSETAIL. 394 Myoſurus.

EMPAL. *Cup* five leaves; half fpear-fhaped; blunt, reflected; coloured; deciduous; joined together
   above the bafe.
BLOSS. *Petals* five; very fmall; ſhorter than the cup;
   tubular at the bafe; opening obliquely inwards.
CHIVES. *Threads* five (or more); as long as the cup.
   Tips oblong; upright.
POINT *Seedbuds* numerous; fitting upon the receptacle
   in an oblong conical figure. *Shafts* none. *Summits*
   fimple.
S. VESS. None. *Receptacle* very long. ſhaped like a pillar; containing the feeds which are layed one over
   another like tiles, and are covered.
SEEDS. Numerous; oblong: tapering.

   OBS. *The number of chives varies greatly in this genus. It is
nearly allied to the* Crowfoot: *the* Petals *are a ſort of* Honey-cup.
       O   MOUSETAIL

Little
Minimus

MOUSETAIL. As there is only one species known Linnæus gives no defcription of it—*The* Stem *fimple. Root-leaves narrow; ftrap-fhaped.* Bloffoms *greenifh.*

Myofurus. *Ray's Syn.* 251.

Holofteo affinis cauda muris. *Pauh. pin.* 190.

Holofteum loniceri, cauda muris vocatum. *Park.* 500.

Cauda muris. *Gerard* 426.

Gravelly meadows. A. April—May.

The whole plant is acrid.

C L A S S VI.

S I X C H I V E S.

THE flowers of this clafs contain fix chives all of the *fame length*, whereas in the 15th clafs the chives though fix in number are *unequal in length,* four of them being long and two of them fhort.

The Bulbous Roots in this clafs are fome of them noxious, as thofe of the Daffodil, the Hyacinth and the Fritillary; others are corrofive, as Garlic; but by roafting or boiling they lofe a great part of their acrimony.

CLASS VI.

(196)

CLASS VI.

SIX CHIVES.

ORDER I. ONE POINTAL.

* Flowers with a cup and a blossom.

144. FRANKWORT. *Blossom* five petals. *Cup* one leaf; beneath. *Capsule* one cell, with many seeds.

145. BARBERRY. - *Blossom* six petals. *Cup* six leaves; beneath. *Berry* with two seeds.

† *Six-chived Grasspoly.*

* * Flowers with a sheath, or husk.

146. DAFFODIL. - - *Blossom* superior. *Petals* six. *Honey-cup*, bell-shaped; on the outside of the chives.

147. GARLIC. - - - *Blossom* beneath. *Petals* six egg-shaped; sitting.

148. BELLWORT. - - *Blossom* beneath; with six clefts *Berry* with three seeds.

149. HYACINTH. - - *Blossom* beneath; with six clefts. Three honey-cup pores at the top of the seedbud.

150. ASPHODEL. - - *Blossom* beneath; with six flat petals.

151. BETHLEMSTAR. - - *Blossom* beneath; with six petals. *Threads* alternately broader at the base.

150. SQUILL.

152. SQUILL. - - - *Blossom* beneath. *Petals* six; deciduous. *Threads* cylindrical; of equal thickness.
153. SPARAGUS. - - - *Blossom* beneath. *Petals* six. *Berry* with six seeds.
154. FRITILLARY. - - *Blossom* beneath. *Petals* six; egg-shaped; with a *Honey-cup* pore at the base.

† *Rush.*

* * * *Flowers imperfect.*

155. SPICEWORT. - - *Sheath* containing several flowers. *Capsule* with three cells.
156. RUSH. - - - - *Cup* six leaves. *Capsule* with one cell.
157. PURSLAIN. - - *Cup* with twelve clefts. *Capsule* with two cells.

Order II. Two Pointals.

† *Round-leaved Dock.*
† *Pepper, and Spotted Snakeweed.*

Order III. Three Pointals.

158. TUBEROOT. - - *Empalement* a sheath. *Blossom* with six divisions resembling petals.
159. ARROWGRASS. - *Cup* three leaves. *Blossom* three petals. *Capsule* opening at the base.
160. DOCK. - - - *Cup* three leaves. *Blossom* three petals. *Seed* single; three cornered.

† *Fenced Asphodel.*

Order V. Many Pointals.

161. WATERPLANTAIN. *Cup* three leaves- *Blossom* three petals. *Seed-vessels* many.

144 FRANKWORT.　445 Frankenia.

EMPAL. *Cup* one leaf; funnel-fhaped; ten cornered; permanent. *Rim* with five fharp teeth; ftanding out.

BLOSS. *Petals* five; the *Claws* as long as the cup. *Border*
flat: limbs nearly round and expanding. *Honeycup*
a channelled tapering claw, fixed to each claw of
the petals.

CHIVES. *Threads* fix; as long as the cup. *Tips* roundifh;
double.

POINT. *Seedbud* oblong, *Shaft* fimple; as long as the
chives. *Summit* divided into fix fegments.

S. VESS. *Capfule* oval; of one cell and three valves.

SEEDS. Many; egg-fhaped; fmall.

Smooth　　　FRANKWORT. The leaves ftrap-fhaped; crowded; fring-
Lævis　　at the bafe—*Stems numerous; branched; trailing.* Bloffoms
purple.
　　Lychnis fupina maritima ericæ facie. *Ray's Syn.* 338.
　　Erica fupina maritima. *Park.* 1484.
　　Polygonum maritimum minus, foliolis ferpylli. *Bauh pin.*
281.
　　Polygonum pufillo vermiculato ferpylli folio. *Gerard.* 567.
　　Smooth fea heath.
　　Marfhes on the fea-coaft. P. Auguft.

Mealy　　　FRANKWORT. The leaves inverfely egg-fhaped; dented
Pulverulenta　at the end; dufty on the under fide
　　Alfine maritima fupina, foliis chamæfices. *Ray's Syn.* 352.
　　Anthyllis maritima chamæficæ fimilis. *Bauh. pin.* 282.
　　Broad-leaved fea heath.
　　Sea coaft. A. July.

145 BARBERRY. 442 Berberis.

EMPAL. *Cup* of fix leaves; ftanding wide. *Leaves* egg-fhaped, narroweft at the bafe; concave; coloured; deciduous: alternately fmaller.

BLOSS. *Petals* fix; roundifh, concave; not quite upright; fcarcely larger than the cup. *Honeycups* two roundifh coloured fubftances, growing to the bafe of each petal.

CHIVES. *Threads* fix; upright; compreffed; blunt. *Tips* two; adhering to each fide of the thread, at the end.

POINT. *Seedbud* cylindrical; as long as the chives. *Shaft* none. *Summit* round and flat; broader than the feedbud, encompaffed by a fharp border.

S. VESS. *Berry* cylindrical; blunt; dimpled, with one cell.

SEEDS. Two; oblong; cylindrical; blunt.

BARBERRY. The fruit-ftalks fupporting bunches of flow-ers—*In fearching for the honey-cup at the bafe of the petals when the flower is fully expanded, if you happen to touch the threads tho' ever fo flightly, the tips burft and throw out the duft with a confiderable expanfive force.* The Leaves *change into thorns.* Bloffom *yellow; fometimes ftreaked with orange.* Berries *red.*
Common Vulgaris

Berberis dumetorum. *Bauh pin.* 454. *Ray's Syn.* 465.
Berberis. *Park.* 1559.
Spina acida five oxyacantha. *Gerard.* 1325.
1. There is one variety with white berries;
2. And another in which the berries have no feeds.
Pipperidge bufh.
Woods and hedges. S. June.
The leaves are gratefully acid. The flowers are offenfive to the fmell when near, but at a proper diftance their odour is extremely fine. The berries are fo very acid that birds will not eat them, but boiled with fugar they form a moft agreeable rob or jelly. They are ufed likewife as a dry fweetmeat and in fugarplumbs. An Infufion of the Bark in white wine is purgative. The roots boiled in lye dye wool yellow. In Poland they dye Leather of a moft beautiful yellow with the Bark of the root. The inner Bark of the ftems dyes linnen of a fine yellow, with the affiftance of Alum. This fhrub fhould never be permitted to grow in corn lands, for the ears of wheat that grow near it never fill, and its influence in this refpect has been known to extend as far as three or four hundred yards acrofs a field.
Cows, Sheep, and Goats eat it; Horfes and Swine refufe it.

SIX CHIVES.

146 DAFFODIL. Narciſſus 403.

EMPAL. *Sheath* oblong; blunt; compreſſed; opening upon the flat ſide: ſhrivelling.

BLOSS. *Petals* ſix; egg-ſhaped; tapering; flat; fixed outwardly above the baſe of the tube of the honey-cup. *Honey-cup,* one leaf; funnel-ſhaped, but rather cylindrical; with a coloured border.

CHIVES. *Threads* ſix; awl-ſhaped; fixed to the tube of the honeycup, but ſhorter than the honeycup. *Tips* rather long.

POINT. *Seedbud* beneath: roundiſh; with three blunt corners. *Shaft* thread-ſhaped; longer than the chives: *Summit* with three clefts; concave; blunt.

S.VESS. *Capſule* roundiſh; bluntly three cornered; with three cells and three valves.

SEEDS. Numerous; globular; with little appendages.

Pale
Poeticus

DAFFODIL. The ſheath with only one flower. Honeycup wheel-ſhaped; very ſhort; ſkinny; ſomewhat ſcolloped—*Bloſſom pale yellow.*

Narciſſus medio luteus. *Gerard.* 124. vulgaris. *Park.* 74. *Ray's Syn.* 371.

Narciſſus pallidus circulo luteo. *Bauh. pin.* 51.

Primroſe peerleſs.

In meadows, not common. P. April.

Wild
Pſeudo-
Narciſſus

DAFFODIL. The ſheath with only one flower. Honeycup bell-ſhaped; upright; curled; as long as the petals, which are egg-ſhaped—*And pale yellow.*

Narciſſus ſylveſtris pallidus; calyce luteo. *Bauh. pin.* 52. *Ray's Syn.* 371.

Pſeudo-Narciſſus anglicus *Gerard.* 138. vulgaris. *Park.* 74. Woods and hedges. P. March—April.

The duſt in the microſcope appears kidney-ſhaped. The freſh roots are acrid.

147 GARLIC

147 GARLIC. Allium. 409

EMPAL. *Sheath* common to several flowers; roundish; shrivelling.

BLOSS. *Petals* six; oblong.

CHIVES. *Threads* six; awl-shaped; generally as long as the blossom. *Tips* oblong; upright.

POINT. *Seedbud* superior: short; somewhat three cornered; the corners marked by a line. *Shaft* simple. *Summit* sharp.

S.VESS. *Capsule* very short; broad; of three lobes, three cells and three valves.

SEEDS. Many; roundish.

OBS. In some species every other chive alternately is broad, and the tip is fixed in a fork at the end of the thread.

* Stem leaves flat. Rundles bearing capsules.

GARLIC. The stem with flat leaves. Rundle globular; Round-headed chives three pointed; petals rough upon the back, which is Ampelopra-somewhat keel shaped—*Chives a little longer than the blossom.* Blos-sum soms *pale purple. Lateral root solid.*
Allium holmense sphærico capite. *Ray's Syn.* 370.
Allium sphærico capite, folio latiore seu scorodoprasum alterum. *Bauh pin.* 74
Scotodoprasum primum clusii. *Gerard.* 180.
In the island of Holms at the mouth of the Severn. P.
This is eaten along with other pot-herbs. It communicates its flavour to the milk and butter of cows that graze where it grows.

* * Stem leaves flat ; rundles bearing bulbs.

GARLIC. The stem with flat leaves, and bulbs. Sheath of Broad-leaved the leaves cylindrical; sheath of the flower without any sharp Arenarium point; chives with three points—*Bulbs and* Blossoms *blue.* Chives *a little longer than the blossom.* Leaves *narrow, entire at the edges.*
Allium sylvestre amphycarpon, foliis porraceis, floribus et nucleis purpureis. *Ray's Syn.* 370.
Broad-leaved mountain Garlic.
On mountains in the north. P. July.

*** STEM

*** *Stem leaves nearly cylindrical. Rundle bearing Bulbs.*

Crow
Vineale

GARLIC. The ſtem with cylindrical leaves, and bulbs. Chives three pointed—*Bloſſoms ſmall, violet coloured. Two long briſtles from the chives project out of the bloſſoms.*

Allium ſylveſtre. *Park.* 870. *Gerard.* 179. *Ray's Syn.* 369.

Allium campeſtre juncifolium capitatum purpuraſcens majus. *Bauh. pin.* 74.

Allium ſylveſtre vinearum. *Bauh. pin.* 72.

In paſtures. P. June.

The young ſhoots are eaten in ſallads, or boiled as a pot-herb.

Wild
Oleraceum

GARLIC. The ſtem with cylindrical leaves, and bulbs. Leaves rough; ſemi-cylindrical furrowed on the under ſide. Chives ſimple—*Root a ſolid bulb.* Stem *two or three feet high; upright, or only a little bent towards the top; ſmooth; not ſcored; ſolid. Bulbs egg-ſhaped, forming a roundiſh knob; from betwixt theſe ariſe ſeveral thread ſhaped fruit-ſtalks, each ſupporting a ſingle flower which is nodding; cylindrical, but ſomewhat bell-ſhaped, very minute white dots hardly viſible to the naked eye, are ſcattered over the ſurface of the whole plant.* Bloſſom *whitiſh green, with three dark purple ſtreaks on each petal.*

Allium ſylveſtre bicorne, flore exherbaceo albicante, cum triplici in ſingulis petalis ſtria atro-purpurea. *Ray's Syn.* 370.

Allium montanum bicorne, flore ex-albido. *Bauh. pin.* 75.

Wild Garlic, with an herbaceous ſtriated flower.

Cornfields and paſtures. P. July.

The tender leaves are very commonly boiled in ſoups, or fried with other herbs.—Cows, Goats, Sheep and Swine eat it.

**** *Leaves growing from the roots. Stalk naked.*

Ramſon
Urſinum

GARLIC with naked, ſemi-cylindrical ſtalks. Leaves ſpearſhaped; on leaf ſtalks. Rundle flat at the top—*Leaves as long as the ſtalk.* Threads *fixed to the claws of the petals; ſhorter than the ſhaft.* Bloſſoms *white.*

Allium ſylveſtre latifolium. *Bauh. pin.* 74. *Ray's Syn.* 370.

Allium urſinum. *Gerard.* 180. *Park.* 871.

In woods and hedges. P. May.

An infuſion of this plant in brandy is eſteemed a good remedy for the gravel. Pennant's Tour. 1772, p. 175, other plants that grow near it will not flouriſh.—Cows eat it; but it communicates its flavour to the milk.

148 BELLWORT. 425 Convallaria.

EMPAL. *Cup* none.

BLOSS. One petal; bell-fhaped; glofly. *Border* with fix
clefts: *Segments* blunt; expanding and reflected.

CHIVES. *Threads* fix; awl-fhaped; ftanding on the petal;
fhorter than the bloffom. *Tips* oblong; upright.

POINT. *Seedbud* globular. *Shaft*, thread-fhaped; longer
than the chives. *Summit* blunt; three cornered.

S. VESS. *Berry* globular; with three cells, fpotted
before it is ripe.

SEED. Solitary; roundifh.

BELLWORT with a naked ftalk, and bell-fhaped bloffoms— Lii
white; fometimes red; or purple and white by cultivation, which Majalis
will likewife make the bloffoms double.
 Lilium convallium. *Gerard.* 410. *Ray's Syn.* 265.
 Lilium convallium album. *Bauh. pin.* 304.
 Lilium convallium flore albo. *Park.* 349.
 Lilly of the Valley. May—Lily.
 Woods and heaths. P. May.
The flowers though fragrant when frefh, have a narcotic fmell
when dry. Reduced to powder they excite fneezing. An extract prepared from the flowers or from the roots partakes of the
bitternefs as well as of the purgative properties of aloes. The
dofe from twenty to thirty grains. A beautiful and durable green
colour may be prepared from the leaves by the affiftance of lime.
Sheep and Goats eat it. Horfes, Cows and Swine refufe it.

BELLWORT with funnel-fhaped bloffoms, and alternate Odoriferous
leaves, embracing the ftem. Stem two edged: fruit-ftalks at the Polygonatum
bafe of the leaves; feldom fupporting more than one flower—
from a fpan to a foot high. Bloffoms *white. By cultivation the*
bloffoms will become double and the ftalks purple.
 Polygonatum floribus ex fingularibus pediculis. *Ray's Syn.* 263.
 Polygonatum latifolium flore majore odoro. *Bauh. pin.* 303.
 Polygonatum majus flore majore. *Park.* 696.
 Polygonatum latifolium 2. Clufii. *Gerard.* 904.
 Sweet fmelling Solomon's Seal.
 In fiffures of rocks in Yorkfhire. P. May—June.
The roots are very nutritious, and the young fhoots are eaten
at Conftantinople as Sparagus,—Sheep and Goats eat it. Horfes,
Cows and Swine refufe it.

BELLWORT.

Solomons
Multiflora

BELLWORT with funnel-shaped bloffoms, and alternate leaves, embracing the ftem. Stem cylindrical. Fruit-ftalks at the bafe of the leaves fupporting many flowers—*Bloffoms white.*
Polygonatum. *Gerard.* 903. *Ray's Syn.* 263.
Polygonatum vulgare. *Park.* 696.
Polygonatum latifolium vulgare. *Bauh. pin.* 303.
1. By cultivation the leaves are fometimes ftriped.
Solomon's Seal.
In woods. P. May—June.
The roots and the young fhoots are nutritious.—Cows, Goats and Sheep eat it.

149 HYACINTH. 427 Hyacinthus.

EMPAL. *Cup* none.
BLOSS. One petal; bell-fhaped. *Border* with fix clefts; reflected. *Honey-cups;* three pores filled with honey, at the point of the feedbud.
CHIVES. *Threads* fix; awl-fhaped; rather fhort. *Tips* approaching.
POINT. *Seedbud* roundifh; but with three edges, and three furrows. *Shaft* fimple; fhorter than the bloffom. *Summit* blunt.
S.VESS. *Capfule* nearly round, but with three corners, three cells and three valves.
SEEDS. Two (for the moft part), roundifh.

Harebell
Non-fcriptus

HYACINTH. Bloffoms bell-fhaped with fix divifions, rolled back at the ends—*Floral* Leaves *two, and generally longer than the bloffoms.* Segments *of the petal rolled back.* Summit *moift.* Bloffoms *blue; fometimes white.*
Hyacinthus oblongo flore cæruleus major. *Bauh. pin.* 43.
Hyacinthus Anglicus. *Gerard.* 110. *Ray's Syn.* 373.
Hyacinthus Anglicus, belgicus feu Hifpanicus. *Park.* 122.
Englifh Hyacinth or Harebells.
Woods and hedges. P. May.
The frefh roots are poifonous. they may be converted into ftarch.—It furnifhes food for the Wood Tyger Moth, *Phalæna Plantaginis.*

ONE POINTAL. 205

150 ASPHODEL. 422 Anthericum.

EMPAL. *Cup* none.
BLOSS. *Petals* fix; oblong; blunt; greatly expanded.
CHIVES. *Threads* fix; awl-fhaped; upright. *Tips* fmall; with four furrows; fixed fide-ways to the threads.
POINT. *Seedbud* with three corners, but flightly marked. *Shaft* fimple; as long as the chives. *Summit* blunt; three cornered.
S. VESS. *Capfule* egg-fhaped; fmooth; with three furrows; three cells and three valves.
SEEDS. Numerous; angular.

OBS. *The* Fenced Afphodel *hath a* Cup *with three teeth, and three diftinct* Summits *without any fhaft. In the* Baftard Afphodel *the petals fland upright and are betwixt ftrap and fpear-fhaped. The* Threads *are woolly; the* Seeds *fpindle-fhaped; and the* Shaft *very fhort.*

ASPHODEL. The leaves nearly flat; ftalk fupporting a fingle flower—*furnifhed with three or four floral leaves; fpear-fhaped, alternate, fheathing.* Bloffom *beneath; white on the infide; pale red without.* Saffron Serotinum
Bulbocodium alpinum juncifolium, flore unico; intus albo, extus fqualide rubente. *Ray's Syn.* 374. Tab. 17. fig. 1.
Pfeudo narciffus gramineo folio. *Bauh. pin.* 51.
Bulbocodium ferotinum. *Hudfon.* 122.
Mountain Saffron.
On Snowdon and other high mountains in Wales. P. Auguft.

ASPHODEL. The leaves fword fhaped; threads woolly— *The flowers appear very remarkable from the woolly threads.* Bloffoms *greenifh.* Baftard Offifragum
Phalangium Anglicum paluftre iridis folio. *Ray's Syn.* 375.
Afphodelus Lancaftriæ verus. *Gerard.* 97.
Pfeudo-Afphodelus paluftris Anglicus. *Bauh. pin.* 29.
Baftard or Lancafhire Afphodel.
In turf bogs. P. Auguft.
This plant is fuppofed to foften the bones of animals that eat it, but this opinion wants confirmation.—Cows and Horfes eat it. Sheep and Swine refufe it.

ASPHODEL.

Fenced
Calyculatum

ASPHODEL. The leaves fword-fhaped; cup with three lobes. Threads not woolly. Flowers with three fummits—*Bloſſoms greeniſh.*

Phalangium Scoticum paluſtre minimum, iridis folio. *Ray's Syn.* 375.

Pſeudo-Aſphodelus alpinum. *Bauh. pin.* 29.

Scottiſh Aſphodel.

In marſhes. P. September.

Sheep and Goats refuſe it.

151 BETHLEMSTAR. 418 Ornithogalum.

EMPAL. *Cup* none.

BLOSS. *Petals* ſix; fpear-fhaped; upright below the middle, but expanding and flat above; permanent, but fading,

CHIVES. *Threads* ſix; upright; broadeſt at the baſe; ſhorter than the bloſſom. *Tips* ſimple.

POINT. *Seedbud* angular. *Shaft* awl-ſhaped; permanent. *Summit* blunt.

S.VESS. *Capſule* roundiſh, angular; with three cells and three valves.

SEEDS. Many; roundiſh.

OBS.. *The* Threads *in ſome ſpecies are upright and flat; every other thread having three points and the* Tip *fixed upon the middle point.*

Yellow
Luteum

BETHLEMSTAR. The ſtalk angular; with two leaves. Fruit-ſtalks forming ſimple rundles—*Bloſſoms green on the outſide, but yellow within.*

Ornithogalum luteum. *Bauh. pin.* 71. *Park.* 140. *Ray's Syn.* 372.

Ornithogalum luteum ſeu cepe agraria. *Gerard.* 165.

In the northern counties. P. April—May.

The bulbous roots of all the ſpecies are nutritious and wholeſome.—Horſes, Goats and Sheep eat it. Swine are not fond of it. Cows refuſe it.

BETHLEMSTAR. The bunch of flowers very long. Spiked Threads fpear-fhaped. Fruit-ftalks equal; expanding when in Pyrenaicum flower, but afterwards approaching the ftalk—*Bloffoms greenifh and ftreaked on the outfide; white within.*
Ornithogalum anguftifolium majus, floribus exalbo virefcentibus. *Bauh. pin.* 70. *Ray's Syn.* 373.
Afphodelus bulbofus. *Gerard.* 97.
Afphodelus bulbofus galeni, five Ornithogalum majus flore fubvirente. *Park.* 136.
Spiked Star of Bethlem with a greenifh flower.
In meadows. P. May.

BETHLEMSTAR. The flowers in broad-topped fpikes; Common fruit-ftalks taller than the ftalk. Chives notched at the end— Umbellatum *Bloffoms white; with a green ribon on the outfide.*
Ornithogalum umbellatum medium anguftifolium. *Bauh. pin.* 70.
Ornithogalum vulgare et verius, majus et minus. *Ray's Syn.* 372.
Ornithogalum vulgare. *Gerard.* 165. *Park.* 136.
In woods and paftures. P. May.

152 SQUILL. 419 Scilla.

EMPAL. *Cup* none.
BLOSS. *Petals* fix; egg-fhaped; greatly expanding; deciduous.
CHIVES. *Threads* fix; awl-fhaped; half as long as the petals. *Tips* oblong; fixed fideways to the threads.
POINT. *Seedbud* roundifh. *Shaft* fimple, as long as the chives; falling off. *Summit* fimple.
S.VESS. *Capfule* nearly egg-fhaped; fmooth; with three furrows, three cells, and three valves.
SEEDS. Several; roundifh.

SQUILL. With few flowers, and thofe nearly upright— Vernal *Flowers generally four, and all of a height. Root folid.* Bloffoms Bifolia blue.
Hyacinthus ftellaris bifolius germanicus. *Bauh. pin.* 45.
Hyacinthus ftellatus fuchfii. *Gerard.* 106. *Ray's Syn.* 372.
Hyacinthus ftellatus vulgaris, feu bifolius fuchfii. *Park.* 126.
Vernal Star Hyacinth.
Sea-coaft. P. May.

SQUILL

Autnmnal   SQUILL. Leaves betwixt thread and ſtrap-ſhaped. Flowers
Autumnalis   in broad topped ſpikes. Fruit-ſtalks naked; aſcending; as long
as the flowers—*The Bloſſom is as large as a Pea, and as long as
the fruit-ſtalk :* blue.

    Hyacinthus autumnalis minor. *Gerard.* 110. *Park.* 132. *Ray's
Syn.* 373

    Hyacinthus ſtellaris autumnalis minor. *Bauh. pin.* 47.

    Leſſer Autumnal Star Hyacinth.

    Dry paſtures. P. September.

153 SPARAGUS.   424 Aſparagus.

EMPAL. *Cup* none.

BLOSS. *Petals* ſix; oblong; permanent: connected by
the claws into an upright tube. The three inner
petals alternate; reflected at the top.

CHIVES. *Threads* ſix; thread-ſhaped; ſtanding on the
petals; upright; ſhorter than the bloſſom. *Tips*
roundiſh.

POINT. *Seedbud* turban-ſhaped. with three corners. *Shaft*
very ſhort. *Summit* a prominent point.

S. VESS. *Berry* globular; with three cells and a dot at
the end.

SEEDS. Two; ſmooth; roundiſh; but angular on the
inſide.

    OBS. *It is not eaſy to ſay whether the bloſſom is compoſed of one
or of ſix petals; but the figure of it differs in different ſpecies.*

Cultivated   SPARAGUS. The ſtem herbaceous; cylindrical; upright.
Officinalis   Props in pairs. Leaves like briſtles—*The outer props ſolitary; the
inner ones ſmaller and in pairs.* Fruit-ſtalks *in pairs; limber; pendant; each ſupporting a ſingle bell-ſhaped bloſſom. The inner petals
longer than the others.* Bloſſoms *yellowiſh green.* Berries *red.*

    Aſparagus. *Park.* 454. *Gerard.* 1110. *Ray's Syn.* 267.

    Aſparagus ſativus. *Bauh. pin.* 489.

    Sperage.

    1. There is a variety with thicker leaves.

    In fields near the ſea. P. July.

    It delights in a ſandy ſoil.

    The young ſhoots of this plant in its cultivated ſtate, are very
univerſally eſteemed for their flavour and nutritious qualities.
They impart a fœtid ſmell to the urine.

    The *Sparagus Chryſomela* lives upon it.

154 FRITILLARY. 411 Fritillaria.

EMPAL. *Cup* none.

BLOSS. Bell-shaped, expanding at the base. Petals six; oblong; parallel. *Honey-cup* a hollow in the base of each petal.

CHIVES. *Threads* six; awl-shaped; approaching the shaft. *Tips* four-cornered; oblong; upright.

POINT. *Seedbud* oblong, three-sided; blunt. *Shaft* simple; longer than the chives. *Summit* with three clefts; expanding; blunt.

S. VESS. *Capsule* oblong; blunt; with three lobes, three cells, and three valves.

SEEDS. Several; flat; outwardly half circular; standing in two rows.

FRITILLARY, with all the leaves alternate, and one flower Chequered only on a stem—*Fruit-stalks slender*; Blossom *nodding*; *chequered* Meleagris *with purple, and greenish yellow. The* Stalk *grows considerably longer after the flowering is over.*

Fritillaria præcox purpurea variegata. *Bauh. pin.* 64.
Fritilliaria vulgaris. *Park.* 40.
Fritillaria variegata. *Gerard.* 149.
Common chequered Daffodil or Fritillary.
Meadows and pastures. P. April—May.

155 SPICEWORT. 434 Acorus.

EMPAL. *Stalk* cylindrical; undivided; covered by the florets. *Sheath* none. *Cup* none.

BLOSS. *Petals* six; blunt; concave; flexible; thicker upwards and generally lopped.

CHIVES. *Threads* six; rather thick; something longer than the petals. *Tips* thick; terminating; double; connected.

POINT. *Seedbud* hunched; oblong; as long as the chives. *Shaft* none. *Summit* a prominent point.

S. VESS. *Capsule* short; triangular; tapering each way; blunt. Cells three.

SEEDS. Several; oblong: egg-shaped.

Flag
Calamus

SPICEWORT. As there is only one fpecies known Linnæus gives no defcription of it—*Leaves thick; narrow; two edged. Flowering ftalk thicker than the leaves.*

Acorus verus, feu calamus officinalis. *Bauh. pin.* 34.
Acorus verus feu calamus officinarum. *Park.* 140. *Ray's Syn.* 437.
Acorus verus, officinis falfo calamus. *Gerard.* 62.
Sweet Smelling Flag, or Calamus.
In rivulets and marfhes. P. May.

The roots have a ftrong aromatic fmell, and a warm pungent, bitterifh tafte. The flavour is greatly improved by drying. They are commonly imported from the Levant, but thofe of our own growth are full as good, and might fupply the place of fome of the foreign fpices. The Turks candy the roots, and think they are a prefervative againft contagion.—Neither Horfes, Cows, Goats Sheep or Swine will eat it.

156 RUSH. 437 Juncus.

EMPAL. *Hufk* two valves.
Cup fix leaves; oblong; tapering; permanent.
BLOSS. None: unlefs you call the leaves of the coloured cup petals.
CHIVES. *Threads* fix; hair-like; very fhort. *Tips* oblong; upright; as long as the cup.
POINT. *Seedbud* three-cornered, tapering. *Shaft* fhort; thread-fhaped. *Summits* three; long; thread-fhaped; woolly; bent inwards.
S.VESS. *Capfule* covered; three-cornered; with one cell and three valves.
SEEDS. Several, roundifh.

OBS. *There is no particular mention of the colour of the bloffoms in this genus, for they are all brown or approaching to blacknefs.*

* Straws naked.

Marine
Acutus

RUSH. The ftraw nearly naked; cylindrical; fharp pointed. Panicle terminating: fence of two leaves; thorny—
Juncus acutus capitulis forghi. *Bauh. pin.* 10. *Ray's Syn.* 431.
Juncus maritimus capitulis forghi. *Park.* 1192.
Sea Hard Rufh.
In marfhes on the fea-coaft. P. July—Auguft.

RUSH.

RUSH. The ſtraw naked; ſtiff and ſtraight. Flowers in a Roundheaded
lateral knob.— Conglomeratus
Juncus lævis vulgaris panicula compactiore. *Ray's Syn.* 432.
Juncus lævis panicula non ſparſa. *Bauh. pin.* 12.
Juncus lævis glomerato flore. *Park.* 1192.
In boggy paſtures. P. June—Auguſt.

RUSH. The ſtraw naked, ſtiff and ſtraight. Flowers in a Soft
lateral panicle.— Effuſus
Juncus lævis vulgaris panicula non ſparſa noſtras. *Ray's Syn.*
452.
Juncus lævis panicula ſparſa major. *Bauh. pin.* 12. *Park.* 1191.
Juncus lævis. *Gerard.* 39.
Seaves.
Wet paſtures. P. May—Auguſt.
Ruſhes are ſometimes uſed to make little baſkets. The pith of
this and the preceding ſpecies is uſed inſtead of Cotton to make
the wick of Ruſh-lights.—Horſes and goats eat it.

RUSH. The ſtraw naked; terminated by a membranaceous Hard
ſubſtance which is bent inwards. Flowers in a lateral panicle.— Inflexus
Juncus acutus panicula ſparſa. *Bauh. pin.* 11.
Juncus acutus. *Gerard.* 39. *Ray's Syn.* 432.
Juncus acutus vulgatius. *Park.* 1193.
In paſtures and road ſides. P. June.

RUSH. The ſtraw naked; thread-ſhaped; nodding. Flowers Little
in a lateral panicle—*The panicle not always expanded; placed about* Filiformis
the middle of the ſtraw, which is ſo ſlender as hardly to ſupport itſelf.
Juncus parvus, calamo ſupra paniculam longius productio.
Ray's Syn. 432.
Juncus lævis panicula ſparſa minor. *Bauh. pin.* 12.
Leaſt Soft Ruſh.
Boggy mountains. P. Auguſt.

RUSH. The ſtraw naked; leaves briſtle-ſhaped. Flowers in Moſs
little congregated heads, without leaves.— Squarroſus
Juncus montanus paluſtris. *Ray's Syn.* 432.
Gramen junceum maritimum majus. *Gerard.* 35. *Park.* 1270.
Gramen foliis et ſpica junci. *Bauh. pin.* 5.
Gooſe Corn.
On heaths. P. June.
Horſes eat it.

SIX CHIVES.

Jointed
Articulatus

RUSH. The leaves with knotted joints. Petals blunt.—*In watery situations the leaves are compressed, but in the woods they are cylindrical. The leaves are not so properly jointed as separated transversely by valves, which become visible, when pressed by the fingers or held against the light.*

Juncus foliis articulosis, floribus umbellatis. *Ray's Syn.* 433.
Gramen Junceum folio articulato aquaticum. *Bauh. pin.* 1270.
Jointed leaved Rush.
Woods and wet pastures. P. July—August.

Bulbous
Bulbosus

RUSH. The leaves strap-shaped and chanelled. Capsules blunt—*Roots creeping* ; *thick.* Straw *thread-shaped* ; *a little compressed.* Floral Leaves *as long as the broad-topped spike of flowers. General and partial broad-topped spikes irregular.* Capsules *egg-shaped* ; *brown* ; *shining.*

Gramen junceum minimum capsulis triangularibus. *Ray's Syn.* 434.
Graminis juncei varietas minor. *Gerard.* 4.
Wet commons. P. August.
Cows, Goats, Sheep and Horses eat it.

Toad
Bufonius

RUSH. The straw forked ; leaves angular. Flowers solitary ; sitting—*Capsules brown* ; *shining.*
Gramen juncoides mininum, anglo-britannicum. *Ray's Syn.* 434.
Juncus palustris humilior erectior. *Ray's Syn.* 434.
Gramen Junceum. *Gerard.* 4.
Gramen junceum parvum, seu Holosteum Matthioli, et gramen Bufonium Flandrorum. *Park.* 1190.
Gramen nemorosum calyculis paleaceis. *Bauh. pin.* 7.
1. It varies in being tall ; or short ; or narrow-leaved.
Wet sandy ground. A. July—August.
Horses eat it.

Hairy
Pilosus

RUSH. The leaves flat and hairy. Flowers in a branching broad-topped spike.—
Gramen nemorosum hirsutum vulgare. *Ray's Syn.* 416.
Gramen nemorosum hirsutum. *Gerard.* 16.
Gramen nemorosum hirsutum majus. *Park.* 1184.
Gramen nemorosum hirsutum majus latifolium. *Bauh. pin.* 7.
Common hairy Wood Rush, or Grass.
1. There is a variety which is larger ; with broader leaves, viz. juncus Sylvaticus. *Huason* 132.
In thick woods. P. April—May.
Goats, Sheep and Horses eat it. Cows refuse it.

RUSH.

RUSH. The leaves flat and a little hairy. Flowers in spikes; Field
some fitting and some on fruit-stalks.— Campestris

Gramen hirfutum capitulis pfyllii. *Bauh. pin.*

Gramen nemorofum hirfutum minus anguftifolium. *Park.*
1185.

Gramen exile hirfutum. *Gerard.* 16. *Ray's Syn.* 416.

1. There is a variety with a more compact fpike. *Ray's Syn.* 416.
Small hairy Wood-rufh.

In paftures. The variety in turfy bogs. P. April.

Sheep, Goats and horfes eat it.

157. PURSLANE. 446 Peplis.

EMPAL. *Cup* one leaf; bell-fhaped; large; permanent.
Rim with twelve teeth; every other tooth reflected.

BLOSS. *Petals* fix; egg-fhaped; very minute; growing
from the mouth of the cup.

CHIVES. *Threads* fix; awl-fhaped, fhort. *Tips* roundifh.

POINT. *Seedbud* egg-fhaped. *Shaft* very fhort. *Summit*
round and flat.

S.VESS. *Capfule* heart-fhaped. *Cells* two, with an oppofite partition.

SEEDS. Many; three cornered; fmall.

OBS. *In several flowers, upon one and the fame plant, the bloffom
is altogether wanting.*

PURSLANE. The flowers have rarely any petals—*Stems* Water
numerous; creeping. Leaves *oppofite; two at each joint.* Blof- Portulaca
foms *fitting at the bafe of the leaves; purplifh.*

Portulaca. *Ray's Syn.* 368.

Alfine paluftris minor, ferpylli folia. *Bauh. pin.* 120.

Alfine rotundifolia, five Portulaca aquatica. *Gerard.* 614.

Alfine aquatica minor folio oblongo, feu Portulaca aquatica.
Park. 1260.

Marfhes and fhallow ftagnant waters. P. September.

SIX CHIVES.

Order III. Three Pointals.

158 TUBEROOT.   457. Colchicum.

EMPAL. None; except sometimes a sort of scattered sheathes.

BLOSS. With six divisions. *Tube* angular; extending down to the root. *Segments* of the border betwixt spear and egg-shaped; concave, upright.

CHIVES. *Threads* six; awl-shaped; shorter than the blossom. *Tips* oblong, with four valves; fixed sideways to the threads.

POINT. *Seedbud* buried within the root. *Shafts* three; thread-shaped; as long as the chives. *Summits* reflected; channelled.

S. VESS. *Capsule* of three lobes; connected on the inside by a seam; blunt; with three cells, opening inwards at the seams.

SEEDS. Many; nearly globular; wrinkled.

Saffron   TUBEROOT. With flat, spear-shaped, upright leaves—
Autumnale *Blossoms pale purple. Cultivation produces a great variety of colours and makes the blossoms double.*

Colchicum commune. *Bauh pin.* 67. *Ray's Syn.* 373.

Colchicum Anglicum purpureum et anglicum album, *Park,* 153. *Gerard.* 157.

Meadow Saffron,

In rich pastures.   P. September.

This is one of those plants that upon the concurrent testimony of ages was condemned as poisonous; but Dr. *Storck* of *Vienna* hath taught us that it is an useful Medicine. The roots have a good deal of acrimony. An infusion of them in vinegar, formed into a syrup by the addition of sugar or honey, is found to be a very useful pectoral and diuretic. It seems in its virtues very much to resemble the Squill, but it is less nauseous and less acrimonious.

159 ARROWGRASS. 453 Triglochin.

EMPAL. *Cup* three leaves; nearly round; blunt, concave; deciduous.

BLOSS. *Petals* three; egg-fhaped; concave; blunt; refembling the leaves of the cup.

CHIVES. *Threads* fix; very fhort. *Tips* fix; fhorter than the petals.

POINT. *Seedbud* large. *Shafts* none. *Summits* three, or fix; reflected; downy.

S.VESS. *Capfule* oblong egg-fhaped; blunt; with as many cells as there were fummits. Valves fharp; opening at the bafe.

SEEDS. Solitary; oblong.

ARROWGRASS. The capfule with three cells; fomewhat Marfh ftrap-fhaped—*Bloffoms in a long terminating fpike*; *greenifh.* Paluftre
    Juncajo paluftris et vulgaris. *Ray's Syn.* 435.
    Gramen junceum fpicatum feu triglochin. *Bauh. pin.* 6.
    Gramen marinum fpicatum. *Gerard.* 18. alterum. *Park.* 1279.
    Arrow-headed Grafs.
    In wet ditches and marfhes. P. July—Auguft.
    Cows are extremely fond of it; Horfes, Sheep, Goats and Swine eat it.

ARROWGRASS. The capfules with fix cells; egg-fhaped— Spiked
Leaves femicylindrical. Bloffoms *purple.* Maritimum
    Gramen marinum et fpicatum. *Park.* 1270. *Ray's Syn.* 435.
    Gramen junceum et fpicatum alterum. *Bauh. pin.* 6.
    Sea fpiked Grafs.
    In fields on the fea-coaft. P. June.
    It is falt to the tafte, but Horfes, Cows, Sheep, Goats and Swine are very fond of it,

SIX CHIVES.

160 DOCK. 451 Rumex.

EMPAL. *Cup* three leaves; blunt; reflected; permanent.
BLOSS. *Petals* three; egg shaped; not unlike the cup, but
larger; approaching: permanent.
CHIVES. *Threads* six; hair-like; very short. *Tips* upright; double.
POINT. *Seedbud* turban-shaped: three cornered. *Shafts*
three; hair-like; reflected; standing out betwixt
the chinks of the approaching petals. *Summits* large;
jagged.
S.VESS. None. The petals closing in a three cornered
form contain the seed.
SEEDS. Single; three sided.

* Chives and pointals in the same flower. Valves beaded.

Bleeding
Sanguineus

DOCK. The valves very entire; only one beaded. Leaves
betwixt heart and spear-shaped—*The outer valve of the flower is
marked with a large red globular substance. The other valves have
seldom any globules, and if they have they are very small.* Blossoms
reddish.
Lapathum folio acuto rubente. *Bauh pin.* 115. *Ray's Syn.*
142.
Lapathum sativum sanguineum. *Gerard.* 390.
Bloodwort.
In Woods. B. July.

Curled
Crispus

DOCK. The valves entire; beaded: leaves spear-shaped;
waved at the edges; sharp—
Lapathum folio acuto crispo. *Bauh. pin.* 115. *Ray's Syn.* 141.
Lapathum acutum minus. *Park.* 226.
Lapathi acuti varietas folio crispo. *Gerard.* 387.
Fields and road-sides. P. June—July.
The fresh roots bruised and made into an ointment or decocti-
on cure the Itch. The seeds have been given with advantage in
the dysentery.
Cows and Goats refuse it.

Golden
Maritimus

DOCK. The valves toothed and beaded. Leaves strap-shaped
—*Blossoms dirty yellow.*
Lapathum minimum. *Bauh pin.* 115.
1. With larger seeds and fewer whorls of flowers.
Lapathum folio acuto, flore aureo. *Bauh pin.* 115. *Rays Syn.*
142
Lapathum aureum. *Ray's Syn.* 142.
Near the sea-coast. 1. In ditches and amongst rubbish. P.
July—August.

DOCK.

DOCK. The valves toothed and beaded. Leaves oblong heart-Sharp-pointed
shaped; tapering to a point—   Acutus
    Lapathum acutum. *Gerard.* 388, *Ray's Syn.* 142.
    Lapathum acutum majus. *Park,* 1224.
    Lapathum folio acuto plano, *Bauh pin.* 115.
    Uncultivated and wet fituations. P. June.
    Cows and Horfes refufe it.
    It is infefted by the Dock-loufe, *Aphis Rumicis.*

DOCK. The valves toothed and beaded; leaves oblong heart- Broad-leaved
shaped; a little fcolloped, and rather blunt—   Obtufifolius
    Lapathum vulgare folio obtufo. *Ray's Syn.* 141.
    Lapathum fylveftre folio fubrotundo. *Bauh. pin* 115.
    Lapathum fylveftre vulgatius. *Park.* 1225.
    Lapathum fylveftre folio minus acuto. *Gerard.* 388.
    Road-fides and wet places. P. July—Auguft.

DOCK. The valves toothed; one or two beaded. Root-leaves Fiddle
fiddle-fhaped—*Stem Leaves oblong egg-fhaped, entire; not indent-* Pulcher
ed at the fides. Flowering Branches *crooked. The outer valve of
the cup beaded.*.
    Lapathum pulchrum Bononienfe finuatum. *Ray's Syn.* 142.
    Dry fields. P. June.

   * **Chives and pointals in the fame flower. Valves not beaded.*

DOCK. The valves very entire; not beaded. Leaves heart- Water
shaped; fmooth; fharp.—   Aquaticus
    Lapathum aquaticum folio cubitali. *Bauh pin.* 116.
    Lapathum maximum aquaticum, five hydrolapathum. *Ray's
Syn.* 140.
    Hydrolapathum magnum. *Gerard.* 389.
    Hydrolapathum majus. *Park.* 1225.
    Lapathum longifolium nigrum paluftre, five Britannica, an-
tiquorum vera, vel hydrolapathum nigrum. *Munting.*
    Great Water-dock.
    In wet ditches, marfhes and flow ftreams. P. July—Aug.
    There is fome confufion amongft Botanical Writers in afcer-
taining which is the fpecies of Dock, fo famous for medicinal
purpofes. Linnæus in his Materia Medica makes the *Herba
Britannica* of the Difpenfatory Writers to be the *Lapathum
aquaticum, folio cubitali* of Bauhine; *pin.* 116; and fays the
*valves are very entire, and not beaded: and the leaves betwixt
heart and fpear-fhaped.* In the *Flora Suecica,* ed. 2. he calls the
leaves heart-fhaped. In the 3d. edition of the *Species Plantarum*
he adds to the laft mentioned character of the leaves, the epi-
thet *fharp*; and afterwards in the 12th edit. of the *Syftema Naturæ* he fays the *leaves are heart-fhaped, fmooth and fharp.* Mr.
Hudfon in his *Flora Anglica,* does not feem to confider the
                                  RUMEX

Rumex *aquaticus* as a British species, but takes the Rumex *Britannica* of Linnæus to be our *great Water Dock*; and after giving the specific character of that from the *species plantarum*, adds, the Synonyms of *Gerard, Parkinson, Bauhine, Munting*, and *Ray*. Now in the Rumex *Britannica* all the *Valves* are *beaded*, and the *Leaves* are *spear-shaped*. Linnæus too seems to consider this species as a native of *Virginia*, and as unknown to *Bauhine*; he accordingly quotes the Synonyms of *Colden* and *Gronovius*; writers upon the North American Plants. Dr. Martyn in his *Catalog. Horti. Cantab.* considers the *Great Water Dock* as the Rumex *aquaticus* of Linnæus, and takes notice of Hudson calling it the Rumex *Britannica*. Dr. Lewis too in his *Materia Medica*, describing the *Great Water Dock* says the *Valves* are *not beaded*.

But then how happens it that a North American plant should obtain the trivial name *Britannica?* Munting, who wrote a volume to prove the Rumex *Britannica* to be the true *Herba Britannica* of the ancients, (see *Pliny*) says the word *Britannica* had no reference to the island of Great-Britain, but is composed of three Teutonic words, signifying, to *fasten loose teeth*. His description of the root does not agree with that of *Bauhine*. In a dried specimen of my own, the *Valves* are *not beaded*; but a friend of mine hath one in his collection in which the *Valves* are *all beaded*. How shall we reconcile these differences? Is the plant subject to these varieties? or do both the species grow wild in Great-Britain? Now the difficulty is pointed out, future observations will soon determine the matter. The virtues are probably the same in both. It is a medicine of considerable efficacy both externally applied as a wash for putrid spongy gums, and internally in some species of Scurvy. In Rheumatic pains and in Chronical diseases, owing to obstructed viscera, it is said to be useful. The powdered root is one of the best things for cleaning the teeth. The *Root* of our species is blackish on the out-side, white within; sometimes with a reddish tinge, but soon changing to a yellowish brown when exposed to the air.

The black and white Weevil, *Curculio Lapathi*, is found upon the leaves.

Of the *American Species* there are two varieties; in one the root is yellow both internally and externally: in the other it is black on the outer surface, with circular rings, and orange coloured within; brittle and spongy, but not fibrous: seldom branched, but furnished with slender lateral threads. It grows in deep black muddy places. This is the root used by the native Indians of North America with such success in foul ill-conditioned ulcers, and which they refused to discover to the Europeans for any price; but a lucky circumstance revealed the secret, fecret. I faw an ill-conditioned ulcer in the mouth, which had deftroyed the palate, cured with it, by wafhing the mouth with a decoction of the root, and drinking a fmall quantity of the fame decoction daily. *Cold. Pl. Coldfhing.* u. 68.

DOCK. The flowers with only two pointals—*Bloffom with* Round-leaved *four clefts.* Cup *of two leaves.* Seedbud *compreffed.* Stem *trailing.* Digynus
  Acetofa cambro-britannica montana. *Park.* 745.
  Round-leaved Mountain Sorrell.
  On Mountains. P. June.
  Goats eat it.

  *** *Chives and pointals on diftinct plants.*

DOCK. The leaves oblong; arrow-fhaped—*On purplifh leaf-* Sorrel *ftalks.* Bloffoms *reddifh.* Acetofa
  Lapathum acetofum vulgare. *Ray's Syn.* 143.
  Acetofa vulgaris. *Park.* 742.
  Acetofa pratenfis. *Bauh. pin.* 114.
  Oxalis feu acetofa. *Gerard.* 396.
  Cuckow-meat.
  In paftures. P. June.
1. There is a variety that is larger found upon Rocks near the Sea.
  The leaves are eaten in fauces and in fallads. The Laplanders ufe them to turn their milk four. In France they are cultivated for the ufe of the table, being introduced in foups, ragouts and fricafies. In fome parts of Ireland they eat them plentifully with milk, alternately biting and fupping. The Irifh alfo eat them with fifh and other alcalefcent food. The dried root gives out a beautiful red colour when boiled.
  Horfes, Cows, Goats, Sheep and Swine eat it.—The Sorrel Loufe, *Aphis Acetofæ*, feeds upon it.

DOCK. The leaves betwixt fpear and halberd-fhaped—*Roots* Little *creeping.* Bloffoms *yellowifh.* Acetofella
  Lapathum acetofum repens lanceolatum. *Ray's Syn.* 143.
  Acetofa arvenfis lanceolata. *Bauh. pin.* 114.
  Acetofa minor lanceolata. *Park.* 744.
  Oxalis tenuifolia. *Gerard.* 397.
  Sheeps Sorrel.
  Sandy cornfields and paftures. P. May—July.
  Horfes, Cows, Goats, Sheep and Swine eat it.

  Obs. *The fpotted red and white Underwing Moth,* Phalæna Fuliginofa—*the wild arrach Moth,* Ph. Atriplicis—*the bramble Moth,* Ph. Rumicis, *and the* Meloe Profcarabæus, *are found upon the different fpecies of this genus.*

Order

Order V.　Many Pointals.

161 THRUMWORT.　460 Alifma.

EMPAL. *Cup* three leaves; egg-fhaped; concave; permanent.

BLOSS. *Petals* three; circular; large; flat; greatly expanded.

CHIVES. *Threads* fix; awl fhaped; fhorter than the bloffom. *Tips* roundifh.

POINT. *Seedbuds* more than five. *Shafts* fimple. *Summits* blunt.

S. VESS. *Capfule* compreffed.

SEEDS. Solitary; fmall.

Great
Plantago

THRUMWORT. The leaves egg-fhaped; fharp. Fruit with three flatted corners—*Bloffoms reddifh white.*
　　Plantago aquatica. *Ray's Syn.* 257.
　　Plantago aquatica major. *Gerard.* 417. *Park.* 1245.
　　Plantago aquatica latifolia. *Bauh. pin.* 190.
　　Great Water Plantain.
　　In fhallow waters. P. June—July.
　　Goats eat it. Horfes, Cows, Sheep and Swine refufe it.

Star-headed
Damafonium

THRUMWORT. The leaves oblong heart-fhaped. Flowers with fix pointals; capfules awl-fhaped—*Bloffoms white.*
　　Plantago aquatica minor ftellata. *Gerard.* 417.
　　Plantago aquatica ftellata. *Bauh pin.* 190.
　　Plantago aquatica minor muricata. *Park.* 1245.
　　Damafonium ftellatum Dalechampii. *Ray's Syn.* 272.
　　Starry-headed Water Plantain.
　　Ditches and ftagnant waters. P. June—Auguft.

Floating
Natans

THRUMWORT. The leaves egg-fhaped; blunt. Fruit-ftalks folitary.—
　　Creeping Water Plantain.
　　Lakes. P. Auguft.

Small
Ranunculoides

THRUMWORT. The leaves betwix fpear and ftrap-fhaped. Fruit globular; fcurfy—*The flower opens about noon, and that is the beft time to look for it.* Bloffoms *bluifh white.*
　　Plantago aquatica minor. *Park.* 1245. *Ray's Syn.* 257.
　　Plantago aquatica humilis. *Gerard.* 417.
　　Plantago aquatica anguftifolia. *Bauh pin.* 190.
　　Leffer Water Plantain.
　　In water, but not common. P. June—Auguft.

C L A S S

C L A S S   VII.

S E V E N   C H I V E S.

O R D E R I. ONE POINTAL.

162 WINTERGREEN. *Cup* feven leaves. *Bloff.* with feven divifions ; flat. *Berry* one cell : dry.

162 WINTER.

SEVEN CHIVES.

162 WINTERGREEN. 461 Trientalis.

EMPAL. *Cup* seven leaves; spear-shaped; tapering; expanding; permanent.

BLOSS. Starry; flat; of *one Petal* with seven divisions, slightly adhering at the base. *Segments* betwixt egg and spear-shaped.

CHIVES. *Threads* seven; hair-like; growing on the claws of the blossom; standing wide; as long as the cup. *Tips* simple.

POINT. *Seedbud* globular. *Shaft* thread-shaped; as long as the chives. *Summit* a knob.

S. VESS. *Berry* not unlike a capsule; dry; globular of one cell. Coat very thin; opening by various seams.

SEEDS. Several; angular. *Receptacle* large; hollowed out to receive the seeds.

OBS. *Although seven is. the general number in this plant it is not invariably so. The fruit is a dry berry, not opening with valves like a capsule.*

Chickweed   WINTERGREEN. The leaves spear-shaped; very entire—
Europæa   *The petals close when rain approaches, and the flowers hang down.*
Blossoms *white*; *on long fruit stalks.*
Alsinanthemos. *Ray's Syn.* 286.
Pyrola alsines flore Europæa. *Bauh. pin.* 191. *Park* 509.
Woods and heaths   P. June.
Horses, Goats and Sheep eat it. Cows refuse it.

CLASS.

CLASS VIII.

EIGHT CHIVES.

ORDER I. ONE POINTAL.

* *Flowers perfect.*

† *Maple.*

163 WILLOWHERB. *Bloss.* four petals. *Cup* four leaves : superior. *Caps.* four cells. *Seeds* feathered.

164 CENTAURY. *Bloss.* with eight clefts. *Cup* eight leaves : beneath. *Caps.* one cell. Two valves. *Seeds* many.

165 WHORTLE. - *Bloss.* one petal. *Cup* with four teeth : superior. *Chives* growing on the feat. *Fruit* a berry.

166 HEATH. - *Bloss.* one petal. *Cup* four leaves : beneath. *Chives* growing on the feat. *Fruit* a capsule.

† *Heath Andromeda.*

* * *Flowers imperfect.*

† *Primrose Birds-nest.*

167 MEZEREON. *Cup* with four equal clefts ; resembling a blossom and inclosing the chives. *Berry* pulpy.

Order

Order II. Two Pointals.

† *Saxifrage.*   † *Pale Snakeweed.*

Order III. Three Pointals.

168 SNAKEWEED.   *Bloss.* none. *Cup* with five divisions. *Seed* single; naked.

Order IV. Four Pointals.

169 MOSCHATEL.   *Bloss.* with four or five clefts: superior. *Cup* two leaves. *Berry* with four or five seeds.

170 WATERPINE.   *Bloss.* four petals. *Cup* four leaves. *Capsule* four cells.

171 TRUELOVE.   *Bloss.* four petals; awl-shaped. *Cup* four leaves. *Berry* four cells.

† *Whorled Featherwort.*

163 WILLOW.

163 WILLOWHERB. 471 Epilobium.

EMPAL. *Cup* four leaves; fuperior. *Leaves* oblong; tapering; coloured; deciduous.

BLOSS. *Petals* four; circular; expanding; broadeft on the outer part, and notched at the end.

CHIVES. *Threads* eight: awl-fhaped; alternately fhorter. *Tips* oval; compreffed; blunt.

POINT. *Seedbud* beneath; cylindrical; very long. *Shaft* thread-fhaped. *Summit* with four clefts; thick; blunt; rolled backwards.

S. VESS. *Capfule* very long; cylindrical; fcored; of four cells and four valves.

SEEDS. Numerous; oblong; crowned with a feather. *Receptacle* very long; four cornered; loofe; limber; coloured.

OBS. *In fome fpecies the chives and pointals are upright, in others they lean towards the lower fide of the bloffom.*

* Chives declining.

WILLOWHERB with leaves betwixt ftrap and fpear-fhaped; Rofebay fcattering. Flowers unequal—*Chives declining.* Bloffoms *red.* Anguftifo-*Receptacle replete with honey. Cultivation produces bloffoms of diffe-* lium *rent colours and purple ftreaks in the leaves.*

Lyfimachia chamænerion dicta anguftifolia. *Bauh. pin.* 245.

Lyfimachia fpecioſa, quibufdam onagra dicta, filiquofa. *Ray's Syn.* 310.

Chamænerion flore delphinii. *Park.* 548.

Goats are extremely fond of it; Cows and Sheep eat it. Horfes and Swine refufe it.

Woods and hedges. P. July—Auguft.

The Elephant Moth, *Sphinx Elpenor,* is found upon this plant.

* * Chives upright, irregular. Petals cloven.

WILLOWHERB with oppofite, fpear-fhaped, woolly, toothed Smallflowere leaves. Seed-veffels fmooth—*Bloffoms pale purple.* Parviflorum

Lyfimachia filiquofa hirfuta, parvo flore. *Bauh. pin.* 245. *Ray's Syn.* 311.

Lyfimachia fylvatica. *Gerard.* 476.

Lyfimachia filiquofa fylveftris hirfuta. *Park.* 549.

Epilobium hirfutum. *Hudfon.* 140. *Sp. Plant.* 494.

Small flowered hairy Willowherb.

Banks of rivers. P. July.

Q WILLOWHERB.

*** *Chives upright, regular; petals cloven.*

Hairy
Hirſutum

WILLOWHERB. The leaves oppoſite; ſpear-ſhaped; ſer-rated; running along and embracing the ſtem—*The whole plant is a little hairy.* Stem *cylindrical*; *hairs expanding.* *Lower* Branches *oppoſite.* Flowers *in bunches*; *terminating*; *alternate*; *one upon each little fruit-ſtalk.* Seedbud *woolly.* Petals *cloven half way down.* Chives *upright*; *regular.* Bloſſoms *red.*
    Epilobium ramoſum. *Hudſon.* 141.
    Lyſimachia ſiliquoſa hirſuta magno flore. *Bauh. pin.* 245.
Ray's Syn. 311.
    Great flowered Willowherb, or Codlings and Cream.
    Wet hedges and rivulets. P. July.
    The top ſhoots have a very delicate fragrancy, but ſo tranſi-tory that before they have been gathered five minutes it is no longer perceptible.—Horſes, Sheep and Goats eat it. Cows are not fond of it; Swine refuſe it.

Smooth-leaved
Montanum

WILLOWHERB. The leaves oppoſite; egg-ſhaped; toothed —*Stem very ſoft*; *cylindrical*; *upright*; *reddiſh.* Leaves *covered with, very fine ſoft down eſpecially on the under ſurface.* Petals *cloven*; *red*; *ſometimes white.*
    Lyſimachia campeſtris. *Gerard.* 478.
    Lyſimachia ſiliquoſa glabra major. *Bauh. pin.* 245.
    Lyſimachia ſiliquoſa major. *Park.* 548.
    In wet gravelly ſoil. B. June.
    Goats eat it. Horſes are not fond of it.

Narrowleaved
Tetragonum

WILLOWHERB. The leaves ſpear-ſhaped; ſet with little teeth. The lower ones oppoſite. Stem four cornered—*The young leaves have livid ſpots, and the tender top of the plant hangs down.* Bloſſoms *red.*
    Lyſimachia ſiliquoſa glabra media ſeu minor. *Gerard.* 479.
Ray's Syn. 311.
    Lyſimachia ſiliquoſa glabra minor. *Bauh. pin.* 245.
    Ditches and rivulets. P. July.

Marſh
Paluſtre

WILLOWHERB. The leaves oppoſite; ſpear-ſhaped: very entire. Petals notched at the end. Stem upright—*Leaves in-diſtinctly toothed.* Petals *reddiſh.* Pods *on fruit-ſtalks.*
    Lyſimachia ſiliquoſa glabra minor anguſtifolia. *Gerard.* 479.
Ray's Syn. 311.
    Lyſimachia ſiliquoſa glabra anguſtifolia. *Bauh. pin.* 245.
    In marſhes. P. July.
    Horſes, Sheep and Goats eat it. Swine refuſe it.

WILLOWHERB.

WILLOWHERB. The leaves oppofite; betwixt egg and Mountain fpear-fhaped; very entire. Pods fitting; Stem creeping—*hardly* Alpinum *a fpan long.* Bloffoms *pale red.*
Lyfimachia filiquofa glabra minor latifolia. *Ray's Syn.* 311.
Near rivulets on the fides of hills. P. July.

164 CENTAURY. 1258 Chlora.

EMPAL. *Cup* of eight fharp leaves: permanent.
BLOSS. One petal; funnel-fhaped. The *Border* divided into eight fpear-fhaped *Segments*, lapping over each other.
CHIVES. *Threads* eight awl-fhaped: inferted into the upper part of the tube. *Tips* fimple.
POINT. *Seedbud* oblong; four cornered; longer than the tube. *Shaft* cylindrical; cloven. *Summits* two; horfe fhoe-fhaped.
S.VESS. *Capfule* oblong; four cornered; of one cell, and two valves.
SEEDS. Numerous; fmall; egg-fhaped.

OBS. *This generic defcription was taken from nature, before the publication of the* Mantiffa Plantarum. *Linnæus had confidered it as a fpecies of* GENTIAN, *and in the laft edition of the* Sp. Pl. *arranged it as fuch; but Mr. Hudfon, convinced of the impropriety of this arrangement, confidered it as a new genus and called it* BLACKSTONIA; *his defcription however differs confiderably both from that of Linnæus and from mine; but perhaps the plant itfelf is fubject to variations.*

CENTAURY with perforated leaves—*Bloffoms yellow; in loofe* Yellow *clufters; terminating.* Perfoliata
Centaureum luteum perfoliatum. *Bauh. pin.* 278. *Ray's Syn.* 287.
Centaureum parvum luteum lobelii. *Gerard.* 547.
Centaureum minus luteum et perfoliatum non ramofum. *Park.* 271.
Blackftonia perfoliata. *Hudfon.* 146.
In paftures. A. July.

165 WHORTLE. 483 Vaccinium.

EMPAL. *Cup* very fmall; fuperior: permanent.

BLOSS. One petal; bell-fhaped; with four clefts. *Segments* rolled backwards.

CHIVES. *Threads* eight; fimple. *Tips* with two horns; opening at the point, and furnifhed with two expanding awns which are fixed to the back.

POINT. *Seedbud* beneath. *Shaft* fimple; longer than the chives. *Summit* blunt.

S.VESS. *Berry* with four cells; globular, with a hollow dimple.

SEEDS. Few; fmall.

OBS. *The* Biberry WHORTLE *and feveral foreign fpecies have frequently one fourth more in number in the parts of the flower. In this fpecies the cup is very entire, but in others it hath four clefts. In the* Craneberry WHORTLE *the new blown bloffom is almoft entire and is rolled backwards towards its bafe.*

* Leaves deciduous.

Bilberry
Myrtillus

WHORTLE. The fruit-ftalks fupporting only one flower. Leaves ferrated; egg fhaped; deciduous. Stem angular—*Chives ten.* Bloffom *with five clefts*; *reddifh white.* Berries *bluifh black.*

Vitis idœa, foliis oblongis crenatis, fructu nigricante. *Bauh. pin.* 470.

Vitis idœa angulofa. *Ray's Syn.* 457.

Vaccinia nigra. *Gerard.* 1415. vulgaria. *Park.* 1455.

Black-worts. Whortle-berries. Biberries. Wind-berry.

Woods and heaths. P. April—May.

The berries are very acceptable to children, either eaten by themfelves, or with milk, or in tarts. The moor game live upon them in the autumn. The juice ftains paper or linen purple—Goats eat it. Sheep are not fond of it. Horfes and Cows refufe it.

Great
Uligi nofum

WHORTLE. The fruit-ftalks fupporting only one flower; leaves very entire; oval; blunt; fmooth—*fringed at the bafe*; *deciduous.* Bloffoms *purplifh white.* Berry *bluifh black.*

Vitis idœa foliis fubrotundis exalbidis. *Bauh. pin.* 470.

Vitis idœa magna quibufdam, five myrtillus grandis. *Ray's Syn.* 457.

Vitis idœa foliis fubrotundis major. *Gerard.* 1416.

Vaccinia nigra fructu majore. *Park.* 1455.

Great Bilberry Bufh.

Northern counties. S. April—May.

Children fometimes eat the berries; but in large quantities they occafion giddinefs and a flight head-ach, efpecially when full grown

grown and quite ripe. Brookes fays many vintners in France make ufe of the juice to colour their white wines red.—Horfes, Cows, Sheep and Goats eat it. Swine refufe it.

* * *Leaves Evergreen.*

WHORTLE with flowers in bunches; terminating; nod- Red ding. Leaves inverfely egg-fhaped; very entire; rolled back at Vitis idæa the edges; dotted on the under fide—*Bloffoms reddifh white.*

Vaccinia rubra. *Gerard.* 1415.

Vaccinia rubra buxeis foliis. *Park.* 1458.

Vitis idœa foliis fubrotundis non crenatis, baccis rubris. *Bauh. pin.* 470.

Vitis idœa fempervirens fructu rubro. *Ray's Syn.* 457.

Red Whortle-berries.

High commons and heaths. S. April—May.

The berries are eaten largely by the country people, as nature prefents them. They may be made into tarts or jelly.—Goats eat it. Cows, Sheep and Horfes refufe it.

WHORTLE. The leaves egg-fhaped; very entire; rolled Craneberry back at the edges. Stems creeping; thread-fhaped; naked— Oxycoccos *Flowers in a very fhort bunch; upon very long fruit-ftalks.* Floral Leaves *two; alternate.* Bloffoms *rolled back; reddifh white.* Berry *red, mottled with purple.*

Vaccinium paluftre. *Park.* 1229.

Vaccinia paluftria. *Gerard.* 1425.

Vitis idœa paluftris. *Bauh. pin.* 471.

Oxycoccos feu vaccinia paluftria. *Ray's Syn.* 267.

Cran-berries. Mofs-berries. Moor-berries.

In wet moffy and peaty land. S. May—June.

The berries are hardly eatable alone, but when made into tarts fome people prefer them to all other fweetmeats. They may be kept for years, by drying them a little in the fun, and then ftopping them up in dry bottles.—Goats and Swine eat it. Horfes, Cows and Sheep refufe it.

166 HEATH. 484 Erica

EMPAL. *Cup* with four leaves; egg shaped; upright; coloured; permanent.

BLOSS. One petal, bell-shaped; with four clefts, often distended.

CHIVES. *Threads* eight; hair-like; standing on the receptacle. *Tips* cloven at the point.

POINT. *Seedbud* roundish. *Shaft* thread-shaped; straight; longer than the chives. *Summits* resembling a little crown; with four clefts and four edges.

S.VESS. *Capsule* roundish; inclosed; smaller than the cup; with four cells and four valves.

SEEDS. Numerous; very small.

OBS. *In some species the cup is double; in the* Firleaved HEATH *it consists of one leaf with four divisions. The figure of the blossom varies betwixt egg-shaped and oblong. The chives in some species are longer and in others shorter than the blossom.*

* Tips with two horns.

Common
Vulgaris

HEATH. The tips inclosed within the blossom. Blossoms unequal; bell-shaped; rather shorter than the cup. Leaves opposite; arrow-shaped—*Stems brown*; *woody*. Leaves *opposite; somewhat three edged; the lower edge channelled; fixed to the stem above the base, which is cloven and sharp.* Blossom *with four clefts going more than half way down; purplish.* Shaft *ascending.*

Erica vulgaris. *Park.* 1480. *Ray's Syn.* 470.

Erica vulgaris seu pumila. *Gerard.* 1380.

Erica vulgaris glabra. *Bauh. pin.* 485.

Ling. Grig.

1. There is a variety with white blossoms.

On commons and moors. S. June—August.

This plant, but little regarded in happier climates, is made subservient to a great variety of purposes, in the bleak and barren Highlands of Scotland. The poorer inhabitants make walls for their cottages, with alternate layers of heath, and a kind of mortar made of black earth and straw. The woody roots of the heath being placed in the center, the tops externally and internally. They make their beds of it, by placing the roots downwards, and the tops only being uppermost, they are sufficiently soft to sleep upon. Cabbins are thatched with it.

In the island of Ilay, ale is frequently made by brewing one part malt, and two parts of the young tops of heath; sometimes they add hops. Boethius relates that this liquor was much used by the *Picts. Pennant's Tour.* 1772, p. 229.

Woollen

Woollen cloth boiled in alum water, and afterwards in a ftrong decoction of the tops of heath, comes out a fine orange colour. The ftalks and tops will tan leather. In England befoms are made of it, and faggots to burn in ovens, or to fill up drains that are to be covered over. Sheep and Goats will fometimes eat the tender fhoots, but they are not fond of them.

Cattle not accuftomed to browfe on heath, give bloody milk, but are foon cured by drinking plentifully of water. *Pennant's Tour.* 1772. p. 229. Horfes will eat it. Bees extract a great deal of honey from the flowers ; and where heath abounds the honey has a reddifh caft ; otherwife it is white.

HEATH. The tips inclofed within the bloffoms ; which are egg-fhaped and grow in bunches. Leaves growing by threes ; ftrap-fhaped, fmooth—*flefhy.* Bloffoms *purplifh pink.* Stem *afh-coloured, woody.*   Fine-leaved Cinerea

Erica tenuifolia. *Gerard.* 1380. *Ray's Syn.* 471.
Erica humilis cortice cinereo, arbuti flore. *Bauh. pin.* 486.
Erica virgata, feu VI. Clufii. *Park.* 1483.
1. Little branches in diftant whorls, three in each whorl.   *Whorled*
Erica ternis per intervalla ramulis. *Bauh. pin.* 486.
On heaths and in very dry woods. S. June—Auguft.

HEATH. The tips inclofed within the bloffoms ; which are nearly globular ; incorporated ; longer than the leaves. Leaves growing by fours, fringed, expanding—*Bloffoms purple.*   Crofs-leaved Tetralix

Erica major flore purpureo. *Gerard.* 1382.
Erica ex rubro nigricans fcoparia. *Bauh. pin.* 486.
Erica brabantica folio coridis hirfuto quaterno. *Ray's Syn.* 471.
On moift commons. S. July—Auguft.
Goats eat it.

* * *Tips fimple, blunt ; notched at the end.*

HEATH. The tips inclofed within the bloffoms which are egg-fhaped, but irregular and collected into bunches. Leaves growing by threes, fringed, expanding—*The whole plant is fome-what woody, and about two feet high.* Leaves *oblong egg-fhaped, fharp ; rolled back at the edges.* Bloffoms *large, egg-fhaped, contracted and irregular at the mouth ; pale pink or purple.*   Rough-leaved Ciliaris

Erica vulgaris hirfuta. *Gerard.* 1382. *Ray's Syn.* 471.
Erica hirfuta anglica. *Ray's Syn.* 602.
Erica vulgaris hirfutior. *Park.* 1480.
On Commons. S. June—July.

HEATH.

Fir-leaved
Multiflora

HEATH. The tips appearing above the bloſſoms, which are cylindrical, and longer than the cup. Leaves ſtrap-ſhaped, expanding; growing by fives—*Bloſſoms numerous, purple.* Tips *red; ſo much divided as to appear like two on each thread.*
Erica foliis corios multiflora. *Ray's Syn.* 471.
On commons. S. June—July.
The Heath Fritillary Butterfly, *Papilio Maturna,* is found upon the different ſpecies of this genus.

167 MEZEREON.   485 Daphne.

EMPAL. *Cup* none.
BLOSS. One petal; funnel-ſhaped. *Tube* cylindrical; cloſed at the baſe, longer than the *Border* which hath four clefts; the *Segments* egg-ſhaped; ſharp; flat; expanding.
CHIVES. *Threads* eight; ſhort; ſtanding in the tube; four of them alternately lower than the other four. *Tips* upright; roundiſh; with two cells.
POINT. *Seedbud* egg-ſhaped. *Shaft* very ſhort. *Summit* a knob; flat, but ſomewhat depreſſed.
S. VESS. *Berry* of one cell; roundiſh.
SEED. Single; nearly globular; fleſhy.

Olive
Mezereum

MEZEREON with flowers growing by threes; ſitting upon the Stem. Leaves ſpear-ſhaped; deciduous—*The* terminating Buds *produce leaves; the* lateral Buds, *flowers; which open very early in the ſpri*ng*; often in the winter, and are ſo thick ſet as to make the branches appear of a beautiful red.*
Laureola folio deciduo, flore purpureo. *Bauh. pin.* 426.
Camælæa Germanica, ſeu Mezereon. *Gerard.* 1403. vulgo. *Park.* 205.
Spurge Olive. Widow-wail.
In woods near Andover. S. February—March.
Varieties are
1. Leaves with yellow ſtripes.
2. Bloſſoms white; Berries yellowiſh.
3. Bloſſoms pale red.
An ointment prepared from the bark or the berries, hath been ſuceſsfully applied to ill conditioned ulcers. The whole plant is very corroſive. Six of the berries will kill a wolf. A woman gave twelve grains of the berries to her daughter who had a quartan ague: ſhe vomitted blood, and died immediately. A decoction made of two drams of the cortical part of the root, boiled in three pints of water till one pint is waſted; and this quantity drank daily, is found very efficacious in reſolving vene-

venereal

real nodes and other indurations of the periofteum. See Dr.
Ruffel's paper in the *Med. Obf. and Inquiries.* vol. 3. p. 189.

The confiderable and long continued heat and irritation that
it produces in the throat when chewed, made me firft think of
giving it in a cafe of a difficulty in fwallowing, feemingly occafioned by a paralytic affection. The patient was directed to chew
a thin flice of the root as often as fhe could bear to do it ; and in
about two months fhe recovered her power of fwallowing. This
woman bore the difagreeable irritation, and the ulcerations its
acrimony occafioned in her mouth, with great refolution ; for
fhe was reduced to fkin and bone, and for three years before had
fuffered extremely from hunger without being able to fatisfy her
appetite ; for fhe fwallowed liquids very imperfectly and folids
not at all. The complaint came on after lying-in.—The plant
is eaten by Sheep and Goats, but Cows and Horfes refufe it.

MEZEREON with five flowers forming a bunch at the bafe Laurel
of the leaves ; which are fmooth and fpear-fhaped—*Bloffoms yel-* Laureola
lowifh green. The five flowers terminating the bunches are collected
into a little rundle and are furnifhed with concave floral leaves.

Laureola. *Gerard.* 1405. *Park.* 205. *Ray's Syn.* 465.
Laureola fempervirens flore viridi, quibufdam laureolamas.
Bauh. pin. 462.
1. There is a variety with variegated leaves.
Spurge Laurel.
Woods and hedges. S. March.
Very happy effects have been experienced from this plant in
rheumatic fevers. It operates as a brifk and rather fevere purgative. It is an efficacious medicine in worm cafes ; and upon
many accounts deferves to be better known to phyficians ; but in
lefs fkilful hands it would be dangerous, as it is poffeffed of confiderable acrimony. The whole plant hath the fame qualities,
but the bark of the root is the ftrongeft. Dr. Alfton fixes the
outfide dofe at ten grains,

EIGHT CHIVES.

Order III. Three Pointals.

168 SNAKEWEED. 495 Polygonum.

EMPAL. *Cup* turban-shaped; with five divisions: coloured within. *Segments* egg-shaped; blunt: permanent.
BLOSS. None; unless you call the cup the blossom.
CHIVES. *Threads* generally eight; awl-shaped; very short. *Tips* roundish; fixed side-ways to the threads.
POINT. *Seedbud* three cornered. *Shafts* generally three; thread shaped; very short. *Summits* simple.
S. VESS. None. The *Cup* laps round the seed.
SEED. Single; three cornered; sharp.

OBS. *In the fourth and fifth species there are only six, and in the third species only five chives. In the third, fourth, fifth and sixth species there are only two pointals. In the* Pepper SNAKEWEED *the seedbud and the seed are egg-shaped and compressed.*

* Flowers in a single spike.

Biftort
Biftorta

SNAKEWEED. The stem undivided; supporting a single spike of flowers. Leaves egg-shaped, running down the leaf-stalks.—*Spike pale red.*
Biftorta major. *Gerard.* 399. *Ray's Syn.* 147. vulgaris. *Park.* 391.
Biftorta major, radice minus intorta. *Bauh. pin.* 192.
In wet meadowes. P. May—June.
The root is one of the strongest vegetable astringents.

Smaller
Viviparum

SNAKEWEED. The stem undivided; supporting a single spike of flowers; leaves spear-shaped—*Blossoms whitish. The lower flowers of the spike are frequently changed into vegetating bulbs*
Biftorta minor. *Gerard.* 399. *Ray's Syn.* 147. noftras. *Park.* 392.
Biftorta alpina minor. *Bauh. pin.* 192.
1. There is a variety in which the lower leaves are almost circular and very minutely serrated. *Ray's Syn.* 147.
Small Biftort.
High pastures. P. June.
The roots dried and ground to powder are nutritious.—Cows, Goats and Swine eat it. Sheep are not fond of it. Horses refuse it.

* Pointal cloven or divided. Chives fewer than eight.

SNAKEWEED. The flowers with five chives; pointal clo- Perennial
ven half way down, fpike egg fhaped—*reddifh. When it grows in* Amphibium
the water the chives are fhorter than the bloffom; but when it grows
upon dry land they are longer then the bloffom. In the former fituati-
on the plant floats upon the water, is fmooth and of a pleafant green;
but in the latter it is rough and dark.
Perficaria falicis folio perennis, potamogiton anguftifolium
dicta. *Ray's Syn.* 145
   Potamogiton anguftifolium. *Gerard.* 821.
   Potamogiton falicis folio. *Bauh. pin.* 131.
   Fontalis major longifolia. *Park.* 1254.
   Perennial arfmart. Red-fhanks.
   Ditches and marfhes. P. July.
Horfes Goats, Sheep and Swine eat it. Cows refufe it.

SNAKEWEED. The flowers generally with fix chives; Pepper
pointals deeply divided. Leaves fpear-fhaped; props but little Hydropiper
fringed—*Leaves fmooth on both furfaces, with very minute briftles*
at the edges. Bloffoms *purple or white; in long terminating fpikes.*
Summits *globular.*
   Perficaria urens, five hydropiper. *Bauh. pin.* 101.
   Perficaria vulgaris acris, feu hydropiper. *Ray's Syn.* 144.
   Perficaria vulgaris acris, feu minor. *Park.* 856.
   Hydropiper. *Gerard.* 445.
   Water-pepper. Arfmart. Lakeweed.
   Rivulets, ditches. A. July—Auguft.
The whole plant has an acrid burning tafte. It cures little
aphthous Ulcers in the mouth. It dyes wool yellow. The afhes
of this plant mixed with foft foap is a noftrum, in a few hands,
for diffolving the ftone in the Bladder; but it may be reafonably
queftioned whether it has any advantage over other cauftic pre-
parations of the vegetable Alcali.
Horfes, Cows, Goats, Sheep and Swine refufe it.

Spotted
Perficaria

SNAKEWEED. The flowers generally with fix chives and
two pointals. Spikes oblong egg-fhaped ; leaves fpear-fhaped;
props fringed—*Leaves downy on the under furface ; and fometimes
with a dark fpot in the fhape of a crefcent on the upper furface.* Bloffoms *in terminating fpikes ; reddifh.*
Perficaria maculofa. *Gerard.* 445. *Ray's Syn.* 145.
Perficaria mitis maculofa et non maculofa. *Bauh. pin.* 101.
Perficaria vulgaris mitis feu maculofa. *Park.* 856.
1. There is one variety with narrower leaves, fending forth
flowers from each joint. *Bauh. pin.* 101. *Ray's Syn.* 145.
2. And another in which the leaves are downy on the under furface. *Ray's Syn.* 145. In fome inftances there are feven
chives in each flower, and the pointal is fingle, but the fhaft
divided halfway down.
Dead or fpotted Arfmart.
Common. A. Auguft.
Its tafte is flightly acid and aftringent. Woollen Cloth dipped in a folution of Alum obtains a yellow colour from this plant.
Goats, Sheep and Horfes eat it. Cows and Swine refufe it.

* * * Florets with eight chives.

Pale
Penfylvani-
cum

SNAKEWEED. The flowers with eight chives and two
pointals. Fruit-ftalks covered with rough hairs. Leaves fpear-
fhaped ; props without awns.—*The whole plant is not much unlike
the preceding fpecies, but more ftiff and ftraight.* Stem *branched ; angular.* Leaves *very rough along the middle rib on the under furface.
The* Prickles *on the fruit-ftalks fecrete a glutinous liquor.* Flowers *upon little foot-ftalks, forming a fpike fomewhat branched;* Bloffoms *pale-red.*
Perficaria mitis major, foliis pallidioribus. *Ray's Syn.* 145.
Pale Arfmart.
Cornfields and among rubbifh. A. Auguft.

Marine
Maritimum

SNAKEWEED. The flowers with eight chives and three
pointals, growing at the bafe of the leaves ; which are betwixt
oval and fpear-fhaped ; evergreen. Stem fomewhat woody—
The Leaves *become blue when dried. The whole plant refembles the*
Knotgrafs SNAKEWEED *but is larger and more firm.* Bloffoms
white ; four in a place.
Polygonum marinum. *Ray's Syn.* 147.
Polygonum marinum maximum. *Gerard.* 564.
Pylygonum maritimum latifolium. *Bauh. pin.* 281.
Polygonum marinum majus. *Park.* 444.
Sea Knotgrafs.
Sea coaft. P. July.

SNAKEWEED.

SNAKEWEED. The flowers with eight chives, and three Knotgraſs pointals; growing at the baſe of the leaves; which are ſpear- Aviculare ſhaped. Stem trailing; herbaceous—*Bloſſoms pale pink.* Shafts *hardly perceptible.*

Polygonum mas vulgare. *Gerard.* 565. *Ray's Syn.* 146.
Polygonum mas vulgare majus. *Park.* 443.
Polygonum latifolium. *Bauh. pin.* 281.

1. The figure of the leaves varies a good deal in different ſituations.
Knotgraſs.
Roadſides and Cornfields. A. June—September.

The ſeeds are uſeful for every purpoſe in which thoſe of the next ſpecies are employed. Great numbers of ſmall Birds feed upon them.
Cows, Goats, Sheep, Horſes and Swine eat it.
It affords nouriſhment to the *Chryſomela Polygoni.*

* * * * *Leaves nearly heart-ſhaped*

SNAKEWEED. The leaves betwixt heart and arrow-ſhap- Buckwheat ed: ſtem unarmed; nearly upright; the angles of the ſeeds Fagopyrum equal—*Bloſſoms purpliſh white; in long looſe ſpikes.*

Fagopyrum. *Ray's Syn.* 144. *Gerard.* 89. *Park.* 1141.
Eryſimum Theophraſti folio hederaceo. *Bauh. pin.* 27.
Buckwheat. Branks. Frenchwheat. Crap.
Cornfields. A. July—Auguſt.

This plant is very impatient of cold, dying at the very firſt attack of froſt. The ſeeds furniſh a nutritious meal, which is not apt to turn acid upon the Stomach. It is made into thin cakes in ſome parts of England, called Crumpits. It is uſual with Farmers to ſow a crop of Buckwheat and to plough it under when fully grown, as a manure to the land. The ſeeds are excellent food for poultry. Sheep that eat this plant become unhealthful. As it flowers late in the ſummer, M. Du Hamel, in his obſervations upon the management of Bees, adviſes to move the hives in the autumn to a ſituation where plenty of this plant is ſown.
Cows, Goats and Sheep eat it. Swine and Horſes refuſe it.

238 EIGHT CHIVES.

Binding SNAKEWEED. The leaves heart-shaped; stem angular;
Convolvulus twining. Flowers blunt—*Tips violet colour.* Flowers *in bunches
on fruit-stalks arising from the base of the leaves. The whole fruit-
stalk is covered by the flowers.* Bloſſoms *greeniſh white.* Leaves
ſometimes arrow-ſhaped with the two angles at the baſe lopped.

Convolvulus minor ſemine triangulo. *Bauh. pin.* 295.
Convolvulus minor atriplicis folio. *Park.* 171.
Volubilis nigra. *Gerard.* 863.
Fagopyrum ſcandens ſylveſtre. *Ray's Syn.* 144.
Black Bindweed.
Cornfields and garden-hedges. A. June—September.

The ſeeds are quite as good for uſe as thoſe of the preceding
ſpecies.

Cows and Goats eat it. Sheep, Swine and Horſes refuſe it.

OBS. *The ſpotted Buff Moth,* Phalæna Lubricipeda, *is found
upon ſeveral ſpecies of this genus.*

Order IV. Four Pointals.

169 MOSCHATEL. 501 Adoxa.

EMPAL. *Cup* beneath: cloven; flat; permanent.
BLOSS. One petal; with four clefts; flat. *Segments* egg-
ſhaped; ſharp; longer than the cup.
CHIVES. *Threads* eight; awl-ſhaped; as long as the cup.
Tips roundiſh.
POINT. *Seedbud* beneath the receptacle of the bloſſom.
Shafts four; ſimple; upright; as long as the chives:
permanent. *Summits* ſimple.
S. VESS. *Berry* globular; betwixt the cup and the bloſ-
ſom. the cup connected with the underſide of
the berry, which hath four cells, and a hollow
dimple.
SEEDS. Solitary; compreſſed.

OBS. *Such are the characters of the terminating flowers; but the
lateral flowers have bloſſoms with five clefts; ten threads, and five
pointals.*

MOSCHATEL.

MOSCHATEL. Lateral bloſſoms with five clefts—*Bloſſoms* Tuberous
green, with yellow chives. Fruitſtalk *quadrangular; naked; ter-* Moſchatellina
minating. *From the baſe of the buds and leaves, ſpring out ſolitary*
Runners, *which deſcend to the ground and take root.* Berries
reddiſh.

 Moſchatellina foliis fumariæ bulbofæ. *Ray's Syn.* 261.
 Ranunculus nemoroſus moſchatella dictus. *Park.* 226.
 Ranunculus nemorum moſchatellina dictus. *Bauh. pin.* 178.
 Radix cava minima viridi flore. *Gerard.* 1090.
 Woods and ſhady places. P. March—April.
 Goats eat it. Cows refuſe it.

170 WATERPINE. 502 Elatine.

EMPAL. *Cup* four leaves; circular; flat; as large as the
 bloſſom; permanent.
BLOSS. *Petals* four; egg-ſhaped; blunt; ſitting; expand-
 ing.
CHIVES. *Threads* eight; as long as the bloſſom. *Tips*
 ſimple.
POINT. *Seedbud* large; round; globular but depreſſed.
 Shafts four; upright; parallel; as long as the chives.
 Summits ſimple.
S. VESS. *Capſule* large; round; globular but depreſſed:
 with four cells and four valves.
SEEDS. Several; creſcent-ſhaped; upright; ſurrounding
 the receptacle like a wheel.

 WATERPINE, with leaves growing in whorls—*long and* Boggy
narrow. Alſinaſtrum
 Equiſetum paluſtre linariæ ſcopariæ folio. *Bauh. pin.* 15.
 Alſinaſtrum gratiolæ folio. *Ray's Syn.* 346.
 Water-wort.
 In muddy ditches and bogs. P. Auguſt.

171 TRUE-

171 TRUELOVE. 500 Paris.

EMPAL. *Cup* of four leaves; permanent. *Leaves* spear-
shaped, sharp; as large as the blossom; expanding.

BLOSS. *Petals* four; expanding; awl-shaped; resembling
the cup; permanent.

CHIVES. *Threads* eight; awl-shaped; short beneath the
tips. *Tips* long, growing to the middle of the
threads, and on each side of them.

POINT. *Seedbud* roundish, but with four angles. *Shafts*
four; expanding; shorter than the chives. *Summits*
simple.

S. VESS. *Berry* globular; with four angles and four cells.

SEEDS. Several; lying in a double range.

Four-leaved
Quadrifolia

TRUELOVE. As there is only one species known, Linnæus
gives no description of it.—*Stem single, naked.* Blossom *pale
green.* Berry *bluish black. Leaves of the cup longer than the petals;
petals longer than the chives; chives longer than the shafts, and these
longer than the berry.*

Herba Paris. *Gerard.* 405. *Park.* 390. *Ray's Syn.* 264.
Solanum quadrifolium bacciferum. *Bauh. pin.* 167.
Herb Paris. One-berry.
Woods and shady places. P. May—June.

The leaves and berries are said to partake of the properties of
Opium. The juice of the berries is useful in Inflammations of
the eyes. Linnæus says the root will vomit as well as Ipecacuanha, but it must be given in a double quantity.—Goats and
Sheep eat it. Cows, Horses and Swine refuse it.

CLASS.

C L A S S IX.

N I N E C H I V E S.

ORDER III. SIX POINTALS.

172 G LADIOLE. *Cup* none. *Bloff.* fix petals.
Capfules fix. *Seeds* many.

† *Water Frogbit.*

172 GLADIOLE. 507 Butomus.

EMPAL. *Fence* fimple ; of three leaves ; fhort.

BLOSS. *Petals* fix ; circular ; concave ; fhrivelling ; every other petal ftanding on the out-fide, fmaller and fharper.

CHIVES. *Threads* nine ; awl-fhaped. Six of them ftand on the out-fide the others. *Tips* compofed of two plates.

POINT. *Seedbuds* fix; oblong ; tapering ; ending in *Shafts*. *Summits* fimple.

S. VESS. *Capfules* fix; oblong; gradually tapering; up-right; of one valve which opens inwards.

SEEDS. Many ; oblong; cylindrical ; blunt at each end.

Water Umbellatus

GLADIOLE. As there is only one fpecies known Linnæus gives no defcription of it.—*The root leaves long, narrow.* Stem *cylindrical, naked.* Bloffoms *purple and white ; terminating ; fome-times quite white.*

Butomus. *Ray's Syn.* 273.
Juncus floridus. *Park.* 1197.
Juncus floridus major. *Bauh. pin.* 12.
Gladiolus paluftris cordi. *Gerard.* 29.
Flowering Rufh.
In muddy ditches. P. June.
Neither Cows, Horfes, Sheep, Swine or Goats will eat it.

C L A S S X.

T E N C H I V E S.

ORDER I. ONE POINTAL.

* Flowers of many equal Petals.

173 BIRDSNEST.　*Cup* refembling a bloffom; hunched at the bafe. *Capf.* with five
cells. *Seeds* many.

174 PEARLEAF.　-　*Tips* with two horns pointing upwards. *Capf.* with five cells.
Seeds many.

† *Geraniums.*

** Flowers of one regular Petal.

175 GARDROBE.　-　*Bloff.* bell-fhaped; roundifh; *Capf.*
with five cells.

176 STRAWBERRYTREE. *Bloff.* egg-fhaped; tranfparent at
the bafe. *Berry* with five cells.

† *Whortle Bilberry.*

Order II. Two Pointals.

177 KNAWEL.　-　-　*Bloff.* none. *Cup* with five clefts;
fuperior. *Seeds* two.

R 2　　　　　178 SAX

178 SAXIFRAGE.   -   *Bloſſ.* none. *Cup* ſuperior. *Capſ.* with two cells and two bills.

179 SENGREEN.   -   *Bloſſ.* five petals. *Cup* with five diviſions. *Capſ.* with one cell and two bills.

180 SOAPWORT.   -   *Bloſſ.* five petals. *Cup* tubular; naked at the baſe. *Capſ.* with one cell; oblong.

181 PINK.   - - -   *Bloſſ.* five petals. *Cup* tubular; ſcaly at the baſe. *Capſ.* with one cell; oblong.

Order III.   Three Pointals.

182 SANDWORT.   -   *Capſ.* with one cell. *Petals* entire; expanding.

183 STITCHWORT.   -   *Capſ.* with one cell. *Petals* divided almoſt to the baſe; expanding.

184 CAMPION.   -   *Capſ.* with three cells. *Petals* cloven. *Mouth* naked.

185 CATCHFLY.   -   *Capſ.* with three cells. *Petals* cloven. *Mouth* crowned.

† *Dwarf Elder.*

Order IV.   Five Pointals.

186 NAVELWORT.   -   *Capſ.* five, with a honey-cup to each. *Bloſſ.* one petal.

187 STONECROP.   -   *Capſ.* five, with a honey-cup to each. *Bloſſ.* five petals.

188 SPURREY.   - -   *Capſ.* with one cell. *Petals* entire. *Cup* five leaves.

189 MOUSE-EAR.   -   *Capſ.* with one cell. *Petals* cloven. *Cup* five leaves.

190 COCKLE.   -   *Capſ.* with one cell; oblong; *Cup* tubular; like leather.

191 CUCKOWFLOWER.   *Capſ.* with three cells: oblong; *Cup* tubular; membranaceous.

192 WOODSORREL.   *Capſ.* with five cells, angular. *Bloſſ.* somewhat connected at the baſe.

† *Geraniums.*   † *Tuberous Moſchatel.*

173 BIRDS.

173 BIRDSNEST. 536 Monotropa.

EMPAL. None: unless you call the five outermost co-loured petals the cup.

BLOSS. *Petals* ten; oblong; nearly parallel and upright; serrated towards the point; deciduous. The five outermost are hunched at the base, and have a cavity for honey on the inside.

CHIVES. *Threads* ten; awl-shaped; upright; simple. *Tips* simple.

POINT. *Seedbud* roundish, tapering to a point. *Shaft* cylindrical; as long as the chives. *Summit* a blunt knob.

S. VESS. *Capsule* egg-shaped; blunt; with five angles and five valves.

SEEDS. Numerous; chaffy.

OBS. *Such are the generic characters of the terminating flower. But if there are any lateral flowers they contain one fifth part less in number.*

BIRDSNEST. The lateral flowers with eight; but the terminating flower with ten chives—*The whole plant smells sweet, and is of a pale yellow colour, which peculiarity is generally confined to parasitical plants and those that grow in very shady situations.* Primrose Hypopithys

Hypopitys lutea. *Ray's Syn.* 317.
Orobanche hypopitys. *Bauh. pin.* 88.
Birds-nest, smelling like Primrose roots,
In woods; not common. P. July.

The country people in Sweden give the dried plant to cattle that have a Cough.

174 PEARLEAF. 554 Pyrola.

EMPAL. *Cup* with five divisions; very small; permanent.

BLOSS. *Petals* five; circular; concave; expanding.

CHIVES. *Threads* ten; awl-shaped; shorter than the blossom. *Tips* large; nodding; with two horns pointing upwards.

POINT. *Seedbud* roundish, angular. *Shaft* thread-shaped; longer than the chives; permanent. *Summit* rather thick.

S. VESS. *Capsule* roundish; depressed; with five angles and five cells; opening at the angles.

SEEDS. Numerous; chaffy.

OBS. *In some species the threads and shaft are upright, in others declining to one side, and in others again expanding. The shape of the summit is different in different species.*

K 3　　　　　　　PEARLEAF.

Winter-green
Rotundifolia

PEARLEAF. The chives afcending; the pointal declining.
—*Bloffoms white ; on flender fruit ftalks.*
Pyrola. *Gerard.* 408. *Ray's Syn.* 363.
Pyrola noftras vulgaris. *Park.* 508.
Pyrola rotundifolia major. *Bauh. pin.* 191.
Common Wintergreen.
Woods in the North. P. June—July.
Goats eat it; Cows, Horfes, Sheep and Swine refufe it.

Leffer
Minor

PEARLEAF. The flowers difpofed in fcattering bunches.
Chives and pointal ftraight—*This fpecies much refembles the preceding
but is eafily diftinguifhed by attending to the direction of the chives
and pointal. The ftalk in both is three-cornered. It is very probable
that this plant was firft produced by the duft of the* Indented PEAR-
LEAF *impregnating the feedbud of the firft fptcies.* Blossoms *reddifh
white.*
Pyrola minor. *Ray's Syn.* 363.
Pyrola folio minore et duriore. *Bauh. pin.* 191.
Leffer Winter-green.
Woods in Yorkfhire. P. Auguft.

Indented
Secunda

PEARLEAF. The flowers in bunches pointing one way
—*The chives project beyond the bloffom and the fummit beyond the
chives.* Blossoms *white.*
Pyrola folio mucronato ferrato. *Bauh. pin.* 181. *Ray's Syn.*
363.
Pyrola tenerior. *Park.* 509.
Pyrola fecunda tenerior. *Gerard.* 408.
Dented leaved Winter-green.
In woods. P. June.
Goats eat it ; fheep refufe it.

ONE POINTAL. 247

175 GARDROBE. 549 Andromeda.

EMPAL. *Cup* with five divisions; sharp; very small; coloured; permanent.

BLOSS. One petal; bell-shaped; with five clefts. *Segments* reflected.

CHIVES. *Threads* ten; awl-shaped; shorter than the blossom, to which they slightly adhere. *Tips* with two horns; nodding.

POINT. *Seedbud* roundish. *Shaft* cylindrical; longer than the chives; permanent. *Summit* blunt.

S. VESS. *Capsule* roundish; with five angles, five cells, and five valves; opening at the angles.

SEEDS. Numerous; roundish; shining.

OBS. The blossom in some species is egg-shaped, but in others truly bell-shaped.

GARDROBE. The fruit-stalks incorporated; blossoms egg- Rosemary-shaped; leaves alternate; spear-shaped; rolled back at the edges leaved —*Blossoms tinged with red.* Polifolia
Ledum palustre nostras arbutiflore. *Ray's Syn.* 472.
Rosmarinum sylvestre minus nostras. *Park.* 76.
Marsh Cistus. Wild Rosemary.
In Turf Bogs. P. March—April.
Goats and Sheep eat it; Cows and Horses refuse it.

GARDROBE. The flowers in bunches, all pointing one Heath way. Blossoms with four clefts, egg-shaped. Leaves alternate, Daboæcia spear-shaped, rolled back at the edges—*Flowers in a simple bunch, terminating; placed alternately and all pointing one way. One flower only upon each little fruit-stalk, which is somewhat clammy and hath a strap-shaped floral leaf beneath it. Cup of four leaves; awl-shaped; upright; purplish; deciduous; only a fourth part as long as the blossom. Blossom betwixt cylindrical and egg-shaped; violet coloured; twice as large as the first species. The mouth a little contracted; with four clefts, and the segments rolled back. Chives eight; threads white; Tips as long as the threads, and but a little shorter than the blossom; arrow-shaped, brown; lopped at the end; with two perforations. Shaft thread-shaped; as long as the blossom. Summit blunt; with four slight clefts. Fruit a Capsule with four cells and four valves.* It hath all the appearance of the GARDROBE, but the numbers of the HEATH.
Erica cantabrica flore maximo, foliis Myrti, subtus incanis. *Ray's Syn.* 472.
Vaccinium cantabricum. *Hudson.* 143.
Irish Worts.
In spongy wet and uncultivated soil. S.

R 4 176 STRAW-

176 STRAWBERRYTREE. 552 Arbutus.

EMPAL *Cup* with five divisions ; blunt ; very small ; permanent.

BLOSS. One petal ; egg-shaped ; flattish at the base. *Mouth* with five clefts ; *Segments* blunt ; rolled back ; small.

CHIVES. *Threads* ten, awl-shaped but distended ; very slender at the base ; half as long as the blossom, and fixed to the margin of its base. *Tips* slightly cloven ; nodding.

POINT. *Seedbud* nearly globular ; fitting upon the receptacle, which is marked with ten dots. *Shaft* cylindrical ; as long as the blossom. *Summit* rather thick and blunt.

S. VESS. *Berry* roundish ; with five cells.

SEEDS. Small, and hard as bone.

Common Unedo

STRAWBERRYTREE. The stem woody ; leaves smooth ; serrated. Berry containing many seeds—*Blossoms white.* Berries *red.*
Arbutus. *Gerard.* 1496. *Park.* 1489. *Ray's Syn.* 464.
Arbutus folio serrato. *Bauh. pin.* 460.
In the West of Ireland ; particularly near the Lake of Killarney, on barren lime-stone rocks. S.
1. Blossoms oblong : berries oval.
2. Blossoms red.
3. Blossoms double.
It hath obtained a place in our gardens upon account of the beautiful appearance of its fruit ; which the country people in Ireland eat, but they always drink water after it.

Mountain Alpina

STRAWBERRYTREE. The stems trailing ; leaves wrinkled, serrated—*Berries black ; globular ; fitting upon a very small red cup.*
Vaccinia rubra, foliis myrtinis crispis. *Ray's Syn.* 457.
Vitis idæa foliis oblongis albentibus. *Bauh. pin.* 470.
On dry mountains. S.
The berries have something of the flavour of Black Currants, but they are not so good.—Goats refuse it.

STRAWBERRY-

STRAWBERRYTREE. The Stems trailing; leaves very Perennial-entire—*Cups purple.* Bloſſoms *white.* leaved.

In the Highlands of Scotland, and in Wales, upon the moun- Uva-urſi tains. S

The berries are inſipid, pulpy and mealy. The plant is much uſed in Sweden to dye an aſh-colour, and to tan leather. Half a dram of the powdered leaves given every, or every other day, hath been found uſeful in çalculous caſes. It was firſt uſed for this purpoſe at Montpelier, and afterwards Dr. de Haen at Vienna relates ſeveral caſes in which it proved of the greateſt ſervice. Its ſucceſs in England has been uncertain. Sometimes the patients found no relief, but thought their complaints rather aggravated than alleviated; whilſt in other calculous and nephritic caſes the ſymptoms have been almoſt entirely removed. Perhaps upon the whole we ſhall find it no better than other vegetable aſtringents; ſome of which have long been uſed by the country people in gravelly complaints, and with very great advantage; though hitherto unnoticed by the regular practitioners.—Horſes, Cows, Goats and Sheep refuſe it.

Order II. Two Pointals.

177 KNAWEL. 562 Scleranthus.

EMPAL. *Cup* one leaf; tubular; with five ſhallow clefts; ſharp; permanent; narrow at the neck.

BLOSS. None.

CHIVES. *Threads* ten; awl-ſhaped: upright; very ſmall, fixed to the cup. *Tips* roundiſh.

POINT. *Seedbud* roundiſh. *Shafts* two; upright; hair-like; as long as the chives. *Summits* ſimple.

S.VESS. *Capſule* egg ſhaped: very thin; in the bottom of the cup, which cloſes at the neck.

SEEDS. Two; convex on one ſide and flat on the other.

KNAWEL. The cup open at the neck when the fruit Knotgraſs ripens—*Leaves oppoſite.* Bloſſoms *at the baſe of the leaves;* greeniſh. Annuus
Knawel. *Ray's Syn.* 159.
Polygonum germanicum vel Knawell germanorum. *Park.* 747.
Polygonum ſelinoides ſive Knawel. *Gerard.* 453.
Polygonum gramineo folio majus erectum. *Bauh. ſin.* 281.
German Knot-graſs. Annual Knawel.
Sandy ground and corn fields. A. Auguſt.
The Swedes and the Germans receive the vapour ariſing from a decoction of it, into their mouths to cure the Tooth-ach.— Goats and Sheep eat it. Cows refuſe it.

KNAWEL.

Perennial   **KNAWEL.** The cup clofed at the neck when the fruit ripe s
Perennis   —*Leaves hoary.* Bloſſoms *greeniſh*; *terminating*.

Knawel incanum flore majore perenne. *Ray's Syn.* 160. Tab. 5. fig. 1.

Gravelly foil and cornfields. P. Auguſt.

The Poliſh cochineal, *Coccus polonicus*, is found upon the roots, in the ſummer months.

178 SAXIFRAGE.   558 Chryſoſplenium.

EMPAL. *Cup* with four or five diviſions; expanded; coloured; permanent. *Segments* egg-ſhaped; the oppoſite ones narroweſt.

BLOSS. None; unleſs you call the cup ſo becauſe it is coloured.

CHIVES. *Threads* eight or ten; awl ſhaped; upright; very ſhort; ſtanding upon the angular receptacle. *Tips* ſimple.

POINT. *Seedbud* beneath: terminated by two awl-ſhaped *Shafts* as long as the chives. *Summits* blunt.

S.VESS. *Capſule* with two bills and two diviſions; one cell and two valves; encompaſſed by the cup which becomes green.

SEEDS. Many; very ſmall.

     OBS. *The terminating Bloſſom hath five clefts; the others only four.*

Alternate-   **SAXIFRAGE.** The leaves alternate—*on long leaf-ſtalks.*
leaved   Bloſſoms *bright yellow.*
Alternifolium   Saxifraga aurea foliis pediculis oblongis inſidentibus. *Ray's Syn.* 158.

Alternate leaved Golden Saxifrage.

In ſhady woods near rills of water. P. April.

Golden   **SAXIFRAGE.** The leaves oppoſite—*on ſhort leaf-ſtalks.*
Oppoſitifolium   Bloſſoms *bright yellow.*
Saxifraga aurea. *Gerard.* 841. *Park.* 425. *Ray's Syn.* 158.

Saxifraga rotundifolia aurea. *Bauh pin.* 309.

Common Golden Saxifrage.

In watery lanes. P. April.

179 SENGREEN. 559 Saxifraga.

EMPAL. *Cup* one leaf; with five divisions; short; sharp; permanent.
BLOSS. *Petals* five; expanding; narrow at the base.
CHIVES. *Threads* ten; awl-shaped. *Tips* roundish.
POINT. *Seedbud* roundish, but tapering, and ending in two short *Shafts. Summits* blunt.
S. VESS. *Capsule* somewhat egg shaped; with two bills and one cell; opening betwixt the bills.
SEEDS. Numerous; minute.

OBS. *In some species the* Seedbud *is beneath; in others it is superior. After the flower is open, two of the* Chives *opposite to each other, bend down to the* Summits *and discharge their dust perpendicularly over them. The next day two others bend down; and this is continued until they have all done the same.*

* *Leaves undivided. Stem nearly naked.*

SENGREEN. The leaves serrated; stem naked; branching. Hairy
Petals tapering to a point—*white; with a yellow spot near the* Stellaris
base.
Geum palustre minus, foliis oblongis crenatis. *Ray's Syn.* 354.
Hairy Kidney-wort.
Moist rocks. P. June—July.

SENGREEN. The leaves inversely egg-shaped; scolloped; Mountain
on very short leaf-stalks. Stem naked; flowers collected into a Nivalis
ball—*Blossoms white; spotted.*
Saxifraga foliis oblongo-rotundis dentatis floribus compactis.
Ray's Syn. 354. Tab. 16. fig. 1.
On lofty mountains. P. June.

* * *Leaves undivided; stem leafy.*

SENGREEN. The stem leaves egg-shaped; opposite; tiled. Heath-like
Upper leaves fringed—*Stem thread-shaped; hanging down, or* Oppositifolia
creeping. Terminating Flower *solitary, sitting.* Blossoms *blue.*
Saxifraga alpina ericoides, flore cæruleo. *Ray's Syn.* 353.
Sedum alpinum ericoides cæruleum. *Bauh. pin.* 284.
Mountain Heath-like Sengreen.
On barren mountains. P. March—April.

Autumnal SENGREEN. The ftem leaves ftrap-fhaped; alternate;
Autumnalis fringed. Root leaves incorporated—*Cup much fhorter than the*
bloffom; *and green.* Bloffoms *beneath: yellow, fpotted.*
Geum anguftifolium autumnale, flore luteo guttato. *Ray's*
Syn. 355.
In turf bogs. P. July—Auguft.

Yellow SENGREEN. The ftem leaves betwixt ftrap and awl-fhaped;
Aizoides fcattered; naked; unarmed. Stems drooping—*Bloffoms yellow.*
Saxifraga alpina angufto folio, folio luteo guttato. *Ray's Syn.*
353.
Sedum alpinum flore pallido. *Bauh. pin.* 284.
Yellow Mountain Sengreen.
On mountains. P. Auguft.

*** *Leaves gafhed. Stem upright.*

White SENGREEN. The ftem leaves kidney-fhaped; gafhed.
Granulata Stem branched. Root beaded—*Branches without leaves.* Flowers
fuperior. Cup *a little hairy.* Bloffoms *white.*
Saxifraga rotundifolia alba. *Bauh. pin.* 309. *Ray's Syn.* 354.
Saxifraga alba. *Gerard.* 841.
Saxifraga alba vulgaris, *Park.* 424.
White Saxifrage,
Dry ground. P. May.
Goats eat it. Cows, Sheep, Horfes and Swine refufe it.

Rue-leaved SENGREEN. The ftem leaves wedge-fhaped; cloven into
Tridactylites three parts; alternate. Stem upright; branched—*The whole*
plant is fet with hairs which pour out a clammy liquor at their points.
The truth of the obfervation at the end of the generic character is rea-
dily to be remarked in this fpecies. In very dry fituations the leaves
are not cloven. Bloffoms *white.*
Saxifraga verna annua humilior. *Ray's Syn.* 354,
Sedum tridactylites tectorum. *Bauh. pin.* 285.
Paronychia rutaceo folio. *Gerard.* 624.
Paronychia foliis incifis. *Park.* 556.
Whitlow-grafs.
Walls. Roofs. A. April.

Tufted SENGREEN. The root leaves incorporated; ftrap-fhaped;
Cafpitofa either entire, or elfe cloven into three parts. Stem upright,
nearly naked, fupporting one or two flowers.—*Petals blunt;*
greenifh white: yellow when dry. The plant is fmooth; the upper
branches clammy. Stem *leaves two or three; ftrap-fhaped; undivided.*
Sedum tridactylites alpinum minus. *Bauh. pin.* 284.
Small Mountain Sengreen.
Mountains in Weftmoreland. P. Auguft.

SENGREEN.

*** Leaves gashed; stems trailing.

SENGREEN. The stem leaves strap-shaped; either entire Cloven or cloven into three parts. Suckers trailing. Stem upright; Hypnoides almost naked.—*Cup green*; Chives *yellow*. Blossom *greenish white. The whole plant is clammy, and becomes reddish when fully grown.*

Saxifraga muscosa trifido folio. *Ray's Syn.* 354.
Sedum alpinum trifido folio. *Bauh. pin.* 284.
Sedum alpinum laciniatis Ajugæ foliis. *Park.* 739.
Trifid Sengreen.
On mountains. P. April—May.

180 SOAPWORT.   564 Saponaria.

EMPAL. *Cup* one leaf; tubular; with five teeth; permanent.

BLOSS. *Petals* five. *Claws* narrow; angular; as long as the cup. *Border* flat. *Limbs* broader towards the end; blunt.

CHIVES. *Threads* ten; awl-shaped; as long as the tube of the blossom. Every other chive fixed to the claws of the petals: five of them ripening later than the others. *Tips* oblong; blunt; fixed side-ways to the threads.

POINT. *Seedbud* somewhat cylindrical. *Shafts* two; straight; parallel; as long as the chives. *Summits* sharp.

S. VESS. *Capsule* as long as the cup; cylindrical; of one cell; covered.

SEEDS. Many; small. *Receptacle* loose.

Smooth   SOAPWORT. The cup cylindrical. Leaves betwixt egg
Officinalis   and spear shaped—*Blossoms white or reddish, terminating.*
       Saponaria. *Gerard.* 444. vulgaris. *Park.* 641.
       Saponaria major lævis. *Bauh. pin.* 206.
       Lychnis Saponaria dicta. *Ray's Syn.* 339.
       Bruise-wort.
       Hedges and heaths. P. July—August.

Concave   1. Leaves concave.—This singular variety is noticed by Gerard
Hybrida   p. 435; who found it in a small wood called the Spinie near
the village of Lichbarrow in Northamptonshire; but it is not
now to be found there. It seems to be a mule produced betwixt the Soapwort and the Gentian, the dust of the latter
falling upon the pointals of the former. It does not produce
perfect seeds.
       Lychnis Saponaria dicta folio convoluto. *Ray's Syn.* 339.
       Saponaria concava anglica. *Bauh. pin.* 206.
       The whole plant is bitter. Bruised and agitated with water
it raises a lather like soap, which washes greasy spots out of
cloaths. A decoction of it applied externally cures the itch.
The Germans use it instead of Sarsaparilla in venereal complaints.

181 PINK. 565 Dianthus.

EMPAL. *Cup* cylindrical; tubular; scored; permanent;
     with five teeth at the mouth, and encompassed at
     the base with four scales; two of which are opposite
     and lower than the other two.

BLOSS. *Petals* five; *Claws* as long as the cup; narrow;
     fixed to the receptacle. *Limbs* flat; broadest to-
     wards the end; blunt; scolloped.

CHIVES. *Threads* ten; awl shaped; as long as the cup;
     standing wide towards the top. *Tips* oblong oval;
     compressed; fixed side-ways to the threads.

POINT. *Seedbud* oval. *Shafts* two; awl-shaped; longer
     than the chives. *Summits* curled; taper.

S. VESS. *Capsule* cylindrical; covered; one cell, opening
     at the top in four directions.

SEEDS. Many; compressed; roundish. *Receptacle* loose;
     four cornered; only half as long as the seed-
     vessel.

     OBS. *In some species the* Shafts *are but little longer than the chives;
in others they are exceedingly long, but rolled back.*

                                       * *Flower*

* *Flowers incorporated.*

PINK with flowers incorporated in little bundles. The scales Deptford of the cup spear-shaped; woolly; as long as the tube—*Petals* Armeria *tapering; furnished with one or two teeth; red.*

Caryophyllus latifolius barbatus minor annuus, flore minore. *Ray's Syn.* 337.

Caryophyllus pratensis. *Gerard.* 594.

Caryophyllus pratensis noster major et minor. *Park.* 1338.

Caryophyllus barbatus sylvestris. *Bauh. pin.* 209.

In gravelly soils. A. July.

PINK with flowers incorporated in little heads. The scales Limewort of the cup egg-shaped; blunt; without awns; longer than the Prolifer tube—*Blossoms red; sometimes white. They expand about eight in the morning and close about one in the afternoon.*

Caryophyllus sylvestris prolifer. *Bauh. pin.* 209. *Park.* 1338. *Ray's Syn.* 337.

Viscaria. *Gerard.* 601.

Limewort. Wild childing Sweet William.

Gravelly soil. A. August.

Cows and Sheep eat it.

* * *Flowers solitary. Many upon the same stem.*

PINK. The flowers solitary; scales of the cups spear-shaped; Maiden in pairs. Blossoms scolloped.—*Bright red.* Deltoides

Caryophyllus minor repens nostras. *Ray's Syn.* 335.

Caryophyllus sylvestris vulgaris latifolius. *Bauh. pin.* 209.

On heaths and in pastures. P. June—July.

Cows, Horses, Sheep and Goats eat it. Swine refuse it.

PINK. The flowers for most part solitary; scales of the cup Mountain spear-shaped; growing by fours; short. Blossoms scolloped— Glaucus *white.*

On Chedder rocks in Somersetshire. P. July.

* * * *Stem herbaceous; supporting a single flower.*

PINK. The stem generally supporting but one flower. Scales Stone of the cup egg-shaped, blunt. Petals with many clefts. Leaves Arenarius strap-shaped—*Root leaves very numerous. Stem hardly so long as ones finger. Blossoms red.*

Armeriæ species flore in summo caule singulari. *Ray's Syn.* 336.

On walls and very dry stony places. P. July.

The smell of the flowers is extremely fragrant in the night-time.—Cows, Horses and Sheep eat it.

Order

Order III. Three Pointals.

182 SANDWORT. 569 Arenaria.

EMPAL. *Cup* of five leaves. *Leaves* oblong; tapering; expanding; permanent.

BLOSS. *Petals* five; egg-shaped.

CHIVES. *Threads* ten; awl-shaped. Every other thread placed more inwards. *Tips* roundish.

POINT. *Seedbud* egg-shaped. *Shafts* three; upright, but a little reflected. *Summits* rather thick.

S. VESS. *Capsule* egg-shaped; covered; with one cell; opening at the point in five different directions.

SEEDS. Many; kidney-shaped.

Chickweed Peploides

SANDWORT. The leaves egg-shaped, sharp, fleshy— *Blossoms white.*
Alsine marina foliis portulacæ. *Ray's Syn.* 351.
Alsine littoralis foliis portulacæ, *Bauh. pin* 251.
Anthyllis maritima lentifolia. *Bauh pin.* 282. *Park.* 282.
Anthyllis lentifolia seu alsine cruciata marina. *Gerard.* 622.
Sea Chickweed.
On the sea shore. P. June—July.
Horses eat it; Sheep and Cows refuse it.

Plantain-leaved Trinervia

SANDWORT. The leaves egg-shaped, sharp; supported on leaf-stalks; stringy— *Stem forked;* Capsules *pendant.* Leaves *tapering at the base so as to form the leaf-stalks.* Blossoms *white.*
Alsine Plantaginis folio. *Ray's Syn.* 349.
Plantain leaved Chickweed.
Woods and wet hedges. A. May—June.
Sheep are not fond of it.

Little Serpyllifolia

SANDWORT. The leaves somewhat egg-shaped, sharp; fitting. Blossom shorter than the cup— *Petals entire;* white.
Alsine minor multicaulis. *Bauh. pin.* 250. *Ray's Syn.* 349.
Alsine minima. *Gerard.* 488.
Alsine aquatica minima. *Park.* 1259.
Least Chickweed.
Roofs, walls, and very dry places. A. May.
Sheep refuse it.

SANDWORT.

SANDWORT. The leaves thread-shaped; sheathed with Spurry membranaceous props—*The flowers open about nine or ten in the* Rubra *morning, and close at two or three in the afternoon.* Petals *purple.*

Spergula purpurea. *Ray's Syn.* 351.
Saginæ Spergula minima. *Park.* 561.
Alsine Spergulæ facie minor, seu Spergula minor sub-cæruleo flore. *Bauh. pin.* 251.

1. Leaves strap-shaped; as long as the joints of the stem—*This* Sea *hath sometimes only five chives.* Blossoms *purple; sometimes blue.* Marina
Alsine Spergulæ facie media. *Bauh. pin.* 251. *Ray's Syn.* 351.
Purple flowered Chickweed or Spurry. 1. Sea Spurry.
In gravelly soils. 1. on the sea shore. A. June—July.
Goats refuse it. Sheep are not fond of it.

SANDWORT. The leaves awl-shaped. Stem supporting Mountain panicles of flowers. Cup with egg-shaped blunt leaves—*Petals* Saxatilis *white.*

Alsine pusilla pulchro flore, folio tenuissimo nostras, seu Saxifraga pusilla caryophylloides, flore albo pulchello. *Ray's Syn.* 351.
Mountain Chickweed.
On mountains. P. August.

SANDWORT. The leaves awl-shaped; stem supporting a Fine-leaved panicle of flowers. Capsules upright. Petals spear shaped; Tenuifolia shorter than the cup—*white. Little leaves which form the cup greatly tapering and marked on the underside by two green lines. Petals broad; spearshaped; only half as long as the cup.* Leaves *awl-shaped; connected at the base.*

Alsine tenuifolia. *Ray's Syn.* 350.
Fine-leaved Chickweed.
Dry gravelly places. P. June—July.

SANDWORT. The leaves bristle-shaped. Stem almost Larch-leaved naked towards the top. Cups somewhat hairy—*Blossom terminat-* Laricifolia *ing.*

Alsine alpina junceo folio. *Bauh. pin.* 251.
Larch-leaved Chickweed.
On Mountains. P. August.

183 STITCHWORT. 568 Stellaria.

EMPAL. *Cup* five leaves; betwixt egg and spear-shaped; concave; upright; expanding; permanent.

BLOSS. *Petals* five; deeply divided; flat; oblong; shrivelling.

CHIVES. *Threads* ten; thread-shaped; shorter than the blossom. Every other thread shorter. *Tips* roundish.

POINT. *Seedbud* roundish. *Shafts* three; hair-like; expanding. *Summits* blunt.

S.VESS. *Capsule* egg-shaped; covered; with one cell and six valves.

SEEDS. Many; roundish; compressed.

OBS. *In the second species the tips are long and double.*

Broad-leaved
Nemorum

STITCHWORT. The lower leaves heart-shaped; on leaf-stalks. Flowers in panicles, on branching fruit-stalks.—*About a foot high. Leaves pale green on the under surface. Cups smooth. Petals white.*

Alfine montana folio smilacis instar, flore laciniato. *Ray's Syn.* 347.

Alfine montana latifolia, flore laciniato. *Bauh. pin.* 251.

Alfine hederacea montana maxima. *Park.* 761.

Woods, hedges and banks of rivers. P. July—August.

Greater
Holostea

STITCHWORT. The leaves spear-shaped, finely serrated. *Petals cloven—white.*

Caryophyllus holosteus arvensis glaber, flore majore. *Bauh. pin.* 210. *Ray's Syn.* 346.

Gramen leucanthemum. *Gerard.* 47. *Park.* 1325.

Woods and hedges. P. April—May.

Lesser
Graminea

STITCHWORT. The leaves strap-shaped, very entire. Flowers in panicles—*white.*

Caryophyllus holosteus arvensis glaber, flore minore. *Bauh. pin.* 210. *Ray's Syn.* 346.

Gramen leucanthemum alterum. *Gerard.* 47.

Gramen leucanthemum minus. *Park.* 1325.

1. In wet situations the leaves are larger and longer. These variations have been taken notice of by authors, as

Alfine longifolia uliginosis proveniens locis. *Ray's Syn.* 347.

Alfine aquatica media. *Bauh. pin.* 251.

Alfine fontana. *Gerard.* 613.

Very common. P. July.

Horses, Cows, Goats, Sheep and Swine eat it.

184. CAMPION. 566 Cucubalus.

EMPAL. *Cup* one leaf; tubular; with five teeth; permanent.

BLOSS. *Petals* five. *Claws* as long as the cup. *Border* flat. Limbs generally cloven; not crowned by a honey-cup.

CHIVES. *Threads* ten; awl-shaped. Every other thread fixed to the claws of the petals; five of them ripening later. *Tips* oblong.

POINT. *Seedbud* rather oblong. *Shafts* three; awl-shaped; longer than the chives. *Summits* downy; oblong; bending towards the left

S.VESS. *Capsule* covered; tapering; with three cells, opening at the point in five different directions.

SEEDS. Many; roundish.

OBS. *This genus is distinguished from that of* CATCHFLY *by the blossoms not being crowned with honey-cups. The fourth species hath the chives on one plant and the pointals upon another.*

CAMPION. The cups bell-shaped; petals distant. Seed-vessels coloured. Branches straddling—*Blossoms white.*   Berry-bearing Bacciferus
Cucubalus Plinii. *Ray's Syn.* 267.
Alsine scandens baccifera. *Bauh. pin.* 250.
Berry-bearing Chickweed.
In hedges in the Isle of Anglesea. P. July.

CAMPION. The cups nearly globular; smooth; with a network of veins. Capsules with three cells. Blossoms almost naked—*white.*   Bladder Behen
1. Cultivation produces a double white, a purplish, a narrow-leaved, a smooth narrow-leaved and a hairy round-leaved purplish variety.
Lychnis sylvestris quæ benalbum vulgo. *Bauh. pin.* 205. *Ray's Syn.* 337.
Behen album officinarum. *Gerard.* 678.
Papaver spumeum sive benalbum vulgo. *Park.* 263.
White Corn Campion. Sprattling Poppy. White Behen.
Dry pastures and cornfields. P. July.
The leaves boiled have something of the flavour of pease, and proved of great use to the inhabitants of the island of Minorca in the year 1685, when a swarm of locusts had destroyed the harvest. The Gothlanders apply the leaves to erysipelatous eruptions.—
Horses, Cows, Sheep and Goats eat it.

CAMPION.

Dover
Viſcoſus

CAMPION. The lateral flowers hanging down on every ſide. Stem undivided. Leaves reflected at the baſe—*Fruit-ſtalks op-poſite; ſhort; three flowers upon each.* Cup *cylindrical; with ten angles; clammy.* Petals *white; deeply divided.* Chives *longer than the bloſſom.* Tips *green.* Shafts *three; longer than the bloſ-ſom.* Fruit *egg-ſhaped; with one cell.*
　　Lychnis major noctiflora dubrenſis perennis. *Ray's Syn.* 340
　　On Dover Cliffs. P. July.

Catchfly
Otites

CAMPION. The chives and pointals on different plants. Petals ſtrap-ſhaped; undivided—*Pale green, or white.* Root-leaves *lying in a circle on the ground.*
　　Lychnis viſcoſa flore muſcoſo. *Bauh. pin.* 266.
　　Seſamoides ſalamanticum magnum. *Gerard.* 493.
　　Muſcipula ſalamantica major. *Park.* 636.
　　Spaniſh Campion.
　　In gravelly ſoils. P. July—Auguſt.

185 CATCHFLY. 567 Silene.

EMPAL. *Cup* one leaf; tubular; with five teeth: perma-nent.

BLOSS. *Petals* five. *Claws* narrow; as long as the cup; bordered. Limb flat; blunt; frequently cloven. *Honeycup* compoſed of two little teeth at the neck of each petal, and conſtituting a crown at the mouth of the tube.

CHIVES. *Threads* ten; awl-ſhaped; every other thread fixed to the claws of the petals; and ripening later. *Tips* oblong,

POINT. *Seedbud* cylindrical. *Shafts* three; ſimple; long-er than the chives, *Summits* bending to the left.

S.VESS. *Capſule* cylindrical; covered. cells three; open-ing at the point in five different directions.

SEEDS. Many; kidney-ſhaped.

* Flowers ſolitary, lateral.

Corn
Anglica

CATCHFLY. Hairy. Petals notched at the end. Flowers upright. Fruit on reflected, alternate fruitſtalks—*The Lower Leaves inverſely ſpear-ſhaped and fringed at the baſe.* Cups *not hairy, but furniſhed with ſmall reflected ſharp points at the angles.* Bloſſoms *white.*
　　Lychnis ſylveſtris flore albo minimo. *Ray's Syn.* 339.
　　Small campion.
　　Cornfields. A. June—July.

** *Flowers lateral; crowded.*

CATCHFLY. The petals cloven. lateral flowers pointing Nottingham
all one way; drooping. Panicle nodding—*Stem simple; cylin-* Nutans
*drical; a foot high; with three joints below the panicle of flowers;
beset with clammy hairs.* Leaves *spear-shaped; with short hairs;
root-leaves on short leaf-stalks.* Petals *white: narrow; cloven
more than half way down; rolled inwards in the day-time.* Claws
of the blossom twice as long as the cup. Chives *white; twice as long
as the claws of the petals.* Shafts *three; white; as long as the
chives.*
Lychnis fylveſtris alba IX cluſii. *Gerard.* 470. *Rays Syn.*
340.
Lychnis montana viſcoſa alba latifolia. *Bauh. pin.* 205.
Lychnis fylveſtris alba, ſeu ocymoides minus album. *Park.*
631.
High paſtures, and about Nottingham Caſtle. P. June—
July.
Sheep, Horſes, Goats and Swine eat it. Cows refuſe it.

CATCHFLY. The petals cloven. The honeycup crown Sea
ſlightly jointed. Flowers pointing all one way. Fruitſtalks op- Amœna
poſite; ſupporting three flowers. Branches alternate—*Stems
ſpreading; rather ſmooth and aſcending.* Branches *ſtraddling;
ſhort.* Leaves *ſmooth upon the upper ſurface. The angles of the cup
ten; purple and woolly.* Petals *white; cloven half way down.*
Lychnis marina anglica. *Gerard.* 470. *Rays Syn.* 337.
Lychnis marina repens alba. *Park.* 638.
Sea Campion.
On the ſea-ſhore. P, Auguſt.

*** *Flowers growing from the forks of the ſtem.*

CATCHFLY. The cup incloſing the fruit globular; taper- Campion
ing; with thirty ſcores. Leaves ſmooth; petals entire—*pale* Conoidea
purple.
Lychnis fylveſtris anguſtifolia calyculis turgidis ſtriatis. *Bauh.
pin.* 205. *Rays Syn.* 341.
Greater corn Catchfly or Campion.
In gravelly ſoils. A.

CATCHFLY. The cup with ten angles, and with teeth as Night-flower-
long as the tube. Stem forked; petals cloven—*White tinged* ing
with red. Noctiflora
Lychnis Noctiflora. *Bauh, pin.* 205. *Park.* 632. *Rays Syn.*
340.
Cornfields, P. July.

Broad-leaved Armeria

CATCHFLY. The bundles of flowers flat at the top. Upper leaves heart-shaped; smooth. Petals entire—*Red.*
Lychnis viscosa purpurea latifolia lævis. *Bauh. pin.* 205. *Rays Syn.* 341.
Banks of rivers. A. August.

Moss Acaulis

CATCHFLY. Stemless; depressed. Petals notched at the end—*Purple: sometimes pure white.*
Lychnis alpina minima. *Rays Syn.* 341.
Caryophylleus IX clusii Caryophyllus pumilio alpinus. *Gerard.* 593.
Cucubalis acaulis. *Hudson.* 164.
Lychnis alpina pumila folio gramineo, seu muscus Alpinus Lychnidis flore. *Bauh. pin.* 206.
Moss Campion.
On mountains. P. July.

Order IV. Five Pointals.

186 NAVELWORT. 578 Cotyledon.

EMPAL. *Cup* one leaf with five clefts; sharp; small.
BLOSS. One *Petal*; bell-shaped; with five shallow clefts.
Honeycup a hollow scale at the base of each seedbud.
CHIVES. *Threads* ten; awl-shaped; straight; as long as the blossom. *Tips* upright; with four furrows.
POINT. *Seedbuds* five; oblong; rather thick; ending in awl-shaped *Shafts*, longer than the chives. *Summits* simple; reflected.
S.VESS. *Capsules* five; oblong; distended; tapering, of one valve, opening length-ways on the inside.
SEEDS. Many; small.

Venus Umbilicus

NAVELWORT. The leaves alternate, with central leaf-stalks; hooded: serrated and toothed. Stem branched; spikes of flowers generally upright—*Blossoms yellowish; or greenish white. Cultivation occasions some varieties.*
Cotyledon vera radice tuberosa. *Ray's Syn.* 271.
Cotyledon major. *Bauh. pin.* 289.
Umbilicus veneris. *Gerard.* 529. vulgaris. *Park.* 740.
Kidney-wort. Wall Penny-wort.
On old walls and stony places. P. May—July.

187 STONECROP.

187 STONECROP. 579 Sedum.

EMPAL. *Cup* with five divifions; fharp; upright; permanent.

BLOSS. *Petals* five; fpear-fhaped; taper; flat; expanding. *Honeycups* five; each confifting of a fmall fcale; notched at the end, and fixed on the outfide the bafe of each feedbud.

CHIVES. *Threads* ten; awl-fhaped; as long as the bloffom. *Tips* roundifh.

POINT. *Seedbuds* five; oblong; ending in flender *Shafts*. *Summits* blunt.

S.VESS. *Capfules* five; expanding; taper; compreffed; notched at the bafe; opening inwards along the feam.

SEEDS. Many; very fmall.

* *Leaves flat.*

STONECROP. The leaves nearly flat, ferrated. Stem af- Orpine cending. Flowers in leafy broad topped fpikes—*Bloffoms purple,* Telephium *or white.*

 Telephium vulgare. *Bauh. pin.* 287.
 Telephium feu craffula major vulgaris. *Park.* 726.
 Craffula feu faba inverfa. *Gerard.* 519.
 Anacampferos, vulgo faba craffa. *Ray's Syn.* 269.
 Live-long.
 Roofs and walls. P. Auguft.
 Cows, Goats, Sheep and Swine eat it. Horfes refufe it.

* * *Leaves cylindrical.*

STONECROP. The leaves oppofite; egg-fhaped; blunt; Round-leaved flefhy. Stem weak. Flowers fcattered.—*Petals white.* Dafyphyllum
 Sedum minus, circinato folio. *Bauh. / in.* 283. *Ray's Syn.* 271.
 Walls and roofs. P. July.

Yellow
Reflexum

STONECROP. The leaves awl-shaped; scattered; loose at
the base. Lower leaves much curved backwards—*Leaves green
and slender; the lower leaves bent back at the ends like hooks.* Blos-
soms *yellow*
Sedum minus luteum ramulis reflexis. *Bauh pin.* 283 *Ray's
Syn.* 270.
Aizoon scorpioides. *Gerard.* 513.
Vermicularis scorpioides. *Park.* 733.
1. There is a variety with reddish branches. *Gerard.* 512. *Ray's
Syn.* 269.
Prick-madam.
Walls and roofs. P.

Rock
Rupestre

STONECROP. The leaves awl-shaped; growing in five
rows; crowded; loose at the base. Flowers in tufts—*Leaves
bluish green; thick.* Blossoms *yellow.*
Sedum minus e rupe S. Vincentii. *Ray's Syn.* 271.
St. Vincents Rock Stonecrop.
On St. Vincents Rock near Bristol. P. Augu st.
Both this and the preceding species are cultivated in Holland
and Germany to mix with lettuces in sallads.

White
Album

STONECROP. The leaves oblong; blunt; somewhat cylin-
drical; sitting; expanding. Flowers in branching tufts—*The
whole plant is sometimes purple, except the flowers, which are white.*
Sedum minus teretifolium album. *Bauh. pin.* 283. *Ray's Syn.*
271.
Sedum minus officinarum *Gerard.* 512.
Vermicularis flore albo. *Park.* 733.
Vermicularis seu Crassula minor vulgaris. *Park.* 734.
White flower'd Stone crop.
Walls and roofs. P. June—July.
Goats eat it, Sheep refuse it.

STONECROP. The leaves fomewhat egg-fhaped; alternate; Pepper connected; fitting; hunched; rather upright. Flowers in tufts Acre divided into three parts—*Terminating*. Bloffoms *yellow*.

Sedum parvum acre flore luteo. *Ray's Syn.* 270.
Sempervivum minus vermiculatum acre. *Bauh pin.* 283,
Vermicularis feu Illecebra minor acris. *Gerard.* 517.
Illecebra minor, feu Sedum tertium Diofcoridis. *Park.* 736.
Wall Stonecrop, or Pepper.
Walls, roofs, rocks. P. June.

This plant continues to grow when hung up by the root, which is a proof that it receives its nourifhment principally from the air, as is the cafe with moft of the fucculent plants. It is very acrid. Applied externally it blifters. Taken inwardly it excites vomiting. In fcorbutic cafes, and quartan Agues it is an excellent medicine under proper management.

Goats eat it. Cows, Horfes, Sheep and Swine refufe it.

STONECROP. The leaves fomewhat egg-fhaped; connect-Infipid ed; fitting; hunched; rather upright; in fix rows; tiled—*Re-* Sexangulare *fembles the preceding fpecies but the roots are not matted together and there are only two or three bunches of flowers*. Bloffoms *yellow*.

Sempervivum minus vermiculatum infipidum. *Bauh. pin.* 284.
Dry paftures. B. July.
Goats eat it.

STONECROP. The ftem upright; folitary; annual. Leaves Mountain egg-fhaped; fitting; hunched; alternate. Flowers in a tuft, Annuum on curved branches—*Reddifh or white; in Sweden yellow*. Leaves *fpotted with purple*.

Sedum minimum non acre, flore albo. *Ray's Syn.* 270. Tab. 12. fig. 2.
In barren foils, on dry rocks, and on roofs and walls in mountainous countries. P. Auguft.

STONECROP. The ftem upright; leaves nearly flat. Fruit-Marfh ftalks fomewhat hairy—*Bloffoms pale red, ftreaked with purple*. Villofum

Sedum purpureum pratenfe. *Ray's Syn.* 270.
Sedum paluftre fub-hirfutum purpureum. *Bauh. pin.* 285.
Sedum minus paluftre. *Gerard.* 515.
Sedum arvenfe feu paluftre flore rubente. *Park.* 734.
On moift rocks. A. June.

188 SPURREY. 586 Spergula.

Empal. *Cup* five leaves. *Leaves* egg-shaped; blunt; concave; expanding; permanent.

Bloss. *Petals* five; egg-shaped; concave; expanding; entire: larger than the cup.

Chives. *Threads* ten; awl-shaped; shorter than the blossom. *Tips* roundish.

Point. *Seedbud* egg shaped; *Shafts* five; somewhat upright, but reflected: thread-shaped. *Summits* rather thick.

S.Vess. *Capsule* egg shaped; covered with one cell and five valves.

Seeds. Many; globular but depressed; encompassed by a notched border.

Obs. This genus is distinguished from the Mouse-ear by the petals being entire. In the second species there are only five chives.

Corn
Arvensis

SPURREY. The leaves growing in whorls. Flowers with ten chives—*Leaves three or six in a whorl; thread-shaped; hairy; clammy.* Fruit-stalks *branched.* Blossoms *white.*
Alsine spergula dicta major. *Bauh. pin.* 251. *Ray's Syn.* 351.
Saginæ spergula. *Gerard.* 1125. major. *Park.* 562.
In gravelly soil and cornfields. A. August.
Poultry are fond of the seeds, and the inhabitants of Finland and Norway make bread of them when their crops of corn fail. Experience shews it to be very nutritious to the cattle that eat it. —Horses, Sheep, Goats and Swine eat it; Cows refuse it.

Five-chived
Pentandria

SPURREY. The leaves growing in whorls. Flowers with five chives—*Seeds black; surrounded by a white border.*
Spergula annua semine foliaceonigro, circulo membranaceo albo cincto. *Ray's Syn.* 351.
Small Spurrey.
In sandy places. A. July.

Knotted
Nodosa

SPURREY. The leaves opposite; awl-shaped; smooth. Stems simple—*Blossoms terminating; white.*
Alsine palustris foliis tenuissimis, seu Saxifraga palustris Anglica. *Gerard.* 567. *Ray's Syn.* 350.
Alsine nodosa Germanica. *Bauh. pin.* 251.
Saxifraga palustris Anglica. *Park.* 427.
English Marsh Saxifrage.
Moist ground. P. July—August.

189 MOUSE-

189 MOUSE-EAR.   585 Ceraſtium.

EMPAL. *Cup* five leaves. *Leaves* betwixt egg and ſpear-ſhaped; ſharp; expanding; permanent.

BLOSS. *Petals* five; cloven; blunt; upright, but expanding. As long as the cup.

CHIVES. *Threads* ten; thread-ſhaped; ſhorter than the bloſſom: alternately longer and ſhorter. *Tips* roundiſh.

POINT. *Seedbud* egg-ſhaped. *Shafts* five; hair-like; upright; as long as the chives. *Summits* blunt.

S. VESS. *Capſule* betwixt egg-ſhaped and cylindrical; or globular. Blunt; with one cell, opening at the top, with five teeth.

SEEDS. Many; roundiſh.

OBS. *The third ſpecies hath only five chives.*

* Capſules oblong.

MOUSE-EAR. The leaves egg-ſhaped. Petals as large as Narrowleaved
the leaves of the cup. Stems ſpreading.—*Reſembling the next* Vulgatum
ſpecies, but the fruit-ſtalks are not clammy; the ſtems are more nume-
rous and more drooping; and the outer leaves of the cup are not mem-
branaceous at the edges. Bloſſoms *white.*

Alſine hirſuta magno (parvo potius) flore. *Bauh.* ſin. 251.
Alſine hirſuta myoſotis. *Ray's Syn.* 349.
Narrow-leaved Mouſe-ear Chickweed.
In paſtures. A. June.

MOUSE-EAR. Upright; woolly; clammy—*Bloſſoms white.* Broad-leaved
Alſine hirſuta myoſotis latifolia præcocior. *Ray's Syn.* 348. Viſcoſum
Alſine Viſcoſa. *Park.* 768.
Alſine hirſuta altera viſcoſa, *Bauh.* ſin. 251.
1. There is a variety in which the petals are ſmaller, and the cups three times larger than uſual.
Broad-leaved Mouſe-ear Chick-weed.
Dry paſtures. A. April—May.
Horſes and Goats eat it; Cows and Sheep refuſe it.

MOUSE-EAR.

Five-chived
Semidecan-
drium

MOUSE-EAR. The flowers with five chives. Petals notched at the end—*There are five* Chives *with, and five without tips.* Stems *very short.* Leaves *egg-shaped; opposite; a little channelled; blunt; beset with a few very short hairs.* Fruit-stalks *very short; each supporting a single flower.* Cup *hairy; clammy; membranaceous at the edges.* Petals *five, narrow; white. Five honey-cup dots betwixt the perfect chives and the petals.* Seedbud *egg-shaped.*
Cerastium hirsutum minus parvo flore. *Ray's Syn.* 348. Tab. 15. fig. 1.
Alsine hirsuta minor. *Bauh. pin.* 251.
Least Mouse-ear Chickweed.
In pastures. A. April.

Corn
Arvense

MOUSE-EAR. The leaves betwixt strap and spear-shaped; blunt; smooth. Blossoms larger than the cup—*White.*
Caryophyllus arvensis hirsutus flore majore. *Bauh. pin.* 210. *Ray's Syn.* 348.
Caryophyllus holosteus. *Gerard.* 477. arvensis hirsutus. *Park.* 1339.
Corn Mouse-ear Chick-weed.
On heaths, in cornfields and gravelly pastures. P. May.

Mountain
Alpinum

MOUSE-EAR. The leaves betwixt egg and spear-shaped. Stem divided; capsules oblong—*The leaves are sometimes downy, at other times entirely smooth; oval and pointed.* Blossoms *white.*
Alsine myosotis facie, lychnis alpina, flore amplo niveo repens. *Ray's Syn.* 349. Tab. 15. fig. 2.
Caryophylleus holosteus alpinus latifolius. *Bauh. pin.* 210.
Mountain Mouse-ear Chickweed.
On mountains. P. June.
Cows and Sheep eat it.

* * *Capsules nearly globular.*

Marsh
Aquaticum

MOUSE-EAR. The leaves heart-shaped; sitting. Flowers solitary. Fruit pendant—*Stem slender; cylindrical; smooth; jointed; the joints red at the base.* Branches *few; solitary; alternate; simple; generally as long as the stem; hairy towards the ends.* Blossoms *white; twice as large as the cup.* Petals *deeply divided.* Seeds *yellowish red.*
Alsine aquatica major. *Bauh. pin.* 251.
Alsine major repens perennis. *Ray's Syn.* 347.
Alsine major. *Gerard.* 611. et maxima. *Park.* 759.
Marsh Mouse-ear Chickweed.
Banks of rivers. P. July.

MOUSE-EAR. The leaves oblong; downy. Fruit-ftalks Woolly
branched. Capfules glöbular—*Bloffoms white.*　　　　　Tomentofum
　Caryophvllus holofteus tomentofus latifolius. *Bauh. pin.* 210.
　Alfine myofotis lanuginofa alpina grandiflora, feu auricula
muris villofa flore amplo membranaceo. *Ray's Syn.* 349.
　Woolly Moufe-ear Chick-weed.
1. There is a variety with narrow leaves.
　On mountains. P. May.

190 C O C K L E. 583 Agroftemma.

EMPAL. *Cup* one leaf; like leather; tubular; with five
　　teeth; permanent.
BLOSS. *Petals* five. *Claws* as long as the tube of the cup.
　　Limbs expanding; blunt.
CHIVES. *Threads* ten; awl-fhaped; every other thread
　　ripening later; fixed to the claws of the petals.
　　Tips fimple.
POINT. *Seedbud* egg-fhaped. *Shafts* five; thread-fhaped;
　　upright; as long as the chives. *Summits* fimple.
S.VESS. *Capfule* oblong egg-fhaped; covered, with one
　　cell and five valves.
SEEDS. Many; kidney fhaped; dotted. *Receptacles* equal
　　in number to the feeds; loofe; the inner ones gra-
　　dually longer than the others.

　COCKLE. Hairy-cup as long as the bloffom. Petals entire; Corn
naked—*purple; fometimes white, and by cultivation yellow.*　Githago
　Lychnis fegetum major. *Bauh. pin.* 204. *Ray's Syn.* 338.
　Lychnoides fegetum five nigellaftrum. *Park.* 632.
　In cornfields. A. June.
　Horfes, Goats and Sheep eat it.

191 CUCKOWFLOWER.

191 CUCKOWFLOWER. 584 Lychnis.

EMPAL. *Cup* one leaf; tubular; membranaceous; with five teeth; permanent.

BLOSS. *Petals* five. *Claws* as long as the cup; flat; bordered. *Limbs* flat; frequently cloven.

CHIVES. *Threads* ten; longer than the cup, alternately ripening later and fixed to the claws of the petals. *Tips* fixed side-ways to the threads.

POINT. *Seedbud* nearly egg-shaped. *Shafts* five; awl-shaped; longer than the chives. *Summits* downy; bent towards the left.

S.VESS. *Capsule* nearly egg-shaped; covered. Valves five; cells five.

SEEDS. Many; roundish.

OBS. *In the* Campion CUCKOWFLOWER *the chives are upon one plant and the pointals upon another. The* Clammy CUCKOWFLOWER *hath undivided petals, and capsules with five cells.*

Meadow
Flos-cuculi

CUCKOWFLOWER. The petals with four clefts. Fruit roundish—*Stems trailing; but upright when in flower.* Blossoms red; *by cultivation white; and often double.*

Caryophyllus pratensis laciniato flore simplici, sive flos cuculi. *Bauh. pin.* 210.

Lychnis plumaria sylvestris simplex. *Park.* 253. *Ray's Syn.* 338.

Armerius pratensis mas et fæmina. *Gerard.* 600.

Meadow Pinks. Wild Williams. Cuckowflower. Ragged Robin.

In meadows. P. June.

Horses, Sheep and Goats eat it.

It is the food of a sort of Louse, the *Aphis Cucubali.*

Clammy
Viscaria

CUCKOWFLOWER. The petals nearly entire—*The upper joints of the stem are anointed with a reddish black viscid substance, like tar.* Stem *upright, simple.* Blossoms red; *by cultivation white, and sometimes double.*

Lychnis sylvestris viscosa rubra angustifolia. *Bauh. pin.* 205. *Park.* 636. *Ray's Syn.* 340.

Muscipula angustifolia. *Gerard.* 601.

Red German Catchfly.

High pastures. P. May—June.

The Catchfly Weevil, *Curculio viscaria,* is found upon it.

CUCKOW-FLOWER. The flowers with chives on one Campion.
plant, and flowers with pointals on a different plant—*Red, or* Dioica
white. Sometimes the chives and pointals are in the same flower.
By cultivation the blossoms become double.
Lychnis sylvestris albo flore. *Gerard.* 468. *Park.* 630. *Ray's*
Syn. 339.
Lychnis sylvestris alba simplex. *Bauh. pin.* 204.
1. When the flower is red it hath been considered as a variety
by the following authors.
Lychnis sylvestris rubello flore. *Gerard.* 469. *Ray's Syn.*
339.
Lychnis sylvestris flore rubro. *Park.* 631.
Lychnis sylvestris, seu aquatica purpurea simplex. *Bauh.*
pin. 204.
White or red Campion. Batchelor's Buttons.
Woods and hedges. P. May—July.
The Campion Louse, *Aphis Lychnidis,* lives upon it.

192 WOODSORREL. 582 Oxalis.

EMPAL. *Cup* with five divisions; sharp; very short; permanent.
BLOSS. With five divisions, connected by the claws; upright; blunt; with the margin broken.
CHIVES. *Threads* ten; hair-like; upright; the outermost the shortest. *Tips* roundish; furrowed.
POINT. *Seedbud* with five angles. *Shafts* five; thread-
shaped; as long as the chives. *Summits* blunt.
S. VESS. *Capsule* with five corners, and five cells; opening lengthways at the corners.
SEEDS. Nearly round; bursting out of the feed-vessel.

OBS. *In some species the Capsule is short and the Seeds solitary; in*
others it is long and the Seeds many.

WOODSORREL. The stalk supporting a single flower.
Leaves growing by threes, Root scaly; jointed—*In rainy wea-* Acetous
ther the leaves stand upright, but in dry weather they hang down. Acetosella
Blossoms *pale purple.*
Oxys alba. *Gerard.* 1201. *Ray's Syn.* 281.
Trifolium acetosum vulgare. *Bauh. pin.* 330. *Park.* 746.
Cuckow-bread. Sour Trefoil.
Woods and moist shady lanes. P. April.
It is likewise found upon mountains; and Linnæus somewhere
observes, that the plants which chiefly grow upon mountains are
hardly found any where else but in marshes. Is it because the
clouds resting upon the tops of the mountains, keep the air in
a moist

a moist state, like the fogs in meadows and marshes, which are nothing but clouds in the lower part of the atmosphere?

The juice is gratefully acid. The London College directs a Conserve to be made of the leaves beaten with thrice their weight of fine sugar. The expressed juice depurated properly evapporated, and set in a cool place, affords a chrystalline acid salt in considerable quantity, which may be used wherever vegetable acids are wanted. An Infusion of the leaves is an agreeable liquor, in ardent Fevers.—Sheep, Goats and Swine eat it. Cows are not fond of it; Horses refuse it.

Yellow-flow-   **WOODSORREL.** The fruit-stalks forming rundles. Stem
ered   branched, spreading—*Leaves growing by threes; on leaf-stalks;*
Corniculata   Capsules *long and pointed.* Blossoms *yellow.*

Trifolium acetosum corniculatum.   *Bauh. fin.* 230.

In the country about Exeter. A. May—October.

CLASS XI.

TWELVE CHIVES.

NOTWITHSTANDING the title of this Clafs, the number of Chives is not very certain; fome of the Flowers containing more, and fome of them fewer than twelve; fo that it is neceffary to take in another circumftance, (viz.) that in this Clafs the Chives are fixed to the RECEPTACLE, but in the next Clafs they are fixed to the CUP.

The different fpecies of SPURGE are generally fuppofed difficult to inveftigate, but the young Botanift will foon learn to diftinguifh them by attending to the following circumftances.

Whether the

Root is annual, biennial, or perennial. Whether the plant in queftion is a fhrub, and if fo, whether it is thorny, or prickly, or neither.

Stem is naked, cylindrical, or angular.

Leaves are oppofite or alternate; and of what fhape.

Rundle is general or partial; how divided and how fenced.

Flowers have only chives, or both chives and pointals.

Petals are entire; crefcent-fhaped or hand-fhaped, &c.

Capfules are hairy, warty or fmooth.

T   Order

ORDER I. ONE POINTAL.

193 ASARABACCA.   *Bloss.* none. *Cup* three clefts; superior. *Capsule* with six cells.

194 GRASSPOLY.   *Bloss.* six petals. *Cup* twelve clefts; beneath. *Capsule* with two cells.

Order II. Two Pointals.

195 AGRIMONY.   *Bloss.* five petals. *Cup* with five clefts; *Seeds* one, or two.

Order III. Three Pointals.

196 YELLOW-WEED.   *Bloss.* Petals with many clefts. *Cup* divided. *Capsule* with one cell; gaping.

197 SPURGE. - -   *Bloss.* Petals fixed by the center. *Cup* distended. *Capsules* three berries.

Order IV. Four Pointals.
† *Upright Tormentil.*

Order VII. Twelve Pointals.

198 HOUSESLEEK.   *Bloss.* twelve petals. *Cup* with twelve divisions. *Capsules* twelve.

193 ASARA-

193 ASARABACCA. 589 Afarum.

EMPAL. *Cup* one leaf; bell-fhaped; with three fhallow clefts; like leather; coloured; permanent. *Segments* upright; with the point bent inwards.

BLOSS. None.

CHIVES. *Threads* twelve; awl-fhaped; half as long as the cup. *Tips* oblong; growing to the middle of the threads.

POINT. *Seedbud* either beneath; or elfe hidden within the fubftance of the cup, *Shaft* cylindrical; as long as the chives. *Summit* ftarry; with fix reflected divifions.

S. VESS. *Capfule* like leather; generally with fix cells; inclofed within the fubftance of the cup.

SEEDS. Many; egg-fhaped.

ASARABACCA. The leaves growing in pairs; blunt; kidney-fhaped—*Bloffoms purplifh.* Common
Europæum

Afarum. *Bauh. pin.* 197. *Ray's Syn.* 158.
Afarum vulgare. *Park.* 266. *Gerard.* 836.
In woods. P. May.

The root powdered and taken to the amount of thirty or forty grains excites vomiting. If it is coarfely powdered it generally purges. The powder of the leaves is the bafis of moft of the Cephalic Snuffs, which occafion a confiderable difcharge of mucus from the noftrils without much fneezing.—Cows eat it.

194 GRASSPOLY. 604 Lythrum.

EMPAL. *Cup* one leaf; cylindrical; scored; with twelve teeth; every other tooth smaller.

BLOSS. *Petals* six; oblong; rather blunt; expanding; fixed by the claws to the divisions of the cup.

CHIVES. *Threads* twelve; thread-shaped; as long as the cup. The upper threads shorter than the lower ones. *Tips* simple; rising.

POINT. *Seedbud* oblong. *Shaft* awl-shaped; declining; as long as the chives. *Summit* round and flat; rising.

S. VESS. *Capsule* oblong; taper; covered. Cells two.

SEEDS. Numerous; small.

OBS. *In the second species there are only six chives.*

Purple Salicaria

GRASSPOLY. The leaves opposite; betwixt heart and spear-shaped; flowers with twelve chives; growing in spikes— *Stem four cornered; firm; branched.* Blossoms *purple.*
Salicaria vulgaris purpurea, foliis oblongis. *Ray's Syn.* 367.
Lysimachia purpurea. *Park.* 546. *Gerard.* 476.
Lysimachia spicata purpurea. *Bauh. pin.* 246.
Purple spiked Loosestrife, or Willow-herb.
These are the varieties.
1. Leaves soft, long, and four together.
2. Leaves shorter and broader.
3. Stem with six corners; Leaves three together.
Banks of rivers. P. July.
Horses, Cows, Sheep and Goats eat it.

Six-chived Hyssopifolia

GRASSPOLY. The leaves alternate; strap-shaped. Flowers with six chives—*Blossoms at the base of the leaves; pale blue.*
1. Cultivation gives rise to several varieties, viz. white blossomed; blue; broad-leaved white, and broad-leaved blue.
Salicaria Hyssopifolio latiore et angustiore. *Ray's Syn.* 367.
Hyssopifolia. *Bauh. pin.* 218.
Gratiola angustifolia. *Gerard.* 581. seu minor. *Park.* 220.
Small Hedge Hyssop.
In watery places and shallow ponds. A. August—September.

Order

Order II. Two Pointals.

195 AGRIMONY. 607 Agrimonia.

EMPAL. *Cup* one leaf; with five clefts: sharp; small; superior; permanent: surrounded by another cup.

BLOSS. *Petals* five; flat; imperfect at the margin. *Claws* narrow; growing to the cup.

CHIVES. *Threads* hair-like; shorter than the blossom; fixed to the cup. *Tips* small; double; compressed.

POINT. *Seedbud* beneath. *Shafts* two; simple; as long as the chives. *Summits* blunt.

S. VESS. None. The *Cup* grows hard and closes at the neck.

SEEDS. Two; roundish.

OBS. *The number of Chives is exceedingly uncertain; in some plants there are twelve; sometimes ten; frequently seven.*

AGRIMONY. The stem leaves winged; the odd one at the end supported upon a leaf-stalk. Fruit covered with rough hairs.—*Blossoms on long terminating spikes; yellow.* Common Eupatoria

Agrimonia. *Gerard.* 712. *Ray's Syn.* 202. vulgaris. *Park.* 594.

Eupatorium veterum, seu agrimonia. *Bauh. pin.* 321.

Hedges and ditch-banks. P. June.

The Canadians are said to use an Infusion of the root in burning Fevers, and with great success. An Infusion of six ounces of the crown of the root in a quart of boiling water, sweetened with honey, and half a pint of it drank three times a day, Dr. Hill says is an effectual cure for the Jaundice. He advises to begin with a vomit, afterwards to keep the bowels soluble, and to continue the medicine as long as any symptoms of the disease remain.—Sheep and Goats eat it; Cows, Horses and Swine refuse it.

Order III.   Three Pointals.

196 YELLOW-WEED.   608 Reſeda.

EMPAL. *Cup* one leaf; divided. *Segments* narrow; ſharp; upright; permanent; two of them ſtanding more open upon account of the honey-cup petals.

BLOSS. *Petals* ſeveral; unequal. *Some* with three ſhallow clefts. The uppermoſt hunched at the baſe; as long as the cup and containing the honey.

Honey-cup a flat upright gland; riſing from the receptacle; ſituated betwixt the chives and the uppermoſt petal.

CHIVES. *Threads* from eleven to fifteen; ſhort. *Tips* blunt; upright; as long as the bloſſom.

POINT. *Seedbud* hunched; ending in ſome very ſhort *Shafts. Summits* ſimple.

S. VESS. *Capſule* hunched; angular; tapering to the ſhafts: with one cell; opening betwixt the ſhafts.

SEEDS. Many; kidney-ſhaped; connected to the angles of the capſule.

OBS. *There is hardly any genus ſo difficult to characterize as this; the different ſpecies varying ſo much both in figure and number. The eſſential character conſiſts in the* Petals *with three Clefts, one petal bearing the honey-cup in its baſe, and the capſules not cloſed, but always gaping open.*

In the firſt ſpecies the cup hath four diviſions, the petals are three; the uppermoſt petal incloſing the honey-cup hath ſix ſhallow clefts. The lateral and oppoſite petals have three clefts; and there are ſometimes two other very ſmall and entire petals. Shafts three; Chives many.

YELLOW-WEED. The leaves fpear-fhaped; entire; with Dyers a tooth upon each fide the bafe. Cup with four clefts—*Flowers* Luteola *in a nodding fpike which follows the courfe of the fun through the day; pointing towards the Eaft in the morning, to the South at noon, and Weftward in the afternoon: in the night it points to the North Thefe circumftances take place even in a cloudy fky.* Bloffoms *yellow*.

Luteola. *Gerard.* 494. *Ray's Syn.* 366. vulgaris. *Park.* 602.
Luteola herba falicis folio. *Bauh. pin.* 100.
Wild Woad. Dyers-weed.
On barren ground, and on walls. A. June.

This plant affords a moft beautiful yellow dye for cotton, woollen, mohair, filk and linen, and is conftantly ufed by the dyers for that purpofe. Blue cloths dipped in a decoftion of it become green. The yellow colour of the paint called Dutch Pink, is got from this plant. The tinging quality refides in the ftems and roots, and it is cultivated in fandy foils, rich foil making the ftalk hollow and not fo good.—Cattle will not eat it, but Sheep fometimes browfe it a little.

YELLOW-WEED. The leaves with three divifions. Lower Rocket leaves winged. *Bloffoms yellow.* Lutea
Refeda vulgaris. *Bauh. pin.* 100. *Ray's Syn.* 366.
Refeda minor, feu vulgaris. *Park.* 823. *Gerard.* 277.
1. There is a variety with curled leaves.
Bafe Rocket.
Corn-fields and on chalk hills. A. July.

197 SPURGE. 609 Euphorbia.

EMPAL. *Cup* one leaf; permanent; a little coloured;
distended. *Mouth* with four, (and in a few species
with five) teeth.

BLOSS. *Petals* four; (in a few species five;) turban shaped;
hunched; thick; lopped, irregularly situated; alternating with the teeth of the cup and fixed by
their claws to its edge : permanent.

CHIVES. *Threads* many; (twelve or more;) thread-shaped;
jointed; standing on the receptacle; longer than
the blossom; appearing at different times. *Tips*
double; roundish.

POINT. *Seedbud* roundish : three-cornered; standing on
a little fruit-stalk. *Shafts* three; cloven. *Summits*
blunt.

S. VESS. *Capsule* roundish; consisting of three berries and
three cells opening with a spring.

SEEDS. Solitary; roundish.

OBS. *The Capsule is either smooth; hairy, or warty. In some
species the first flowers have Chives, but no Pointals.*

Sea
Peplis.

SPURGE. Forked. Leaves very entire, the shape of half
a heart. Flowers solitary; at the base of the leaves. Stems
trailing—*Blossoms yellow.*
Tithymalus maritimus supinus annuus, peplis dictus. *Ray's
Syn.* 313.
Peplis. *Gerard.* 503. *Park.* 194.
Peplis maritima folio obtuso. *Bauh. pin.* 293.
Small Purple Sea Spurge.
On the Sea Shore. A. June.

* *Flowers in Rundles, with three divisions.*

Petty
Peplis.

SPURGE. The rundle with three divisions; divisions forked.
Partial fences egg-shaped. Leaves very entire; inversely egg-
shaped; on leaf-stalks—*Alternate. All the flowers fertile.* Petals
with two horns. Fruit *smooth, or only rough at the edge.* Blossoms
yellowish green.
Tithymalus parvus annuus, foliis subrotundis non crenatis,
peplus dictus. *Ray's Syn.* 313.
Peplus sive esula rotunda. *Bauh. pin.* 292. *Gerard.* 503.
In rich cultivated soil. A. July.

SPURGE. The rundle with three divifions; divifions forked. Dwarf
Partial fences fpear-fhaped; leaves ftrap fhaped—*Without lateral* Exigua
branches. Flowers *formed of four horned petals. Little leaves of
the rundle and rundlet not broader than the others.* Fruit *fmooth.*
Bloffoms *yellow.*
    Tithymalus leptophyllus. *Park.* 193. *Ray's Syn.* 313.
    Tithymalus five efula exigua. *Bauh. pin.* 291.
    Efula exigua tragi. *Gerard.* 501.
  1. Leaves blunt.
  2. Small; growing on rocks.
    Corn-fields. A. July.

** *Flowers in Rundles with five divifions.*

SPURGE. The Rundle with five divifions; divifions forked. Portland
Partial fences nearly heart-fhaped; concave. Leaves betwixt Portlandica
ftrap and fpear-fhaped; fmooth; expanding—*Stems a little woody*;
fmooth; cylindrical; reddifh. Leaves *alternate*; *on very fhort
leaf ftalks: red on the under fide the bafe.* Flowers *almoft fitting*;
yellow. The firft and fecond flower without Pointals. Fruit *fmooth*;
angular; *furnifhed with fharp points.*
    Tithymalus maritimus minor. Portlandicus. *Ray's Syn.* 313.
Tab. 24. *fig.* 6.
    Sea fhore. S. Auguft.

SPURGE. The rundle generally with five divifions; cloven. Marine
Partial fences betwixt heart and kidney-fhaped. Leaves tiled Paralias
upwards—*Bloffoms yellowifh green.*
    Tithymalus maritimus. *Bauh. pin.* 291.
    Tithymalus paralius. *Gerard.* 499. *Ray's Syn.* 312.
    Tithymalus paralius feu maritimus. *Park.* 184.
    Sea Spurge.
    Sea fhore. P. July—Auguft.

SPURGE. The rundle with five divifions; forked. Partial Corn
fences heart-fhaped; fharp. Leaves betwixt ftrap and fpear- Segetalis
fhaped. The upper leaves the broadeft—*Stem about a foot high*;
reddifh toward the bottom. Branches *at the lower part of the ftem
fhort.* Leaves *pale green.* Petals *yellow*; *horned. Many alternate
fruit-ftalks bearing rundlets arife from the bafe of the upper Leaves.*
    Tythimalus fegetum longifolius. *Ray's Syn.* 312.
    Tytihmalus maritimo affinis, linariæ folio. *Bauh. pin.* 291.
    Corn-fields. A. July—Auguft.

SPURGE.

282   TWELVE CHIVES.

Sun
Heliofcopia

SPURGE. The rundle with five divifions; divifions thrice
cloven, and thefe again forked. Partial fences inverfely egg-
fhaped. Leaves wedge fhaped; ferrated—*Alternate. All the
flowers fertile; with four entire petals.* Fruit *fmooth.* Rundle
with five clefts, and five leaves. Rundlets *with three clefts and
three leaves.* Bloffoms *greenifh yellow.*
Tithymalus helioscopius. *Gerard.* 498. *Park.* 189. *Bau
pin.* 291. *Ray's Syn.* 313.
Wart-wort. Churn-ftaff.
Cultivated places; gardens. A. July.
If Sheep eat it they are purged by it, and their flefh gets a
bad tafte; but this is not the cafe with Cows.

Rough
Verrucofa

SPURGE. The rundle with five divifions; divifions generally
cloven into three, and thefe again into two. Partial fences
egg-fhaped. Leaves fpear-fhaped; ferrated; woolly. Capfules
warty and woolly—*Stem cylindrical.* Rundle *longer than the
ftem; ftanding wide.* Flowers *of the firft, fecond, and third row,
barren.* Petals *four; entire; yellow.*
Tithymalus myrfinites fructu verrucæ fimili. *Bauh. pin.* 291.
Tithymalus verrucofus. *Park.* 181. *Ray'sSyn.* 312.
Rough-fruited Spurge.
Cornfields. B. Auguft.

Broad-leaved
Platyphyllos

SPURGE. The rundle with five divifions; divifions thrice
cloven: and thefe again forked. Partial fences hairy on the
under rib. Leaves ferrated; fpear-fhaped; capfules warty—
Stem upright; one foot high; fmooth. Leaves *alternate; remote;
expanding; with a very few hairs fometimes upon the middle rib on
the under furface.* Petals *yellow. The upper part of the plant of
a yellowifh green. All the flowers fertile.*
Tithymalus platiphyllos. *Ray's Syn.* 312.
Tithymalus arvenfis latifolius germanicus. *Bauh. pin.* 291.
In corn-fields. A. July.

*** *Flowers in Rundles with many divifions.*

Knotty-rooted
Hyberna

SPURGE. The rundle with fix clefts; forked. Partial
fences oval. Leaves very entire. Stem without branches. Cap-
fules warty—*The leaves oblong; fometimes hairy; fometimes fmooth.*
Bloffoms *yellowifh.*
Tithymalus hibernicus. *Ray's Syn.* 312.
Tithymalus latifolius hifpanicus. *Bauh. pin.* 291.
In corn-fields. P. July.

SPURGE.

SPURGE. The rundle with many divifions; forked. Parti- Wood
al fences circular; perforated. Leaves blunt—*Root-leaves dow-* Amygdaloides
ny on the under furface. Petals *crefcent fhaped.* Bloffoms *yellow.*
It retains its leaves all the year.

Tithymalus characias amygdaloides. *Gerard.* 500. *Baub. pin.*
290. *Ray's Syn* 312.

Tithymalus characias vulgaris. *Park.* 186.

Woods and hedges. P. April.

SPURGE. The rundle with many clefts; forked. Partial Red
fences perforated, notched. Leaves very entire. Stem fome- Characias
what woody—*Stems thick; reddifh; four feet high. Leaves fpear-*
fhaped; like Leather; a little downy; green. Rundles *fmall; crowd-*
ed; terminating; fitting. The firft flowers without pointals, Petals
four; purple; moift. Seedbuds *woolly. The* Branches *of one years*
growth are thickeft towards the top and rather woolly. The flower-
ing branches in the beginning of fpring fhoot out from the ends of the
other branches, and after the fruit is perfect they perifh, and then
other branches fpring out from the fides of the bafe.

Tithymalus characias rubens peregrinus. *Baub. pin.* 290.

Tithymalus characias monfpelienfium. *Gerard.* 499. *Park.*
186. *Ray's Syn.* 312.

1. There is variety with yellow flowers.

Woods and hedges. S. June.

The juice of every fpecies of Spurge is fo acrid that it corrodes
and ulcerates the body wherever it is applied; fo that Phyficians have feldom ventured to ufe it internally. Warts or Corns
anointed with the juice prefently difappear. A drop of it put
into the hollow of a decayed and aching tooth deftroys the nerve
and confequently removes the pain. Some people rub it behind
the ears that it may blifter, and by that means give relief.

Order

Order VII. Twelve Pointals.

198 HOUSELEEK. 612 Sempervivum.

EMPAL. *Cup* from six to twelve divisions; concave; sharp, permanent.

BLOSS. *Petals* six to twelve; oblong; spear-shaped; sharp; concave; a little larger than the cup.

CHIVES. *Threads* six to twelve; awl-shaped; slender. *Tips* roundish.

POINT. *Seedbuds* six to twelve; placed in a circle; upright; each ending in a *Shaft*; expanding. *Summits* sharp.

S. VESS. *Capsules* six to twelve; oblong; compressed; short; placed in a circle; tapering outwards; opening on the inner side.

SEEDS. Many; roundish; small.

OBS. *When in a luxuriant growth the numbers often increase, especially the number of the pointals. This genus is nearly allied to the* STONECROP, *but differs from that in always having more than five petals.*

Common
Tectorum

HOUSELEEK with fringed leaves and expanding offsets— *Blossoms on a crooked stalk; pale red.*
Sempervivum majus. *Gerard.* 508. *Ray's Syn.* 269.
Sedum majus vulgare. *Bauh. pin.* 283. *Park.* 730,
Cyphel.
On roofs and old walls. P. July.
The juice either applied by itself, or mixed with cream, gives present relief in burns and other external inflammations. Mixed with honey it is an useful application in aphthous cases.—Sheep and Goats eat it.

C L A S S XII,

TWENTY CHIVES,

THIS is called the clafs of *Twenty Chives*, becaufe the number of chives is generally about twenty, in moft of the plants it contains ; neverthelefs the claffic character is not to be taken from the number of chives only, but from the following circumftances, which will fufficiently diftinguifh it both from the preceding and from the enfuing clafs.

1. CUP confifting of one concave leaf.

2. PETALS fixed by the claws to the *fides* of the cup.

3. CHIVES more than nineteen ; ftanding upon the *fides* of the cup, or upon the bloffom, but *not* upon the receptacle.

Hardly any of the plants of this clafs are poifonous. The fruits are moftly pulpy and efculent.

Order I.

ORDER I. ONE POINTAL.

199. Prune. - -   *Cup* beneath; with five clefts. *Bloſſom* five petals. *Seed-veſſel* pulpy, including an entire ſtone.

† *White Hawthorn.*

Order II. Two Pointals.

200. Hawthorn.   *Cup* ſuperior; with five clefts. *Bloſſom* five petals. *Berry* with two ſeeds,

Order III. Three Pointals.

201. Service. - - *Cup* ſuperior; with five clefts. *Bloſſom* five petals. *Berry* with three ſeeds.

Order IV. Five Pointals.

202. Medlar.   *Cup* ſuperior; with five clefts. *Bloſſom* five petals. *Berry* with five ſeeds.

203. Apple. - - *Cup* ſuperior; with five clefts. *Bloſſom* five petals. Fruit an *Apple* with five cells and many ſeeds.

204. Meadowsweet. *Cup* beneath, with five clefts. *Bloſſoms* five petals. *Capſules* many, collected into a ball.

Order V. Many Pointals.

205. Rose. - - - *Cup* five clefts. *Bloſſoms* five petals. *Cup* forming a berry; with many ſeeds.

                       206. Bramble.

206. BRAMBLE. - - *Cup* five clefts. *Bloſſoms* five petals. *Berry* compound.

207. TORMENTIL. - *Cup* eight clefts. *Bloſſoms* four petals. *Seeds* eight ; without awns.

208. AVENS. - - *Cup* eight clefts. *Bloſſoms* eight petals. *Seeds* many; with downy awns.

209. STRAWBERRY. *Cup* ten clefts. *Bloſſoms* five petals. *Seeds* many; deciduous; ſituated upon a receptacle reſembling a berry.

210. CINQUEFOIL. - *Cup* ten clefts. *Bloſſoms* five petals. *Seeds* many; without awns.

211. BENNET. - - *Cup* ten clefts. *Bloſſoms* five petals. *Seeds* many with a jointed awn.

212. MARSHLOCKS. *Cup* ten clefts. *Bloſſoms* five petals. *Seeds* many, permanent, upon a fleſhy receptacle-

† *Meadowſweet.*

199 PRUNE

199 PRUNE. 620 Prunus.

EMPAL. *Cup* one leaf; bell-shaped; with five clefts; deciduous. *Segments* blunt; concave.

BLOSS. *Petals* five; circular; concave; large; expanding; fixed to the cup by claws.

CHIVES. *Threads* from twenty to thirty; awl-shaped; nearly as long as the blossom; standing on the cup. *Tips* double; short.

POINT. *Seedbud* roundish. *Shaft* thread-shaped; as long as the chives. *Summit* circular.

S.VESS. Nearly globular; pulpy; including a nut or stone.

SEED. A *Nut*; somewhat globular, but compressed.

OBS. *The inside of the cup in some, if not in all the species, is covered with a number of small glands which make an appearance like a hoar-frost. In the* Bullace PRUNE *there are sometimes two pointals.*

Birds-cherry
Padus

PRUNE. The flowers in bunches. Leaves deciduous; with two glands on the under side of the base—*Petals serrated; white.* Fruit *black.*

Cerasus aviuu nigra et racemosa. *Gerard.* 1504. *Ray's Syn.* 463.

Cerasus racemosa sylvestris fructu non eduli, *Bauh. pin.* 451.

Cerasus avium racemosa. *Park.* 1517.

Wild cluster Cherry. Birds Cherry.

Woods and hedges. S. May.

It grows well in woods, groves or fields, but not in a moist soil. It bears lopping and suffers the grass to grow under it. The fruit is nauseous, but bruised and infused in wine or brandy it gives it an agreeable flavour. A strong decoction of the bark is used by the Finlanders to cure venereal complaints; and a decoction of the berries is sometimes given with success in the dysentery. The wood being smooth and tough is made into handles for knives and whips.—Sheep, Goats and Swine eat it. Cows are not fond of it. Horses refuse it.

Black-cherry
Cerasus

PRUNE. The flowers in rundles on very short fruit-stalks. Leaves betwixt egg and spear-shaped; glossy; doubled together —*Props toothed.* Floral Leaves *cloven into three segments. The* terminating Buds *producing leaves; but the* lateral Buds *producing flowers.* Blossoms *white.* Fruit *black.*

Cerasus sylvestris fructu nigro. *Ray's Syn.* 463.

Cerasus sylvestris ac major, fructu sub-dulci, nigro colore inficiente. *Bauh. pin.* 450.

Black Cherry. Mazzards.

Woods and hedges. S. April.

It

It loves a fandy foil, and an elevated fituation. The gum that exfudes from this tree is equal to gum arabic. Haffelquift relates that more than an hundred men, during a fiege, were kept alive for near two months without any other fuftenance than a little of this gum taken into the mouth fometimes, and fuffered gradually to diffolve. The common people eat the fruit either frefh or dried ; and it is frequently infufed in brandy for the fake of its flavour. The wood is hard and tough. It is ufed by the turner, and is formed into chairs and ftained to imitate mahogany. This tree is the original ftock from which many of the cultivated kinds are derived.

PRUNE. The flowers in rundles; fitting. Leaves betwixt Cherry egg and fpear-fhaped; downy beneath : doubled together— Avium *Bloffoms white.* Fruit red.
Cerafus fylveftris fructu rubro. *Ray's Syn.* 463.
Common wild Cherry-tree.
Woods and hedges. S. April.
It grows beft in a rich foil on the fides of hills, unmixed with other trees. It bears cropping, and fuffers the grafs to grow under it.

PRUNE. The fruit-ftalks either in pairs, or folitary. Leaves Plumb betwixt fpear and egg-fhaped : rolled in a fpiral. Branches with- Domeftica out thorns.—
Prunus. *Bauh. pin.* 443.
1. Fruit red; acerb and ungrateful to the tafte. *Ray's Syn.* 463.
2. Fruit large; white. *Ray's Syn.* 462. White Bullace-tree.
In hedge rows. S. April.
It loves a lofty expofure, and is favourable to pafturage. The varieties have probably originated from the red and white cultivated plumb, either fown by defign or accident. The cultivated garden plumbs are derived from this fpecies.

PRUNE. The fruit-ftalks growing in pairs. Leaves egg- Bullace fhaped; fomewhat woolly : rolled in a fpiral. Branches thorny Infititia —*Branches reddifh brown, fmooth: fome of them terminating in a thorn.* Props *narrow: fringed; fharp; cloven at the bafe. Upper furface of the leaves fmooth.* Cups *fmooth.* Bloffoms *white.* Fruit black.
Prunus fylveftris major. *Ray's Syn.* 462.
Prunus fylveftria præcocia. *Bauh. pin.* 444.
Black Bullace-tree.
In hedges. S. April.
The fruit is acid, but fo tempered by a fweetnefs and roughnefs as not to be unpleafant, particularly after it is mellowed by the frofts. A conferve is prepared by mixing the pulp with

U   thrice

thrice its weight of fugar. The bark of the roots and branches
is confiderably ftyptic. An infufion of the flowers fweetened
with fugar is a mild purgative, not improper for children.

Blackthorn
Spinofa

PRUNE. The fruit-ftalk folitary. Leaves fpear-fhaped;
fmooth. Branches thorny—*Props ftrap-fhaped*; *ferrated and
toothed*; *the ends of the teeth appearing as if dead.* Bloffoms *white.*
Fruit *black.*

Prunus fylveftris. *Gerard.* 1497. *Park.* 1033. *Bauh pin.* 444.
Ray's Syn. 462.

Black-thorn. Sloe-tree. Scrogs.
In hedges. S. March—April.

This is not well adapted to grow in hedges, becaufe it fpreads
its roots wide, and encroaches upon the pafturage; but it makes
good dead fence. The wood is hard and tough, and is formed
into teeth for rakes, and walking fticks. The thorns have fomething of a poifonous nature in the autumn. The tender leaves
dried, are fometimes ufed as a fubftitute for Eaft India tea. The
Fruit bruifed and put into wine gives it a beautiful red colour,
and a pleafant fub-acid roughnefs. An infufion of a handful of
the flowers is a fafe and eafy purge. Letters written upon linen
or woollen with the juice of the fruit will not wafh out.—Sheep,
Goats and Horfes eat it.

The different fpecies of PRUNE furnifh nourifhment to the
following infects.

White black-veined Butterfly. *Papilio Cratægi.*
Great tortoife-fhell Butterfly. *Papilio Polychloros.*
Brown hair-ftreak Butterfly. *Papilio Betulæ.*
Lappit Moth. *Phalæna quercifolia.*
Great Egger Moth. *Phalæna quercus.*
Small Egger Moth. *Phalæna laneftris.*
Black-thorn Moth. *Phalæna cærulocephala.*
Emperor Moth. *Phalæna pavonia.*
Lackey Moth. *Phalæna neuftria.*
Ealings Glory. *Phalæna Oxyacanthæ.*
Sallow Moth. *Phalæna Citrago.*
Orange Moth. *Phalæna prunaria.*
- - - - - *Phalæna brumata.*
Prune Loufe. *Aphis Padi.*
Cherry Curculio. *Curculio cerafi.—Pruni.*
Garden Beetle. *Scarabæus horticola.*

Order II. Two Pointals.

200 HAWTHORN. 622 Cratægus.

EMPAL. *Cup* one leaf; concave; expanding; with five teeth; permanent.

BLOSS. *Petals* five; circular; concave; fitting; fixed to the cup.

CHIVES. *Threads* twenty; awl-fhaped: fixed to the cup. *Tips* roundifh.

POINT. *Seedbud* beneath. *Shafts* two; thread-fhaped; upright. *Summits* with knobs.

S.VESS. *Berry* flefhy; nearly globular; dimpled.

SEEDS. Two; rather oblong; feparate; griftly.

OBS. *The white Hawthorn hath generally only one pointal, and one feed.*

HAWTHORN. The leaves egg-fhaped; jagged; ferrated; Whitebean downy underneath—*Bloffoms white*. Fruit red. Aria

Mefpilus alni folio fubtus incano, Aria Theophrafti dicta. *Ray's Syn.* 453.

Aria Theophrafti. *Gerard.* 1327.

Sorbus fylveftris Aria Theophrafti dicta. *Park.* 1421.

Alni effigie, lanato folio major. *Bauh. pin.* 452.

Whitebean tree.

Woods and hedges in hilly countries. S. May.

It loves dry hills and open expofures, and flourifhes either in gravel or clay. It bears lopping and permits the grafs to grow. The wood, being hard, tough and fmooth, is ufed for axle-trees, wheels, walking-fticks, carpenters and other tools. The fruit is eatable when mellowed by the autumnal frofts. An ardent fpirit may be diftilled from it. It feldom bears a good crop of fruit two years together.

Sheep and Goats eat it.

HAWTHORN. The leaves heart-fhaped; with feven an- Service gles. The lower lobes ftraddling—*Bloffoms white*. Fruit *reddifh* Torminalis *brown*.

Mefpilus apii folio fylveftris non fpinofa, feu forbus torminalis. *Bauh. pin.* 454. *Ray's Syn.* 453.

Sorbus torminalis. *Gerard.* 1471. feu vulgaris. *Park.* 1420.

Common wild fervice-tree, or forb.

Woods and hedges. S. April.

There are varieties with leaves more or lefs round, jagged and ferrated

U 2   HAWTHORN.

White
Oxyacantha

HAWTHORN. The leaves blunt; with about three clefts; serrated—*Bloffoms white; fometimes with a reddifh tinge, as in many parts of Worcefterfhire. There is often but one pointal and one feed in each flower.* Fruit *a coral red.*

Mefpilus apii folio fylveftris fpinofa, feu oxyacantha. *Bauh. pin.* 454. *Ray's Syn.* 453.

Oxyacanthus. *Gerard.* 1325.

Spina appendix vulgaris. *Park.* 1025.

1. There is one variety in which the leaves and fruit are larger;
2. Another with double bloffoms and fmaller fruit: and a third
3. that bloffoms twice in the year, called the Glaftonbury-thorn. White-thorn. Haw-thorn.

Hedges and woods. S. May. 3 At Glaftonbury in Somerfetfhire and about Reading in Berkfhire.

Upon account of the ftiffnefs of its branches, the fharpnefs of its thorns, its roots not fpreading wide, and its capability of bearing the fevereft winters without injury, this plant is univerfally preferred for making hedges, whether to clip or to grow at large. The wood is tough, and is formed into axletrees and handles for tools. The berries are the winter food of Thrufh and many other birds.

This genus fupports the following infects.

White black-veined Butterfly. *Papilio Cratægi.*

Black-thorn Moth. *Phalæna Cærulocephala.*

Ealings Glory. *Phalæna Oxyacantha.*

Brimftone Moth. *Phalæna Cratægata.*

Order III. Three Pointals.

201 SERVICE. 623 Sorbus.

EMPAL. *Cup* one leaf; concave; expanding; with five teeth; permanent.

BLOSS. *Petals* five; circular; concave; fixed to the cup.

CHIVES. *Threads* twenty; awl-fhaped; fixed to the cup. *Tips* roundifh.

POINT. *Seedbud* beneath. *Shafts* three; thread-fhaped; upright. *Summits* roundifh.

S. VESS. *Berry* foft, globular, with a hollow dimple.

SEEDS. Three; rather oblong; feparate; griflly.

SERVICE. The leaves winged; fmooth on both fides—*Blof-* Quicken
foms white. Fruit *red.* Aucuparia

Sorbus fylveftris, foliis domefticæ fimilis. *Bauh. pin.* 415.
Ray's Syn. 452.

Sorbus fylveftris, five Fraxinus bubula. *Gerard.* 1473.

Ornus, five Fraxinus fylveftris. *Park.* 1419.

Quicken-tree. Mountain Afh.

Woods and moift hedges. S. May.

1. Leaves ftriped with yellow.

It grows either in woods or open fields, but beft on the fides
of hills, and in fertile foil. It will not bear lopping. Plants
grow well in its fhade. The wood is foft, tough and folid. It
is converted into tables, fvokes for wheels, fhafts, chairs, &c.
The roots are formed into handles for knives and wooden fpoons.
The berries dried and reduced to powder make wholefome
bread; and an ardent fpirit may be diftilled from them which has
a fine flavour, but it is fmall in quantity. The berries too, under
proper management, make an acid liquor fomewhat like perry.
Virgil fays (Georgica. Lib. 3.)

— — — — — — Pocula læti
Fermento atque acidis imitantur vitea forbis.

Horfes, Cows, Sheep, Goats and Swine eat it.

SERVICE. The leaves winged; woolly underneath -*Blof-* True
foms white. Fruit *brownifh.* Domeftica

Sorbus. *Gerard.* 1471. *Ray's Syn.* 452. fativa. *Bauh. pin.*
415.

Sorbus legitima. *Park.* 1420.

Sorb.

Woods in hilly countries. S. April—May.

The fruit is mealy and auftere, not much unlike the Medlar.

The *Chermes Sorbi* and the *Coccinella Bipuftulata,* a fort of Lady
Cow live upon both fpecies.

Order IV. Five Pointals.

202 MEDLAR. 625 Mefpilus.

EMPAL. *Cup* one leaf; concave; expanding; with five teeth; permanent.

BLOSS. *Petals* five; circular; concave; fixed to the cup.

CHIVES. *Threads* twenty; awl-fhaped; fixed to the cup, *Tips* fimple.

POINT. *Seedbud* beneath. *Shafts* five; fimple; upright. *Summits* roundifh.

S. VESS. *Berry* globular; with a deep hollow, but clofed by the cup.

SEEDS. Five; hunched; hard as bone.

OBS. *From the above defcription it appears that the* HAWTHORN, *the* SERVICE, *and the* MEDLAR *are very nearly allied, fo as hardly to be diftinguifhed othernvife than by the number of pointals. The leaves of the* SERVICE *are generally winged; of the* HAWTHORN *angular, and of the* MEDLAR *entire.*

Common
Germanica
     MEDLAR. Without thorns. Leaves fpear-fhaped; downy underneath. Flowers folitary, fiting—*Branches woolly.* Leaf-ftalks *very fhort; channelled.* Cup *terminating; hairy.* Floral Leaf *as long as the bloffom; which is white.* Fruit *reddifh brown.*

Mefpilus. *Gerard.* 1452. *Ray's Syn.* 453. vulgaris. *Park.* 1422.

Mefpilus germanicus, folio laurino non ferrato, five mefpilus fylveftris, *Bauh. pin.* 453.

In hedges. S. May.

Many people are fond of the fruit when it becomes foft and rotten; it is fomewhat auftere, and binds the bowels.

203 APPLE.

203 APPLE. 626 Pyrus.

EMPAL *Cup* one leaf; concave; with five fhallow clefts; permanent. *Segments* expanding.

BLOSS. *Petals* five; circular; concave; large; fixed to the cup.

CHIVES. *Threads* twenty; awl-fhaped; fhorter than the bloffom; fixed to the cup. *Tips* fimple.

POINT. *Seedbud* beneath; *Shafts* five; thread-fhaped; as long as the chives. *Summits* fimple.

S. VESS. An *Apple*; fomewhat globular; with a hollow dimple; flefhy; with five cells; divifions membranaceous.

SEEDS. Several; oblong; blunt; taper at the bafe; convex on one fide; flat on the other.

APPLE. With ferrated leaves, and flowers in broad-topped Pear fpikes—*Leaves very fmooth*; Petals *white*.　　　　Communis

Pyrus fylveftris. *Gerard.* 1457. *Park.* 1500. *Bauh. fin.* 439.
Pyrafter feu pyrus fylveftris. *Ray's Syn.* 452.
Wild Pear-tree
Woods and hedges. S. April.

1. There is a variety with leaves but little ferrated, and double bloffoms;
2. And another that, flowers in the autumn as well as the fpring.

Cultivation hath produced a very great number of other varieties, which may be feen in *Miller's* Dictionary or *Wefton's* Catalogue. It loves a fertile foil, and floping ground; but will not thrive well in moift bottoms. It ftands the fevereft winters, and does not deftroy the grafs. The wood is light, fmooth, and compact; it is ufed by Turners and to make Joiners tools; and for picture frames to be ftained black The fruit is auftere; but when cultivated highly grateful, as is proved by the great variety of excellent Pears which the induftry of mankind has raifed, for they all originate from this.

Horfes, Cows, Sheep and Goats eat it.

It affords nourifhment to the following infects:
Great Tortoife-fhell Butterfly. *Papilio Iolychloros.*
December Moth. *Phalæna Populi.*
Spotted Buff-moth. *Phalæna Lubricipeda.*
Lappit Moth. *Phalæna Quercifolia.*
Yellow Tuffock Moth. *Phalæna Pudibunda.*
Black-thorn Moth. *Phalæna Cærulocephala.*
- - - - - - *Phalæna Brumata.*
Codling Moth. *Phalæna Pomonella.*
Pear-tree Loufe. *Aphis Pyri.*
Grey Chermes. *Chermes Pyri.*
Black Curculio. *Curculio Pyri.*
Yellow-headed Fly. *Mufca Pyraftri.*

U 4　　　　　　　　APPLE.

Crab
Malus

APPLE. With ferrated leaves and flowers in rundles ; fitting
—*Leaves more circular than thofe of the preceding ſpecies ; petals
tinged with red on the out-ſide.*
Malus ſylveſtris. *Gerard.* 1461. *Park.* 1502. *Ray's Syn.* 452.
Malus ſylveſtria. *Bauh. pin.* 435.
Crab-tree. Wilding.
Woods and hedges. S. May.

The induſtry of the Gardeners never was exerted to greater
advantage than in ſweetening and varying this uſeful fruit ; for
the numerous varieties we muſt refer to *their* Catalogues.
It flouriſhes better on declivities and in ſhady places, than in
open expoſures ; or boggy lands. Graſs grows well beneath it.
It is much uſed as a ſtock, on which to ingraft the better kinds
of apples ; becauſe its roots are neither killed by froſt, nor eaten
by Field-mice. The bark affords a yellow dye. The wood is
tolerably hard : it turns very clean, and when made into cogs for
wheels, obtains a poliſh and wears a long time. The acid juice
of the fruit is called by the country people Verjuice, and is
much uſed in recent Sprains, and in other caſes as an aſtringent
or repellent. With a proper addition of ſugar it is probable that
a very grateful liquor may be made with the juice, but little inferior to Old Hock.—Horſes, Cows, Sheep and Goats eat it.
Swine are very fond of the fruit.

This ſpecies nouriſhes the following inſects.

Gypſie Moth. *Phalæna Diſpar.*
December Moth. *Phalæna Populi.*
Black Tuſſock Moth. *Phalæna Faſcelina.*
Yellow-tail Moth. *Phalæna Chryſorrhœa.*
Dagger Moth. *Phalæni Pſi.*
Plumb-tree Moth. *Phalæna Oporana.*
- - - - - - *Phalæna Brumata.*
Codling Moth. *Phalæna Pomonella.*
Black Arches. *Phalæna Monacha.*
Apple Louſe. *Aphis Mali.*
Garden Beetle. *Scarabæus Horticola.*

204 MEADOWSWEET. 630 Spiræa.

EMPAL. *Cup* one leaf; with five fhallow clefts; flat at the bafe; *Segments* fharp; permanent.

BLOSS. *Petals* five; inverfely egg-fhaped, but oblong; fixed to the cup.

CHIVES. *Threads* more than twenty; thread-fhaped; fhorter than the bloffom; fixed to the cup. *Tips* roundifh.

POINT. *Seedbuds* five or more. *Shafts* the fame number; thread-fhaped; as long as the chives *Summits* fomewhat globular.

S. VESS. *Capfules* oblong; tapering; compreffed. Valves two.

SEEDS. Few; tapering; fmall.

OBS. *In the firft fpecies the Capfules are numerous, and form a circle; and in the fecond they are twifted like a fnail fhell.*

MEADOWSWEET. The leaves winged; little leaves uniform; ferrated; ftem herbaceous; Flowers in tufts—*Bloffoms* reddifh white. Drop-wort Filipendula

Filipendula. *Gerard.* 1058. *Park.* 434. *Ray's Syn.* 259.
Filipendula vulgaris. *Bauh. pin.* 163.
Drop-wort.
High paftures. P. July.

The tuberous pea-like roots dried and reduced to powder make a kind of bread, which in times of fcarcity is not to be defpifed. Hogs are very fond of them.—Cows, Goats, Sheep and Swine eat it; Horfes refufe it.—The Eyed Willow Hawk Moth, *Sphinx Ocellata*; and the Burnet Moth, *Sphinx Filipendula*, feed upon this and the following fpecres.

MEADOWSWEET. The leaves winged; with an odd one at the end larger than the reft; and divided into lobes. Flowers in tufts—*Bloffoms yellowifh white.* Common Ulmaria

Ulmaria. *Ray's Syn.* 259.
Ulmaria vulgaris. *Park.* 592.
Regina prati. *Gerard.* 1043.
Barbara capri, floribus compactis. *Bauh. pin.* 164.
Queen of the Meadows.
Moift meadows. P. June—July.

The flowers infufed in boiling water give it a fine flavour which rifes in diftillation.—Sheep and Swine eat it; Goats are extremely fond of it; Cows and Horfes refufe it.

Order

Order V. Many Pointals.

205 ROSE. 631 Rosa.

EMPAL. *Cup* one leaf. *Tube* diftended ; but narrow at
the neck. *Border* globular ; with five divifions ; expanding. *Segments* long ; fpear-fhaped ; narrow.

BLOSS. *Petals* five ; inverfely heart-fhaped ; as long as
the cup and fixed to its neck.

CHIVES. *Threads* many ; hair-like ; very fhort ; fixed to
the neck of the cup. *Tips* three-edged.

POINT. *Seedbuds* numerous ; at the bottom of the cup.
Shafts as many as there are feedbuds ; clofely compreffed by the neck of the cup ; fixed to the fide of
the feedbud. *Summits* blunt.

S. VESS. *Berry* flefhy ; turban fhaped ; coloured ; foft ;
of one cell ; crowned by imperfect fegments ; clofed
at the neck ; formed by the tube of the cup.

SEEDS. Numerous ; oblong ; hairy ; adhering to the
cup.

OBS. *The cup forming the feed-veffel refembles a berry. In fome
fpecies two of the fegments of the cup have a little appendage on each
fide ; the third only on one fide, and the remaining two have none at
all.*

Sweet-briar
Eglanteria

ROSE. The feedbuds globular ; on fruit-ftalks ; glofly. Stem
befet with ftraight prickles ; leaf-ftalks prickly and rough ; leaves
fharp—*Odoriferous.* Flowers *pale red or pale yellow.* Leaves
rufty on the under furface, and clammy at the ends.
Rofa fylveftris odora. *Gerard.* 1272. *Ray's Syn.* 454.
Rofa fylveftris odora, feu Eglanteria flore fimplici. *Park.*
418.
Rofa fylveftris foliis odoratis. *Bauh. pin.* 483.
Eglantine.
In hedges. S. May—June.
The varieties are,
1. Bloffoms red ; double.
2. Bloffoms full, and red.
3. Bloffoms full and flefh-coloured.
4. Bloffoms full and flefh-coloured ; leaves evergreen.
The fweet fmell of the leaves procures it a place in our gardens.

ROSE

ROSE. The feedbuds globular ; gloffy ; fruit-ftalks covered Burnet with ftrong hairs. Stem and leaf-ftalks thick fet with prickles.— Spinofiffima *Fruit roundifh and almoft black.* Bloffoms *white.*

Rofa pumila fpinofiffima, foliis pimpinellæ glabris, flore albo, *Ray's Syn.* 455.

Rofa fylveftris pomifera minor. *Bauh. pin.* 488.

Rofa pimpinella, feu pomifera minor. *Park.* 1018.

Rofa pimpinellæ folio. *Gerard.* 1270.

1. There is a va iety with fmaller leaves and a variegated bloffom. Sandy heaths. S. June.

The ripe fruit is eaten by Children ; it has a grateful fubacid tafte. The juice of it diluted with water dyes filk and muflin of a peach colour ; with the addition of alum a deep violet ; but it has very little effect on woollen or linnen.

Cows, Sheep, Goats and Swine eat it ; Horfes refufe it.

ROSE. The feed-buds globular ; feed-buds and fruit-ftalks co- Apple vered with ftrong hairs : ftem befet with prickles ; leaf-ftalks Villofa prickly ; leaves downy—*on both furfaces.* Cups *prickly.* Blof- foms *deep red.*

Rofa fylveftris pomifera major noftras. *Ray's Syn.* 455.

Rofa fylveftris pomifera major. *Bauh. pin.* 484.

Rofa pomifera major. *Park.* 418.

On mountains. S. June.

1. There is one variety with double bloffoms, and
2. Another with large feedbuds. *Ray's Syn.* 454.

ROSE. Seedbuds globular fmooth. Fruit-ftalks fmooth. Stem White and leaf-ftalks prickly. Flowers in a fort of a tuft—*Bloffoms* Arvenfis *white.* *Hudf. Flor. Angl.* 192.

Rofa fylveftris altera minor, flore albo noftras. *Ray's Syn.* 455.

Rofa arvenfis candida. *Bauh. pin.* 484.

White flowered Dog-rofe.

In hedges. S. July.

ROSE.

ROSE. The feedbuds egg-fhaped : feedbuds and fruit-ftalks gloffy. Stem and leaf-ftalks prickly—*Bloffoms pale red.* Berry *red.* Leaves *five or feven together.* Petals *compofed of two lobes.* Floral-leaves *two ; oppofite ; fringed.*

Rofa fylveftris inodora, feu canina. *Park.* 1017. *Ray's Syn.* 454.

Rofa canina inodora. *Gerard.* 1270.

Rofa fylveftris vulgaris flore odorato incarnato. *Bauh. pin.* 483.

Red flowered Dogs-rofe. Hep-tree. Wild-briar.

In hedges. S. May—June.

1. The bloffoms fometimes, but not always, have a fweet fmell.
2. The leaves are fometimes variegated.

A perfumed water may be diftilled from the bloffoms. The pulp of the berries beat up with fugar makes the conferve of hepps of the London difpenfatory. Mixed with wine it is an acceptable treat in the north of Europe. Several birds feed upon the berries. The leaves of every fpecies of Rofe, but efpecially of this, are recommended in the *Ephemer. Naturæ curiofor.* As a fubftitute, for Eaft India tea, giving out a fine colour, a fub-aftringent tafte, and a grateful fmell, when dried and infufed in boiling water.

Cows, Sheep, Goats and Swine eat it. Horfes are not fond of it.

It is a difficult matter to fay which are fpecies, and which are varieties only, in this genus; Linnæus feems to think that there are no certain limits prefcribed by nature.

The different fpecies of Roses nourifh the following infects.

Province Rofe Moth.   *Phalæna Salicella.*
Emperor Moth.   *Phalæna Pavonia.*
Furbelow Moth.   *Phalæna Libatrix.*
Spotted Elm Moth.   *Phalæna Retularia.*
- - - - - - - *Tenthredo Rofæ.*
- - - - - - *Tenthredo Cynofvati.*
- - - - - - *Ichneumon Bedeguaris.*
- - - - - - *Cicada Rofæ.*
Rofe Loufe.   *Aphis Rofæ.*
Rofe Chaffer.   *Scarabæus Auratus.*
Rofe Fly.   *Mufca Pellucens.*

And thofe Mofflike prickly excrefcences which are frequently found upon the branches of Rofes, efpecially upon the laft fpecies, are the habitations of the *Cynips Rofæ.* This excrefcence was formely in repute as a medicine, and was kept in the fhops under the name of Bedeguar.

An infufion of the full blown bloffoms of all the Rofes, efpecially the paler kinds, is purgative ; but the petals of the red Rofes gathered before they expand, and dried, are aftringent.

206 BRAMBLE.

206 BRAMBLE. 632 Rubus.

EMPAL. *Cup* one leaf; with five divisions. *Segments* oblong; expanding; permanent.

BLOSS. *Petals* five; circular; as long as the cup; upright, but expanding; fixed to the cup.

CHIVES. *Threads* numerous; shorter than the petals; fixed to the cup. *Tips* roundish; compressed.

POINT. *Seedbuds* numerous. *Shafts* small; hairlike; growing on the sides of the seedbuds. *Summits* simple; permanent.

S. VESS. *Berry* compound; composed of little granulations collected into a knob, which is convex above and concave beneath. Each granulation hath one Cell.

SEEDS. Solitary; oblong. *Receptacle* of the seed-vessels conical.

* Shrubs

BRAMBLE. The leaves growing by fives or by threes; in Rafpberry a winged manner. Stem prickly: leaf-stalks channelled—*Stems* Idæus *nearly upright; two feet high.* Leaves *serrated; white with short down underneath.* Blossoms *white.* Berry *red.*

Rubus idæus spinosus. *Bauh. pin.* 479.

Rubus idæus. *Gerard* 1272. *Park.* 557.

Rubus idæus spinosus fructu rubro. *Ray's Syn.* 467.

Raspberry bush. Framboise. Hindberry.

Woods and hedges in high stony places. S. May—June.

The varieties are,

1. White berried.
2. Without prickles.
3. Flowering in the Autumn as well as in the Spring.

The fruit is extremely grateful as nature presents it, but made into a sweetmeat with sugar, or fermented with wine, the flavour is improved. It is fragrant, sub-acid and cooling. It dissolves the tartarous concretions of the teeth, but for this purpose it is inferior to the Strawberry. The white berries are sweeter than the red, but they are generally contaminated by Insects. The fresh leaves are the favourite food of Kids.

Sheep, Goats and Swine eat it; Cows are not fond of it; Horses refuse it.

BRAMBLE.

Dewberry
Cæsius

BRAMBLE: The leaves growing by threes; almost naked.
The lateral leaves consisting of two lobes; stem cylindrical;
prickly—*Stems three feet high; purplish; branched; with pendant
shoots at the top.* Leaves *serrated: the middle leaf egg-shaped.* Blossoms *white.* Fruit *blaish black.*
Rubus repens fructu cæsio. *Bauh. pin.* 479. *Gerard.* 1271.
Rubus minor fructu cæruleo. *Ray's Syn.* 467.
Rubus minor; chamærubus sive humirubus. *Park.* 1013.
Small Bramble.   Dewberry-bush.
Woods and hedges.   S. June—July.
The berries are pleasant to eat, and put into red wine communicate a fine flavour.
Cows, Goats and Sheep eat it; Horses refuse it.

Blackberry
Fruticosus

BRAMBLE. The leaves growing by fives, or by threes;
fingered. Stem and leaf-stalks prickly—*Stem rather angular, extremely long.   Leaves a little woolly on the under surface. The middlemost little leaf larger than the others; heart-shaped; standing on
a leaf-stalk; but the others are sitting.* Blossoms *white.* Fruit *first
red and afterwards black.*
Rubus vulgaris major. *Park.* 1013. fructu nigro. *Ray's Syn.*
46 .
Rubus vulgaris, seu Rubus fructu nigro. *Bauh. pin.* 479.
Common Bramble.   Bumblekites.
Hedges.   S. May—September.
The varieties are,
1. Fruit white.
2. Blossoms double.
3. Leaves variegated.
4. Leaves jagged.
5. Without prickles.
The berries when ripe are black, and do not eat amiss with
wine. The green twigs are of great use in dying woollen, silk
and mohair black.
Cows and Horses eat it; Sheep are not fond of it.

** *Herbaceous.*

Stone
Saxatilis

BRAMBLE. The leaves growing by threes; naked. Wires
creeping along the ground; Herbaceous.—*The granulations of
the berries distinct, and not united as in the other species.* Blossoms
white, fruit red.
Rubus saxatilis. *Gerard.* 1275.
Rubus alpinus saxatilis. *Park.* 1014.
Chamærubus saxatilis. *Bauh. pin.* 479. *Ray's Syn.* 261.
On stony hills. P. June.
The berries are not very good, but Children eat them.
Cows, Goats, Sheep and Swine eat it; Horses refuse it.

BRAMBLE.

BRAMBLE. The leaves undivided; lobed. Stem without Cloudberry prickles, fupporting one flower.—*Stem hardly a foot high.* Blof- Chamæmorus *fom white, or purple.* Berries *red. In this fpecies the chives are found on one plant and the pointals on another; but* Dr. Solander *hath obferved the roots of the two plants to unite under the furface of the ground.*

Chamæmorus. *Gerard.* 173. *Ray's Syn.* 260.
Chamæmorus anglica. *Park.* 1014.
Chamæmorus foliis ribes anglica. *Bauh. pin.* 480.
Knot-berries. Cloud-berries. Knout-berries.
In peat moffes. P. May—June.

The berries are not unpleafant. The Norwegians pack them up in wooden veffels and fend them to the capital of Sweden, where they are ferved up in deferts. The Laplanders bury them under the fnow and thus preferve them frefh from one year to another—Cows, Sheep and Goats eat it—the green Butterfly, *Papilio Rubi*; the Emperor Moth, *Phalæna Pavonia*; the Fox-coloured Moth, *Phalæna Rubi*; the black Tuffock Moth, *Phalæna Fafcelina*; and the Swallow-tail Moth, *Phalæna Sambucaria*, are nourifhed by the different fpecies of *Bramble.*

207 TORMENTIL. 635 Tormentilla.

EMPAL. *Cup* one leaf; flat; with eight hollow clefts. Every other *Segment* fmaller and fharper.
BLOSS. *Petals* four; inverfely heart-fhaped; flat; expanding; fixed by claws to the cup.
CHIVES. *Threads* fixteen; awl fhaped: half as long as the petals; fixed to the cup. *Tips* fimple.
POINT. *Seedbuds* eight; fmall; approaching fo as to form a knob. *Shafts* thread-fhaped; as long as the chives; fixed to the fides of the feedbuds: *Summits* blunt.
S.VESS. None. *Receptacle of the feeds* very fmall, loaded with feeds and inclofed by the cup.
SEEDS. Eight; oblong; taper, but ending in a blunt point.

Upright
Erecta

TORMENTIL. The ſtem nearly upright; leaves ſitting— *Bloſſoms yellow.*

Tormentilla, *Gerard.* 992. *Ray's Syn.* 257. vulgaris. *Park.* 394.

Tormentilla ſylveſtris. *Bauh. pin.* 326.

Septfoil.

Barren paſtures. P. June.

The roots may rank with the ſtrongeſt vegetable aſtringents, and as ſuch have a place in the modern practice of Phyſic. They are uſed in ſeveral countries to tan leather. Farmers find them very efficacious in the dyſenteries of Cattle.

Cows, Goats, Sheep and Swine eat it; Horſes refuſe it.

Creeping
Reptans

TORMENTIL. The ſtem creeping; leaves on leaf-ſtalks— *Bloſſoms yellow.*

Tormentilla reptans. *Ray's Syn.* 257.

Woods and barren paſtures. P. July.

208 A V E N S. 637 Dryas.

EMPAL. *Cup* one leaf; with five or eight diviſions. *Segments* expanding; ſtrap-ſhaped; blunt; equal; ſomewhat ſhorter than the bloſſom.

BLOSS. *Petals* five, (or eight,) oblong; margin broken; expanding; fixed to the cup.

CHIVES. *Threads* numerous; hair-like; ſhort; fixed to the cup. *Tips* ſmall.

POINT. *Seedbuds* many; ſmall; crowded together. *Shafts* hair-like; fixed to the ſides of the ſeedbuds. *Summits* ſimple.

S. VESS. None.

SEEDS. Numerous; roundiſh; compreſſed; retaining the ſhafts, which grow very long and woolly.

Cinquefoil
Pentapetala -

AVENS, with five petals; and winged leaves—*Seven or nine little leaves upon each winged leaf; the lowermoſt the ſmalleſt. The flowering ſtalk twice or thrice as long as the leaves. Flowers ſingle, terminating. This plant is probably the product of the ſeedbuds of the next ſpecies fertilized by the duſt of one of the* BENNETS. *Bloſſoms yellow; ſometimes white.*

Caryophyllata pentaphyllæa. *Park.* 137. *Ray's Syn.* 254.

Caryophyllata alpina pentaphyllæa. *Gerard.* 995.

Caryophyllata alpina quinquefolia. *Bauh. pin.* 322.

In the den of Bethaick in Scotland. P. June—July.

AVENS.

AVENS. With eight petals; and simple leaves—*Egg-shaped*; Mountain
harsh; *serrated*. Blossoms *white*; *on hairy fruit-stalks.*   Octopetala
  Caryophyllata alpina chamædryos folio. *Ray's Syn.* 252.
  Chamædrys spuria montana cisti flore. *Park.* 106.
  Chamædrys alpinum, cisti flore. *Bauh. pin.* 248.
  Teucrium alpinum, cisti flore. *Gerard.* 659.
  On lofty mountains. P. July.
  Neither Cows, Horses, Sheep, Goats or Swine will eat it.

209 STRAWBERRY. 633 Fragaria.

EMPAL. *Cup* one leaf; flat; with ten shallow clefts. *Seg-
    ments* alternately narrower, the narrow ones placed
    on the outside the others.
BLOSS. *Petals* five; circular; expanding; fixed to the cup.
CHIVES. *Threads* twenty; awl-shaped; shorter than the
    blossom; fixed to the cup. *Tips* in the shape of a
    crescent.
POINT. *Seedbuds* numerous; very small; forming a knob.
    Shafts simple; fixed to the sides of the seedbuds.
    Summits simple.
S. VESS. None. *Receptacle of the feeds* a fort of berry;
    partly globular; partly egg-shaped; pulpy; soft;
    large; coloured; lopped at the base; deciduous.
SEEDS. Numerous; very small; tapering; scattered up-
    on the surface of the receptacle.

  STRAWBERRY. With creeping wires.—*Blossoms white.* Wood
Fruit *red*   Vesca
1. Leaves not hairy; wrinkled. Fruit red, conical. *Wallis's
Nat. Hist. Northumberland.*
  Fragaria. *Gerard.* 998. *Ray's Syn.* 254. vulgaris. *Park.*758.
Bauh. pin. 326.
  In woods and hedges. P. 1. In Northumberland. April—
May.
  Cultivation hath produced the following varieties.
1. Gold-striped leaved.
2. White-fruited.—*This sometimes happens without cultivation.*
3. Double-flowering.
4. Red-bushy, without creeping wires.
5. White-bushy, without creeping wires.
6. Versailles, or one-leaved Strawberry.

The berries either eaten alone, or with fugar, or with milk, are univerfally efteemed a moft delicious fruit. They are grateful, cooling, fubacid, juicy, and have a delightful fmell. Taken in large quantities they feldom difagree. They promote perfpiration, impart a violet fmell to the urine, and diffolve the tartarous incruftations upon the teeth. People afflicted with the Gout or Stone have found great relief by ufing them very largely; and Hoffman fays he hath known confumptive people cured by them. The bark of the root is aftringent.—Sheep and Goats eat it; Cows are not fond of it; Horfes and Swine refufe it.

The Cuckow-fpit Frog-hopper, *Cicada Spumaria*, is very frequently found upon it; and the *Coccus Polonicus*, a kind of Cochineal is found upon the roots.

Barren
Sterilis

STRAWBERRY. The ftems drooping. Flowering branches flexible—*Suckers thick, depreffed; covered with fpear-fhaped props of the colour of rufty iron. Leaves growing by threes, inverfely egg-fhaped; ferrated; dented; flexible; hairy; white underneath. Leaf-ftalks very hairy. Flowering ftems thread-fhaped; drooping; flexible; with a few fmall leaves. Flowers folitary; white; on fruit-ftalks.*

Fragaria fterilis. *Bauh. pin.* 327. *Ray's Syn.* 254.
Fragaria minime vefca. *Park.* 758. feu fterilis. *Gerard.* 998.
Hedges. P. April.

210 CINQUEFOIL. 634 Potentilla.

EMPAL. *Cup* one leaf; fomewhat flat; with ten fhallow clefts. *Segments* alternately fmaller; reflected.

BLOSS. *Petals* five; circular; or heart fhaped; expanding; fixed by claws to the cup.

CHIVES. *Threads* twenty; awl-fhaped; fhorter than the petals; fixed to the cup. *Tips* in the fhape of a long crefcent.

POINT. *Seedbuds* numerous; very fmall; forming a knob. *Shafts* thread-fhaped; as long as the chives; fixed to the fides of the feedbuds. *Summits* blunt.

S.VESS. None. *Receptacle of the feeds* roundifh; dry; very fmall; permanent; covered with feeds; inclofed in the cup.

SEEDS. Numerous; tapering.

* *Leaves*

* Leaves winged.

CINQUEFOIL. The leaves winged. Stem somewhat woody. Shrub
B ssoms yellow. Fruticosa
Pentaphylloides fruticosa. *Ray's Syn.* 256.
Near Mickleforce in Teasdale, and several parts of Yorkshire.
S. June.
The beautiful appearance of its numerous flowers gains it ad-
mittance into low garden hedges. Besoms are made of it.—
Cows, Horses, Goats and Sheep eat it; Swine refuse it.

CINQUEFOIL. The leaves winged; serrated. Stem creep- Silver
ing. Fruit stalks supporting a single flower.—*In clayey soils the* Anserina
leaves are white and silky. Blossoms *yellow.*
Potentilla. *Bauh. pin.* 321. *Park.* 593.
Pentaphylloides argentina dicta. *Ray's Syn.* 256.
Argentina. *Gerard.* 973.
Silvet-weed. Wild Tansey. Goose-grass. Goose-tansey.
In fields near foot-paths. P. June—July.
The leaves are mildly astringent. Dried and powdered they
are given with success in Agues. The usual dose is a meat
spoonful of the powder every three hours betwixt the fits. The
roots in the winter time eat like Parsneps. Swine are fond of
them.—Cows, Horses, Goats and Swine eat it; Sheep refuse it.

CINQUEFOIL. The leaves winged; alternate: little leaves Upright
growing by fives; egg-shaped; scollopped. Stem upright— Rupestris
Blossoms white.
Pentaphylloides erectum. *Ray's Syn.* 255.
Quinquefolium fragiterum. *Bauh. pin.* 326.
Pentaphyllum fragiterum. *Gerard.* 991. *Park.* 397.
Upright Bastard Cinquefoil.
On the sides of mountains. P. July.
Cows, Horses, Goats and Sheep eat it.

* * Leaves fingered.

Tormentil
CINQUEFOIL. The leaves growing by fives; wedge-shaped; Argentea
jagged; downy underneath; stem upright—*Blossoms yellow.*
Pentaphyllum erectum, foliis profunde sectis, subtus argenteis,
flore luteo. *Ray's Syn.* 255.
Pentaphyllum rectum minus. *Park.* 400.
Quinquefolium folio argenteo. *Bauh. pin.* 235.
Quinquefolium tormentillæ facie. *Gerard.* 987.
In gravelly soil. P. June.
Goats and Swine eat it; Sheep, Horses and Cows refuse it.

Hairy
Opaca

CINQUEFOIL. The root-leaves growing by fives ; wedge-shaped ; ferrated. Stem leaves nearly oppofite. Branches thread-shaped ; drooping—*Bloſſoms yellow.*
Pentaphyllum parvum hirſutum   *Ray's Syn.* 255.
Pentaphyllum incanum minus repens. *Gerard.* 989.
Quinquefolium minus repens lanuginoſum *Park.* 399.
Quinquefolium minus repens lanuginoſum luteum. *Bauh. ſin.* 22ʒ.
Small rough Cinquefoil.
In Yorkſhire. P. June.

Spring
Verna

CINQUEFOIL. The root-leaves growing by fives ; ſharply ferrated ; dented. Stem leaves growing by threes. Stem de-clining—*Purpliſh ; much branched ; ſupporting many flowers. At the baſe of the petals there is a ſpot which is ſometimes of a tawny yellow colour. Leaves ſitting ; deeply ſerrated and a little hairy towards the ends. Petals, chives, and pointals yellow.*
Quinquefolium minus repens luteum. *Bauh. ſin.* 325. *Ray's Syn.* 213.
In barren paſtures. P. May—June.
Horſes, Cows, Goats and Sheep eat it.

White
Alba

CINQUEFOIL. The leaves growing by fives ; approach-ing at the ends ; ferrated. Stems thread-ſhaped ; trailing. Re-ceptacle rough with hair—*Bloſſoms white ; ſolitary.*
Quinquefolium album majus. *Bauh. ſin.* 325.
Quinquefolium ſylvaticum majus flore albo. *Gerard.* 987.
In Wales. P. Auguſt.

Common
Reptans

CINQUEFOIL. The leaves growing by fives. Stem creep-ing. Fruit-ſtalks ſupporting a ſingle flower.—*Bloſſoms yellow. Props in pairs ; each prop compoſed of three egg-ſhaped leaves.*
Pentaphyllum vulgatiſſimum. *Park.* 398. *Ray's Syn.* 255.
Quinquefolium majus. *Gerard.* 987. repens. *Bauh ſin.* 32ʒ.
Five-leaved Graſs.
Road-ſides. P. June—Sept.
The red cortical part of the root is mildly aſtringent and antiſeptic. A decoction of it is a good gargle for looſe teeth and ſpongy gums.—Horſes, Cows, Goats and ſheep eat it.

MANY POINTALS. 309

211 BENNET. 636 Geum.

EMPAL. *Cup* one leaf; with ten shallow clefts; nearly upright. *Segments* alternately very small and sharp.

BLOSS. *Petals* five; rounded. *Claws* narrow; as long as the cup; fixed to the cup.

CHIVES. *Threads* numerous; awl-shaped; as long as the cup; fixed to the cup. *Tips* short; rather broad; blunt,

POINT. *Seedbuds* numerous; forming a knob. *Shafts* long; hairy; fixed to the sides of the seedbuds. *Summits* simple.

S.VESS. None. *Receptacle of the seed* oblong; hairy; standing upon the reflected cup,

SEEDS. Numerous; compressed; covered with strong hairs and armed with a long and crooked awn formed of the shaft.

BENNET. Flowers upright; fruit globular; hairy. Awns Avens hooked; naked. Leaves lyre-shaped—*Props toothed.* Blossoms Urbanum *yellow.*

Caryophyllata. *Gerard.* 995. *Ray's Syn.* 253.
Caryophyllata vulgaris. *Bauh. pin.* 321. *Park.* 136.
1. There is a variety with a larger flower; and cultivation produces striped leaves.
Common Avens. Herb-Bennet.
Woods and hedges P June—August.
The roots gathered in the Spring before the stem grows up, and put into ale, give it a pleasant flavour and prevent its going four. Infused in wine it is a good stomachic. Its taste is mildly austere and aromatic; especially when it grows in warm, dry situations, but in shady moist places it has little virtue.—Cows, Goats, Sheep and Swine eat it; Horses are not fond of it.

BENNET. The flowers nodding. Fruit oblong. Awns Water downy; twisted—*Blossoms purplish. By cultivation they become* Rivale *double.*

Caryophyllata montana purpurea. *Gerard.* 995. *Rays Syn.* 253.
Caryophyllata montana seu palustris purpurea. *Park.* 136
Caryophyllata aquatica nutante flore. *Bauh. pin.* 321.
Water Avens.
Mountains in the North, and in bogs. P. July.
The powdered root will cure Tertian Agues, and is daily used for that purpose by the Canadians,—Sheep and Goats eat it; Cows, Horses and Swine are not fond of it,

X 3 212 MARSH-

212 MARSHLOCKS. 638 Comarum.

EMPAL. *Cup* one leaf; with ten shallow clefts; coloured. *Segments* alternately smaller, and placed under the others; permanent.

BLOSS. *Petals* five; oblong; tapering; three times smaller than the cup, to which they are fixed.

CHIVES. *Threads* twenty; awl-shaped; fixed to the cup; as long as the blossom; permanent. *Tips* in the shape of a crescent; deciduous.

POINT. *Seedbuds* numerous; roundish; very small; forming a knob. *Shafts* simple; short; fixed to the sides of the seedbuds. *Summits* simple.

S. VESS. None. *Receptacle of the seeds* shaped like a double purse; fleshy; large; permanent.

SEEDS. Numerous; taper; covering the receptacle.

Purple
Paluftre

MARSHLOCKS. As there is only one species Linnæus gives no description of it—*Leaves winged.* Blossoms *solitary; terminating; purple.*

Pentaphylloides palustre rubrum. *Ray's Syn.* 256.
Pentaphyllum rubrum palustre. *Gerard.* 987.
Quinquefolium palustre rubrum. *Bauh. pin.* 326.
Purple Marsh Cinquefoil.

1. There is a variety with hairy leaves.
In putrid marshes. P. June.

The root dyes a dirty red. The Irish rub their milking-pails with it, and it makes the milk appear thicker and richer. Goats eat it; Cows and Sheep are not fond of it; Horses and Swine refuse it.

C L A S S XIII.

IN this Clafs the Chives are numerous ; and ftand
upon the RECEPTACLE, whereas in the preceding Clafs
they are placed upon the fides of the cup or upon the
PETALS. A regard to this circumftance of the *Situation*,
is of more importance than an attention to the *Number* of
the chives.

Moft of the plants of this Clafs are poifonous

CLASS XIII.

MANY CHIVES.

ORDER I. ONE POINTAL.

* Petals four.

213 POPPY. - - *Cup* two leaves. *Capsule* with one cell; crowned.
214 CELANDINE. *Cup* two leaves. *Seed-veſſel* a Pod.
215 CHRISTOPHER. *Cup* four leaves. *Berry* with one cell. *Seeds* in a double row.

* * Petals five.

216 CISTUS. - - *Capſule* nearly globular. *Cup* five leaves; two ſmaller than the reſt.
217 LIME. - - *Capſule* with five cells; like leather. *Seed* one. *Cup* deciduous.

† Wild Larkſpur.

* * * Petals Many.

218 WATERLILY. *Berry* with many cells; outer coat like bark. *Cup* large.

Order II. Two Pointals.

† Burnet Iron-wort.

Order

Order III. Three Pointals.

219 LARKSPUR. - *Cup* none. *Blossom* five petals ; the upper petal horn-shaped. *Honey-cup* cloven ; sitting.

† *Dyers Yellow-weed.*

Order V. Five Pointals.

220 COLUMBINE. *Cup* none. *Blossom* five petals. *Honey-cups* five ; horned in the lower part.

Order VI. Six Pointals.

221 WATERSOLDIER. *Cup* with three divisions. *Blossom* three petals ; *Berry* with six cells, within a sheath.

Order VII. Many Pointals.

222 TRAVELLERSJOY. *Cup* none. *Bloss.* four petals. *Seeds* many ; with awns.

223 MEADOWRUE. *Cup* none. *Bloss.* four or five petals. *Seeds* many ; without awns ; naked.

224 HELLEBORE. *Cup* none. *Bloss.* five petals, permanent. *Honey-cups* many. *Capsules* with many seeds.

225 MEADOWBOUTS. *Cup* none. *Bloss.* five petals. *Capsules* many. *Honey-cups* none.

226 ANEMONE. - *Cup* none. *Bloss.* six petals. *Seeds* many.

227 GLOBEFLOWER. *Cup* none. *Bloss.* fourteen petals. *Honey-cups* narrow. *Capsules* with one seed.

228 CROWFOOT. *Cup* five or three leaves. *Bloss.* five petals. *Seeds* many. *Petals* with a honey-cup in the claw.

229 PHEASANTEYE. *Cup* five leaves. *Bloss.* five or ten petals. *Seeds* many ; angular ; covered with a thick skin.

† *White Water-lily.*

213 POPPY.

213 POPPY. 648 Papaver.

EMPAL. *Cup* two leaves; egg-shaped; imperfect at the margin. *Little leaves* somewhat egg-shaped; concave; blunt; shedding.

BLOSS. *Petals* four; circular; flat; expanding; large; narrowest at the base. Alternately smaller.

CHIVES. *Threads* numerous; hairlike; much shorter than the blossoms. *Tips* oblong; compressed; upright; blunt,

POINT. *Seedbud* nearly globular; large. *Shaft* none. *Summit* flat; large; radiate; crowning the seedvessel.

S. VESS. *Capsule* of one cell, divided half way into many cells; opening beneath the crown into several holes.

SEEDS. numerous; very small. *Receptacle* consists of as many longitudinal plaits, as there are rays in the summit; connected to the sides of the seedvessel.

OBS. *The* Seedvessel *varies in figure from globular to oblong; and the number of rays in the summit are likewise various.*

* Capsules covered with strong hairs.

Rough-headed Hybridum

POPPY. The capsules somewhat globular, with protuberating knobs; and covered with strong hairs. Stem leafy; supporting many flowers—*Blossoms red.*

Papaver laciniato folio, capitulo hispido rotundiore. *Ray's Syn.* 308.

Argemone capitulo rotundiore. *Park.* 369.

Argemone capitulo breviore. *Bauh. pin.* 172.

Argemone capitulo torulo. *Gerard.* 373.

Round rough headed Poppy.

In corn fields. A. June—July.

Long-headed Argemone

POPPY. The capsules club-shaped; covered with strong hairs. Stem leafy; supporting many flowers—*Leaves with three divisions and winged clefts.* Fruit-stalks *rough.* Cups *set with strong hairs.* Blossoms *red.*

Papaver laciniato folio capitulo hispido longiore. *Ray's Syn.* 308.

Argemone capitulo longiore. *Gerard.* 373. *Bauh. pin.* 172. *Park.* 370.

Long rough-headed Poppy.

Corn-fields. A. June.

* * Capsules

** *Capfules fmooth.*

POPPY. The capfules fmooth; globular; Stem hairy; fup- Corn
porting many flowers. Leaves jagged; with winged clefts— Rhæas
Hairs upon the ftem expanding. Stem *cylindrical: branched.* Summit *with twelve rays.* Bloffoms *fcarlet.*
Papaver laciniato folio, capitulo breviore glabro, annuum
Rhæas dictum. *Ray's Syn.* 308.
    Papaver rhæas. *Gerard.* 401. by miftake: it fhould be 371.
    Papaver erraticum rhæas five fylveftre. *Park.* 367.
    Papaver erraticum majus. *Bauh. pin.* 171.
    Red Poppy. Rouud fmooth-headed Poppy. Cop-rofe. Head-
wark. Corn-rofe.
    Amongft Corn. A. June—July.
The bloffoms give out a fine colour, and an infufion of them
made into a fyrup is kept in the fhops. It partakes in a fmall
degree of the properties of Opium.
Sheep and Goats eat it; Horfes refufe it.

POPPY. The capfules oblong; fmooth. Stem fupporting Smooth-head-
many flowers and fet with briftles which lye clofe to it. Leaves ed
jagged, with winged clefts—*Bloffoms red.* Dubium
Papaver laciniato folio capitulo longiore glabro. *Ray's Syn.*
309.
    Long fmooth-headed Poppy.
    Cows and Goats eat it; Sheep are very fond of it; Horfes
refufe it.

POPPY. The cups and capfules fmooth. Leaves embracing Sleepy
the ftem; jagged.—*Stem and leaves fmooth.* Summits *ten. Blof-* Somniferum
foms white, tinged with purple. Seeds *white.*
    Papaver fylveftre· *Gerard.* 370. *Ray's Syn.* 308.
    Papaver hortenfe femine albo. *Bauh. pin.* 170.
    Wild Poppy.
    Uncultivated places. A. July.
Opium is nothing but the milky juice of this plant infpiffated
by the heat of the Sun. The Edinburgh College direct an extract to be prepared from the heads, i. e. the feed-veffels. This
extract is fuppofed to be milder in its effects than the foreign
Opium, but it requires double the quantity for a dofe. A fyrup
is made with a decoction of the heads, and kept in the fhops under the name of Diacodion. The feeds are fometimes ufed to
make Emulfions, but they have nothing of the narcotic virtues
of the other parts of the plant.

POPPY.

Yellow
Cambricum

POPPY. The capfules fmooth; oblong. Stem fmooth; fupporting many flowers. Leaves winged, jagged—*Summits five or fix; diftinct.* Bloffoms *yellow.*
Papaver luteum perenne, laciniato folio, Cambro-britannicum. *Ray's Syn.* 309.
Argemone lutea Cambro-britannica. *Park.* 369.
In mountainous countries. P. June—Auguft.

214 CELANDINE. 647 Chelidonium.

EMPAL. *Cup* two leaves roundifh; *little leaves* fomewhat egg-fhaped; concave; blunt; fhedding.
BLOSS. *Petals* four; circular; flat; expanding; large; narrower at the bafe.
CHIVES. *Threads* about thirty; flat; broader upwards; fhorter than the bloffom. *Tips* oblong; compreffed; blunt; upright; double.
POINT. *Seedbud* cylindrical; as long as the chives. *Shaft* none. *Summit* a knob; cloven.
S. VESS. *Pod* cylindrical; generally with two valves.
SEEDS. Many; egg-fhaped; fhining; adhering to the little ftalk that connects them with the receptacle.
   Receptacle narrow; fituated betwixt the feams of the valves, and applied clofe to the feams through their whole length. It continues entire.

  OBS. *The feedveffel of the fecond fpecies is a pod like two capfules; that of the third fpecies is a pod with three valves.*

Common
Majus

CELANDINE. With fruit-ftalks forming rundles—*Blof-foms yellow.*
Chelidonium majus. *Gerard.* 1069.
Chelidonium majus vulgare. *Park.* 616. *Bauh. pin.* 144.
Papaver corniculatum luteum, chelidonia dictum. *Ray's Syn.* 309.
1. Jagged leaved.
Chelidonium majus foliis quernis. *Bauh. pin.* 141.
Amongft rubbifh. P. June—July.
Cultivation produces double bloffoms and ftriped leaves.
  The juice of every part of this plant is very acrimonious. It cures Tetters and Ringworms. Diluted with milk it confumes white opake fpots upon the eye. It deftroys warts and cures the Itch. There is no doubt but a medicine of fuch activity will one day be converted to more important purpofes. Horfes, Cows, Goats, Sheep, and Swine refufe it.

CELANDINE.

CELANDINE. With fruit-ſtalk ſupporting a ſingle flower. **Horned** Leaves indented embracing the ſtem. Stem ſmooth—*The whole* **Glaucium** *plant of a bluiſh green colour.* Leaves *rather rough.* Root *leaves with winged clefts.* Flowers *yellow.* Pods *rough ; partitioned into cells.*

Papaver cornutum flore luteo. *Gerard.* 367.
Papaver corniculatum luteum. *Park.* 261. *Bauh. pin.* 171. *Ray's Syn.* 309.
Yellow horned Poppy.
On the ſea coaſt. A. July.

CELANDINE. With fruit-ſtalks ſupporting a ſingle flower. **Violet** Leaves with winged clefts; ſtrap-ſhaped. Stem ſmooth. Pods **Hybridum** with three valves—*This ſeems to have been originally produced by the duſt of a ſpecies of* Celandine *fertilizing the ſeedbud of the rough headed Poppy.* Bloſſoms *purple.*

Papaver corniculatum violaceum. *Bauh. pin.* 172. *Ray's Syn.* 309.
Papaver cornutum flore violaceo. *Gerard,* 367.
Papaver corniculatum flore violaceo. *Park.* 262.
Violet-coloured Horned Poppy.
Amongſt corn. A. Auguſt.

215 CHRISTOPHER. 644 Actæa.

EMPAL. *Cup* four leaves. *Little leaves* circular; blunt; concave; ſhedding.
BLOSS. *Petals* four; tapering each way; larger than the cup; ſhedding.
CHIVES. *Threads* about thirty; hair-like; broader towards the top. *Tips* roundiſh; double; upright.
POINT. *Seedbud* egg-ſhaped. *Shaft* none. *Summit* rather thick; obliquely depreſſed.
S. VESS. *Berry* betwixt oval and globular; ſmooth: with one furrow, and one cell.
SEEDS. Many; ſemi-circular; lopped on the inner ſide; ſtanding in a double row.

CHRISTOPHER. The flowers in egg-ſhaped bunches. **Baneberry** Fruit a pulpy berry—*Stem ſlender; jointed; ſcored.* Bloſſoms **Spicata** *white.* Berries *black.*

Chriſtophoriàna. *Gerard.* 969. *Ray's Syn.* 262.
Chriſtophoriana vulgaris. *Park.* 379.
Aconitum racemoſum. *Bauh. pin.* 182.
Herb Chriſtopher, or Baneberries.
In woods and hedges in Yorkſhire. P. May.—June.

This

This plant is a powerful repellent. The root is useful in some
nervous cafes, but it muft be adminiftered with caution. The
berries are poifonous in a very high degree. It is faid that toads
allured by the fœtid fmell of this plant refort to it; but it grows
in fhady places, and toads are fond of damp and fhady fituations.
—Sheep and Goats eat it ; Cows Horfes and Swine refufe it.

216 CISTUS. 673. Ciftus.

EMPAL. *Cup* five leaves ; permanent. *Little leaves* circu-
lar; concave; three of them large, with two fmall
ones interpofed.
BLOSS. *Petals* five; circular; flat; expanding ; large.
CHIVES. *Threads* numerous; hair like; fhorter than the
bloffom. *Tips* roundifh; fmall.
POINT. *Seedbud* roundifh. *Shaft* fimple; as long as the
chives. *Summit* flat; circular.
S. VESS. *Capfule* roundifh; covered by the cup.
SEEDS. Numerous; roundifh; fmall.

OBS. *The effential character of this genus confifts in the two alternate fmaller leaves of the Cup. In fome fpecies the capfule hath
one cell and three values; but in other fpecies it hath five or ten cells.*

Hoary
Marifolius

CISTUS. The ftem fomewhat woody without props; leaves
oppofite; oblong; on leaf ftalks; flat; hoary on the under, but
fmooth upon the upper furface—*Bloffoms yellow.*
Helianthemum alpinum folio Pilofellæ minoris Fuchfii. *Ray's
Syn.* 342.
Ciftus hirfutus. *Hudfon.* 206.
On rocks in the Nor   P July.
1. There is a variety in which the leaves are oval and a little
hairy on the upper furface. The angle-fhade Moth, *Phalæna me-
ticulofa,* feeds upon it.

Annual
Guttatus

CISTUS. The ftem herbaceous ; without props ; leaves oppofite; fpear-fhaped; three fibred. Flowers in bunches, without
floral leaves—*Bunches of flowers very flender; the two outer and
larger leaves of the cup fringed.* Petals *ftraw couloured with a red
fpot.*
Ciftus flore pallido, punicante macula infignito. *Bauh. pin.*
465. *Ray's Syn.* 343.
Ciftus annuus flore maculato. *Gerard.* 1281.
Ciftus annuus flore guttato. *Park.* 661.
In fandy fields. A. June.

CISTUS. The ftem fomewhat woody ; furnifhed with Narrow-leaved props ; trailing. Leaves oblong egg-fhaped ; a little hairy : Surrejanus Petals fpear-fhaped—*Bloffoms yellow.*

Helianthemum vulgare petalis florum peranguftis. *Ray's Syn.* 341.

In paftures and meadows. P. July.

CISTUS. The ftem fomewhat woody; trailing; with fpear- Sunflower fhaped props. Leaves oblong; a little hairy; rolled back at the Helianthemum edges—*Petals roundifh ; very entire ; yellow.* Flowers *in bunches and hanging down previous to their opening. There is fometimes, but not always, a deep orange coloured fpot at the bafe of the Petals.*

Helianthemum vulgare. *Park.* 656. *Ray's Syn.* 341.
Helianthemum anglicum luteum. *Gerard.* 1282.
Chamæciftus vulgaris flore luteo. *Bauh. pin.* 465.

1. There are two varieties, one with a white and another with
2. A rofe coloured flower. The leaves too are more or lefs broad.

Dwarf Ciftus. Little Sunflower.
Dry hills. P. July.
Goats, Sheep and Horfes eat it, Swine refufe it.

CISTUS. The ftem fomewhat woody ; trailing ; furnifhed Dwarf with props. Leaves oblong egg-fhaped ; downy. Cups fmooth. Polifolius Petals ferrated—*white.*

Chamæciftus montanus polii folio. *Ray's Syn.* 342.
Mountain dwarf ciftus.
On Brent-downs in Somerfetfhire, near the fevern fea. P. July.

217 LIME. 660 Tilia.

EMPAL. *Cup* with five divifions ; concave ; coloured ; almoft as large as the bloffom ; deciduous.

BLOSS. *Petals* five ; oblong ; blunt ; fcolloped at the end.

CHIVES. *Threads* many, (thirty or more,) awl-fhaped ; as long as the bloffom. *Tips* fimple.

POINT. *Seedbud* roundifh. *Shaft* thread-fhaped ; as long as the chives. *Summit* blunt ; with five edges.

S. VESS. *Capfule* like leather ; globular ; with five cells and five valves, opening at the bafe.

SEEDS. Solitary ; roundifh.

OBS. *In general only one feed comes to perfection, and this pufhes afide the others which are barren, fo that an incautious obferver would be apt to pronounce that the capfule hath but one cell.*

LIME.

Linden   LIME. The flowers without a honey-cup—*Leaves heart-*
Europæa   *shaped; serrated.* Blossoms *white.*
          Tilia vulgaris platyphyllos. *Ray's Syn.* 473.
          Tilia fæmina. *Gerard.* 1483. major. *Park.* 1407.
          Tilia fæmina folio majore. *Bauh. pin.* 426.
     1. With smaller leaves. *Bauh. pin.* 426.
     2. With soft hairy leaves, and four sided fruit. *Ray's Syn.* 473.
     3. With six sided seeds.
     4. With variegated leaves.
          Lime tree.   Linden tree.
          Woods and hedges. S. July.

     It flourishes best on the sides of hills, but it will live very well
in meadow grounds.   It is easily transplanted, and grass grows
beneath it; it is useful to form shady walks, and clipped hedges.
The wood is soft, light and smooth; close grained and not subject to the worm.   It makes good charcoal for gunpowder and
for designers.   It is used for leather-cutters boards and for carved
work: It is also employed by the turner.   The leaves are dried
in some countries as winter food for Sheep and Goats.   Cows eat
them in the autumn; but they give a bad taste to the milk. The
bark macerated in water, may be made into ropes, and fishing
nets.   The flowers are fragrant, and afford the best honey for
Bees.   The sap inspissated affords a quantity of sugar.

     It affords nourishment to the following insects.
          Lime-hawk Moth.   *Sphinx Tiliæ.*
          Small Egger Moth.   *Phalæna lanestris.*
          Gipsey Moth.   *Phalæna dispar.*
          White-spot tussock Moth. *Phalæna antiqua.*
          Dagger Moth.   *Phalæna Psi.*
          Buff-tip Moth. *Phalæna Bucephala.*
          Lime Louse.   *Aphis Tiliæ.*
          Tawny Tick.   *Acarus telarius.*

218 WATERLILY. 653 Nymphæa.

EMPAL. *Cup* beneath; with four leaves; large; coloured on the upper furface; permanent.

BLOSS. *Petals* numerous, (often fifteen or more), fitting on the fide of the feedbud; not all in one row.

CHIVES. *Threads* numerous, (often feventy), flat; crooked; blunt; fhort. *Tips* oblong; fixed to the borders of the threads.

POINT. *Seedbud* egg-fhaped; large. *Shaft* none. *Summit* circular; flat; central; fitting; marked with rays; fcolloped at the edge; permanent.

S. VESS. *Berry* hard; egg-fhaped; flefhy; rough; narrow at the neck; crowned at the top; with many cells (ten or fifteen) filled with pulp.

SEEDS. Many; roundifh.

OBS. *The firft fpecies differs from the fecond in having its cup compofed of five circular leaves, and the petals fmall.*

WATERLILY. The leaves heart-fhaped; very entire. Cup of five leaves; large—*Flowers on long fruit-ftalks.* Bloffoms yellow. **Yellow Lutea**

Nymphæa lutea. *Gerard.* 819. *Ray's Syn.* 368.
Nymphæa major lutea. *Bauh. pin.* 131. *Park.* 1252.
Ditches and flow Rivers. P. Auguft.
The roots rubbed with milk deftroy Crickets and Cockroaches.
—Swine eat it; Goats are not fond of it; Cows, Sheep and Horfes refufe it.

WATERLILY. The leaves heart-fhaped; very entire. Cup with four clefts—*The flower opens about feven in the morning; clofes about four in the afternoon and then lyes down upon the furface of the water. The fummits are fifteen or more; placed in a circle; and correfponding with as many cells in the feedveffel.* Bloffoms white. **White Alba**

Nymphæa alba. *Gerard.* 819. *Ray's Syn.* 368. Major. *Bauh. jin.* 193.
Nymphæa alba major vulgaris. *Park.* 1251.
In flow Rivers and Ponds. P. July.
The roots are ufed in Ireland and in the Ifland of Jura to dye a dark brown.—Swine eat it Goats are not fond of it; Cows and Horfes refufe it.
Both the fpecies fupport the Water Loufe, *Aphis aquatilis,* and the Water Wafp Beetle, *Leptura aquatica.*

Y   Order

Order III. Three Pointals.

219 LARKSPUR. 681 Delphinium.

EMPAL. *Cup* none.

BLOSS. *Petals* five; unequal; placed in a circle. The uppermoſt petal more blunt than the reſt before, but extended behind into a ſtraight, tubular, long horn; blunt at the end. The other petals are betwixt egg and ſpear-ſhaped; expanding; nearly equal.

> *Honey-cup* cloven; its front ſtanding in the upper part of the circle of the petals, and its hinder part covered by the tube of the uppermoſt petal.

CHIVES. *Threads* many, (fifteen or thirty) awl-ſhaped; broadeſt at the baſe; very ſmall; leaning towards the uppermoſt petal. *Tips* upright; ſmall.

POINT. *Seedbuds* three, or one; egg-ſhaped; ending in *Shafts* as long as the chives. *Summits* ſimple; reflected.

S. VESS. *Capſules* three, or one; betwixt egg and awl-ſhaped; ſtraight. With one valve, opening inwards.

SEEDS. Many; angular.

Wild
Conſolida

LARKSPUR. The honey-cup of one leaf. Stem ſub-divided.—*Bloſſoms blue; or red, or white, or pale.red, or violet.*
Delphinium ſegetum flore cæruleo. *Ray's Syn.* 273.
Conſolida regalis arvenſis, flore cæruleo. *Bauh. pin.* 142.
Conſolida regalis ſylveſtris. *Gerard.* 1083.
In corn-fields. A. June.
The expreſſed juice of the petals with the addition of a little alum makes a good blue ink. The ſeeds are acrid and poiſonous. When cultivated the bloſſoms often become double.—Sheep and Goats eat it; Horſes are not fond of it; Cows and Swine refuſe it. The Peale-bloſſom Moth, *Phalæna Delphinium*, lives upon it

Order V. Five Pointals.
220 COLUMBINE. 684 Aquilegia.

EMPAL. *Cup* none.

BLOSS. *Petals* five; betwixt fpear and egg fhaped; flat; expanding; equal.

Honey-cups five; equal; alternating with the petals; horned; gradually widening upwards; the mouth afcending obliquely outwards; fixed to the receptacle inwardly. The lower part extended into a long tapering tube, blunt at the end.

CHIVES. *Threads* many, (thirty or forty) awl-fhaped; the outer ones the fhorteft. *Tips* oblong; upright; as tall as the honey-cups

POINT. *Seedbuds* five; egg-fhaped but oblong; ending in awl-fhaped *fhafts*, longer than the chives. *Summits* upright; fimple.

Ten wrinkled, fhort, chaffy fubftances feparate and inclofe the feedbuds.

S. VESS. *Capfules* five; cylindrical; parallel; ftraight; taper; with one valve, opening from the point inwardly.

SEEDS. Many; egg-fhaped and keel-fhaped; fixed to the opening feam.

COLUMBINE. The honey-cups crooked *Bloffoms blue, pendant; fometimes double.*

Common Vulgaris

Aquilegia flore fimplici. *Ray's Syn.* 273.
Aquilegia fylveftris. *Bauh. pin.* 144.
Aquilegia fimplex. *Park.* 1367.
Aquilegia flore cæruleo. *Gerard.* 1093.
Woods and hedges. P. June.

This fpecies is cultivated in our gardens and becomes fufceptible of confiderable variety in the colour of the bloffom; as red; purple; white; light brown; blue and white; brown and white.

Goats eat it; Sheep are not fond of it; Cows, Horfes and Swine refufe it.

Mountain
Alpina

COLUMBINE. The honey-cup ftraight; fhorter than the fpear-fhaped petal—*Leaves growing in double fets; three in each; fmall. Little leaves with many clefts.* Bloffoms *blue.*

Aquilegia montana magno flore. *Bauh. pin.* 144. *Ray's Hift.* 70?

Woody mountains. P. June.

1. There are two varieties; one with fmall flowers, and another with
2. Very fmall flowers and white bloffoms.

Order VI. Six Pointals.

221 WATERSOLDIER. 687 Stratiotes.

Empal. *Sheath* two leaves, inclofing a fingle flower; com-preffed, blunt, approaching, keel-fhaped on each fide; permanent.

Cup one leaf with three divifions; upright; de-ciduous.

Bloss. *Petals* three; inverfely heart-fhaped; upright but expanding; twice as large as the cup.

Chives, *Threads* twenty; as long as the cup; fixed to the receptacle. *Tips* fimple.

Point. *Seedbud* beneath. *Shafts* fix, divided down to the bafe; as long as the chives. *Summits* fimple.

S. Vess. *Berry* covered with a capfule; oval, taper to-wards each end; with fix fides and fix cells.

Seeds. Many; oblong; covered; generally winged.

Common
Aloides

WATERSOLDIER. The leaves fword-fhaped, but triangu-lar; fringed with prickles—*Bloffoms white. The chives and point-als are generally found in the fame flower, but they have been fome-times obferved to be on different plants; and where they are found in the fame flower the Chives are fometimes without tips.*

Stratiotes foliis aloes, femine longo. *Rays Syn.* 290.
Stratiotes five militaris aizoides. *Park.* 1249.
Militaris aizoides. *Gerard.* 826.
Aloe paluftris. *Bauh. fin.* 286.
Water Aloe. Frefh water Soldier.
In flow ftreams and ditches. P. June.

A great variety of infects are nourifhed by this plant: fome of them purfue it down to the bottom of the water and devour the leaves.—Swine eat it; Goats refufe it.

Order.

Order VII. Many Pointals.

222 TRAVELLERSJOY. 696 Clematis.

EMPAL. *Cup* None.

BLOSS. *Petals* four; flexible; oblong.

CHIVES. *Threads* many; awl-fhaped; fhorter than the bloffom. *Tips* fixed to the fides of the threads.

POINT. *Seedbuds* many; roundifh; compreffed; ending in awl fhaped *Shafts.* longer than the chives. *Summits* fimple.

S. VESS. None. *Receptacle* a fmall knob.

SEEDS. Many; roundifh; compreffed; retaining the fhaft, which is varioufly fhaped.

TRAVELLERSJOY.   The leaves winged; little leaves Wild
heart-fhaped. Stem climbing—*Bloffoms white.*   Vitalba
Clematis fylveftris latifolia. *Bauh. pin* 300. feu Viorna. *Park.*
380.
Clematis latifolia feu atragene quibufdam. *Ray's Syn.* 258.
Viorna. *Gerard.* 886.
Great Wild Climber.   Virgin's Bower.
In hedges.   S. July.
1. There is a variety with toothed leaves;
2. And another with entire leaves,
The whole plant is acrid.

223 MEADOWRUE. 697 Thalictrum,

EMPAL. *Cup* none; unlefs you call the bloffom the cup.

BLOSS. *Petals* four; circular; blunt; concave; fhedding.

CHIVES. *Threads* many; broadeft in the upper part; compreffed; longer than the bloffom. *Tips* oblong; upright.

POINT. *Seedbuds* many; roundifh; often ftanding on little foot-ftalks. *Shafts* none. *Summits* thick.

S. VESS. None.

SEEDS. Many; furrowed; egg-fhaped; without awns.

OBS. *In fome fpecies there are long fhafts.*

MEADOW-

MANY CHIVES.

Mountain
Alpinum

MEADOWRUE. The stem undivided; almost naked; flowers in a simple bunch, terminating—*Chives twelve. Pointals eight.* Blossoms *deep red.*

Thalictrum minimum montanum atro-rubens, foliis splendentibus. *Ray's Syn.* 204.

In wet places on mountains. P. June.

Lesser
Minus

MEADOWRUE. The leaves with six divisions. Flowers on crooked fruit-stalks—*Tips of the leaves purple, and the stem clouded with blue.* Blossoms *yellowish white.*

Thalictrum minus. *Gerard.* 1251. *Park.* 264. *Bauh.* ½ in. 337. *Ray's Syn.* 203.

1. There is a variety with broader leaves. *Ray's Syn.* 204.

In high pastures. P. July-August.

Common
Flavum

MEADOWRUE. The stem leafy; furrowed. Flowers upright; forming a panicle with many divisions—*Chives twenty-four. Pointals from ten to sixteen.* Blossoms *whitish.* Root *yellow.*

Thalictrum majus vulgare. *Park.* 263.

Thalictrum seu thalitrum majus. *Gerard.* 1251. *Ray's Syn.* 203.

Thalictrum majus siliqua angulosa aut striata. *Bauh. pin* 336. Moist meadows. P. June.

A Cataplasm made of the leaves has been known to give relief in the Sciatica. The root dyes wool yellow.—Cows, Horses, Goats and Sheep eat it; Swine are not fond of it.

224 HELLEBORE. 702 Helleborus.

EMPAL. *Cup* none: unless you reckon the blossom such which in some species is permanent.

BLOSS. *Petals* five; circular; blunt; large.

Honey-cups many; very short; placed in a circle; consisting of one leaf; tubular; narrowest beneath. *Mouth* with two lips; upright; broken at the margin; the inner lip the shortest.

CHIVES. *Threads* numerous; awl-shaped. *Tips* compressed; narrowest in the lower part; upright.

POINT. *Seedbuds* generally six; compressed. *Shafts* awl-shaped. *Summits* rather thick.

S. VESS. *Capsules* compressed; keel-shaped at both edges; the lower edge the shortest; the upper the most convex; opening.

SEEDS. Several; round; fixed to the seams.

HELLEBORE.

HELLEBORE. The ſtem leafy; ſupporting many flowers. Green
Leaves fingered—*Bloſſoms terminating; green.* Stem *upright; about* Viridis
a foot high. The root leaves often grow longer than the ſtem.
Helleborus niger hortenſis flore viridi. *Bauh. pin.* 185. *Ray's
Syn.* 271.
Helleboraſter minor, flore viridante. *Park.* 212.
Helleboraſtrum. *Gerard,* 976.
Wild black Hellebore.
In woods and high paſtures. P. April.

HELLEBORE. The ſtem leafy; ſupporting many flowers. Setterwort
Leaves like a bird's foot—*Bloſſoms ſomewhat globular, green; ſome-* Fœtidus
times tinged with purple at the edges. Stem *about a yard high.*
Helleborus niger fœtidus· *Bauh. pin.* 185.
Helleboraſter maximus. *Gerard* 976. *Ray's Syn.* 271.
Helleboraſter maximus, ſeu conſiligo. *Park.* 212.
Great baſtard black Hellebore. Bears-foot. Setterwort.
1. Leaves with three diviſions.
Fields and hedges. P. March.
The dried leaves are frequently given to Children to deſtroy
worms, but they muſt be uſed ſparingly being violent in the
operation, and inſtances of their fatal effects are recorded. The
country people put the root into ſetons made through the dew-
laps of Oxen.

225 MEADOWBOUTS. 703 Caltha.

EMPAL. *Cup.* none.
BLOSS. *Petals* five; egg-ſhaped; nearly flat; expanding;
large; ſhedding.
CHIVES. *Threads* numerous; thread-ſhaped; ſhorter than
the petals. *Tips* compreſſed; blunt; upright.
POINT. *Seedbuds* from five to ten; oblong; compreſſed;
upright. *Shafts* none. *Summits* ſimple.
S.VESS. *Capſules* from five to ten; ſhort; tapering; expanding; keel-ſhaped at both edges; opening at
the upper ſeam.
SEEDS. Many; roundiſh, enlarging; fixed to the upper ſeam.

OBS. *This genus differs from the* Crowfoot *in wanting a cup,
from the* Hellebore *in having no Honeycup, and from both in having
a double row of chives.*

Marigold
Paluftris

MEADOWBOUTS. As there is only one fpecies known Linnæus gives no defcription of it—*The inner row of chives with broad tip ; the outer row twice as long, club-fhaped, and the tips compreffed. Leaves kidney-fhaped. Bloffoms yellow. Petals from five to feven ; fomewhat concave. Capfules fometimes more than ten.*

Caltha paluftris flore fimplici. *Bauh. pin.* 276.
Caltha paluftris vulgaris fimplex. *Park.* 1213.
Caltha paluftris major. *Gerard.* 817.
Populago. *Ray's Syn.* 272.

1. There is a variety that is fmaller. The bloffom is fometimes double.

Marfh Marigold. Meadow-bouts.
Wet meadows. P. April.

The flowers gathered before they expand, and preferved in falted vinegar are a good fubftitute for Capers. The juice of the petals boiled with a little alum ftains paper yellow. The remarkable yellownefs of butter in the fpring is fuppofed to be caufed by this plant ; but Cows will not eat it unlefs compelled by extreme hunger, and then Boerhaave fays, it occafions fuch an inflammation that they generally die. Upon May-day the country people ftrew the flowers upon the pavement before their doors.— Goats and Sheep eat it : Cows, Horfes and Swine refufe it,

226 ANEMONE.   694 Anemone.

EMPAL. None.

BLOSS. *Petals* in two or three rows ; three in each row ; rather oblong.

CHIVES. *Threads* numerous ; hair-like ; half as long the bloffom. *Tips* double : upright.

POINT. *Seedbuds* numerous ; forming a knob. *Shafts* taper. *Summits* blunt.

S.VESS. None. *Receptacle* globular or oblong ; with hollow dots.

SEEDS. Many ; taper ; retaining the fhaft.

Pafque
Pulfatilla

ANEMONE. The fruit-ftalks furnifhed with a fort of diftant fence. Petals ftraight. Leaves doubly winged—*Petals fpearfhaped : leaves hairy ; finely divided. Bloffoms folitary ; purple.*

Pulfatilla folio craffiore et majore flore. *Bauh. pin.* 177. *Ray's Syn.* 260.
Pulfatilla vulgaris. *Park.* 341. *Gerard* 385.
Pafque flower.
In high paftures. P. April.

The whole plant is acrid and blifters the fkin. The juice of the petals ftains paper green.

Goats and Sheep eat it ; Horfes, Cows and Swine refufe it.
ANEMONE.

ANEMONE. The feeds fharp; little leaves jagged; ftem Wood supporting a fingle flower—*Bloffoms white, with a tinge of purple.* Nemorofa
The varieties are

1. Double white. 2. Purple. 3. Reddifh purple. 4. Double purple.
Anemone nemorofa flore majore. *Bauh. pin.* 176.
Anemone nemorum album. *Gerard.* 383. *Ray's Syn.* 259,
Ranunculus nemorofus albus fimplex. *Park.* 325.
Woods and hedges. P, April.
The flowers fold up in a curious manner againft rain. The whole plant is acrid. When Sheep that are unaccuftomed to it eat it, it brings on a bloody flux.
Goats and Sheep eat it; Horfes, Cows and Swine refufe it.

ANEMONE. The feeds fharp; little leaves jagged; petals Mountain fpear-fhaped; numerous—*Blue.* Apeñnina
The varieties are

1. Large leaved blue. 2. Double blue. 3. Double violet co-loured. 4. Large leaved white.
Anemone geranii robertiani folio cærulea. *Bauh. pin.* 174.
Ranunculus nemorofus, flore purpureo-cæruleo. *Park.* 325.
Ray's Syn. 259.
Mountain-wood Anemone.
In woods. P. April.

227 GLOBEFLOWER. Trollius 700.

EMPAL. None.
BLOSS. *Petals* about fourteen; nearly egg-fhaped; deci-duous; three in each of the three outer rows, and five in the innermoft row. *Honey-cups* nine; ftrap-fhaped; flat; crooked; perforated on the inner-fide at the bafe.
CHIVES. *Threads* numerous; briftle-fhaped; fhorter than the bloffom. *Tips* upright.
POINT. *Seedbuds* numerous; fitting; like pillars. *Shafts* none. *Summits* fharp-pointed; fhorter than the chives.
S.VESS. *Capfules* numerous; forming a knob; egg-fhap-ed, with a crooked point.
SEEDS. Solitary.

GLOBEFLOWER.

Gowlans
Europæus

GLOBEFLOWER. The petals approaching: the honey-cup as long as the chives—*Bloſſoms globular ; yellow.*

Ranunculus globoſus. *Gerard.* 955. *Park.* 218. *Ray's Syn.* 272.

Ranunculus montanus aconiti folio, flore globoſo. *Bauh. pin.* 182..

Locker-Gowlans.

On the ſides of mountains. P. May—June,

It is cultivated in our flower gardens.

Goats, Sheep and Swine eat it ; Cows and Horſes refuſe it.

228. CROWFOOT.   699. Ranunculus.

EMPAL. *Cup* five leaves ; egg-ſhaped ; concave ; a little coloured ; deciduous.

BLOSS *Petals* five : blunt ; ſhining ; with ſmall claws.

Honey-cup a little cavity, juſt above the claw of each petal.

CHIVES. *Threads* many ; nearly half as long as the petals.
Tips upright ; oblong ; blunt ; double.

POINT. *Seedbuds* numerous, forming a knob. *Shafts* none.
Summits reflected ; very ſmall.

S. VESS. None. *Receptacle* connecting the ſeeds by very ſhort foot-ſtalks.

SEEDS. Many ; irregular ; uncertain ſhaped ; crooked at the point.

OBS. *The eſſential character of this genus conſiſts in the* honey-cup ; *the other parts of the flower are inconſtant. This* honey-cup *is in ſome ſpecies a naked pore ; in others the pore is encompaſſed by a cylindrical border, and in others it is cloſed by a ſcale which is notched at the end.*

In the third ſpecies the cup hath three leaves and the bloſſoms more than five petals. The eleventh ſpecies hath only five chives. The fifth ſpecies hath an awl-ſhaped receptacle, and the fruit in a ſpike.

In ſome ſpecies the ſeeds are roundiſh, in others depreſſed ; ſometimes they are beſet with prickles like a Hedge-hog ; and ſometimes they are but few in number.

Leaves undivided.

CROWFOOT. The leaves betwixt egg and spear-shaped; Spear-leaved on leaf-stalks. Stem declining—*Blossoms deep yellow.* Flammula
Ranunculus flammeus minor. *Gerard.* 961. *Ray's Syn.* 250,
Ranunculus flammeus serratus. *Gerard.* 961. *Park.* 1214.
Ranunculus palustris flammeus minor five angustifolius. *Park.* 1214.
1. Leaves serrated.
Ranunculus longifolius palustris minor, item palustris serratus. *Baub. pin.* 180.
Lesser Spear-wort.
In marshy places. P. June—September.
It is very acrid. Applied externally it inflames and blisters the skin.
Horses eat it; Cows, Sheep, Goats and Swine refuse it.

CROWFOOT. The leaves spear-shaped: stem upright— Great *Blossoms deep yellow; terminating.* Lingua
Ranunculus flammeus major. *Gerard.* 961. *Ray's Syn.* 250.
Ranunculus palustris flammeus major. *Park.* 1215,
Ranunculus longifolius palustris major. *Baub. pin.* 180.
Great Spear-wort.
Moist Meadows. P. May.

CROWFOOT. The leaves heart-shaped; angular; on leaf- Pilewort stalks. Stem supporting a single flower—*Petals generally eight;* Ficaria *spear-shaped.* Cup *composed of three leaves.* Blossoms *yellow.*
Chelidonium minus. *Gerard.* 816. *Park.* 617. *Ray's Syn.* 246.
Chelidonia rotundifolia minor. *Baub. pin,* 309.
Ficaria verna. *Hudson.* 214.
Lesser Celandine.
Meadows and Pastures. P. April.
The young leaves may be eaten in the Spring along with other pot-herbs. It destroys other plants that grow near it.
Goats and Sheep eat; Cows and Horses refuse it.
The Celandine Weevils, *Curculio Dorsalis* is found upon it.

** *Leaves dissected and divided.*

CROWFOOT. The root leaves kidney-shaped; scolloped; Wood jagged. Stem leaves fingered, strap-shaped. Stem supporting se- Auricomus veral flowers.—*Cup united at the base, and scarce perceptibly woolly. The first flowers have sometimes only one or two petals, but as the Spring advances they have always five. Stem once or twice forked.* Blossoms *yellow.*
Ranunculus auricomus. *Gerard.* 954,
Ranunculus nemorosus dulcis. *Park.* 326. *Ray's Syn.* 248.
Ranunculus nemorosus vel sylvaticus folio rotundo. *Baub. pin.* 178.
Sweet Wood Crowfoot. Goldilocks.
Goats and Cows eat it; Horses and Sheep refuse it.

CROW-

Round-leaved
Sceleratus

CROWFOOT. The lower leaves hand-shaped; upper leaves fingered. Fruit oblong—*Bloſſoms yellow.*

Ranunculus paluſtris rotundifolius. *Gerard.* 962. *Ray's Syn.* 249.

Ranunculus paluſtris ſardonius lævis. *Park.* 1215.

Ranunculus paluſtris apii folio lævis. *Bauh. pin.* 180.

Round-leaved Water Crowfoot.

In ſhallow Waters. A. May—June.

The whole plant is very corroſive, and beggars are ſaid to uſe it to ulcerate their feet, which they expoſe in that ſtate to excite compaſſion.

Goats eat it; Cows Horſes and Sheep refuſe it.

Bulbous
Bulboſus

CROWFOOT. The leaves of the cup bent backwards; fruit-ſtalks furrowed; ſtem upright, ſupporting many flowers; leaves compound—*Chives about eighty, the whole plant hairy, more or leſs.* Roots *bulbous.* Bloſſoms *pale yellow.*

Ranunculus bulboſus. *Gerard.* 953. *Park.* 329. *Ray's Syn.* 247.

Ranunculus pratenſis, radice Verticilli modo rotunda. *Bauh. pin.* 179.

1. There is a variety with hairy and paler leaves. *Ray's Syn.* 247.

Butter flower. Gold-cup.

Cows and Horſes have a great averſion to it.

Creeping
Repens

CROWFOOT. The cups open; fruit-ſtalks furrowed; ſuckers creeping; leaves compound—*Leaf-ſtalks flatted at the baſe.* Cups *ſmooth. Bloſſoms cloſe during rain, but do not hang down.* Leaves *ſmooth, ſhining.* Bloſſoms *bright yellow.* Honey-cup *very large.*

Ranunculus pratenſis repens. *Park.* 329. *Ray's Syn.* 247. hirſutus. *Bauh. pin.* 179.

Ranunculus pratenſis. *Gerard.* 951.

Butter-cups.

Paſtures. P. May.

Goats eat it ; Horſes refuſe it.

CROWFOOT. The cups open; fruit-ftalks cylindrical; Upright leaves with three divifions and many clefts. The upper leaves *ftrap-fhaped—Hairs on the ftem preffed clofely down. Leaves of the cup hairy; not joined at the bafe. Segments of the leaves fmooth. Chives about ninety. Threads fhorter than the leaves of the cup. Bloffoms bright yellow.* — in margin: Acris

Ranunculus pratenfis erectus acris. *Bauh. pin.* 178. *Ray's Syn.* 248.

Ranunculus pratenfis erectus acris vulgaris. *Park.* 329.

Ranunculus furrectis cauliculis. *Gerard.* 951.

Upright Meadow Crowfoot.

Paftures. P. June—July.

Cows and Horfes leave this plant untouched though their pafture be ever fo bare. It is very acrid, and its acrimony rifes in diftillation. Some years ago a man travelled in feveral parts of England adminiftering Vomits, which like white Vitriol, operated the inftant they got down into the Stomach. The diftilled water of this plant was his medicine.

Sheep and Goats eat it; Cows, Horfes and Swine refufe it.

CROWFOOT. The feeds prickly; upper leaves doubly compound; ftrap-fhaped--*Chives about fixteen.* Bloffoms *pale yellow.* — in margin: Corn Arvenfis

Ranunculus arvorum. *Park.* 328. *Gerard.* 951. *Ray's Syn.* 249.

Ranunculus arvenfis echinatus. *Bauh. Pin.* 179.

Corn-fields. A. June.

CROWFOOT. The feeds covered with fharp points. Leaves fimple; jagged; fharp; rough with hair. Stem fpreading— *Bloffoms pale yellow.* — in margin: Small-flowerd Parviflorus

Ranunculus hirfutus annuus flore minimo. *Ray's Syn.* 248. Tab. 12. fig. 1.

Paftures and Corn-fields, in gravelly foil. A. May.

CROWFOOT. The leaves nearly circular; with three lobes; very entire. Stem creeping—*Bloffoms white or pale ftraw colour.* — in margin: Ivy leaved Hederaceus

Ranunculus aquatilis hederaceus albus. *Ray's Syn.* 249.

Ranunculus aquatilis hederaceus luteus. *Bauh. pin.* 180.

Ranunculus hederaceus aquaticus. *Park.* 1216.

Ranunculi aquatilis varietas altera. *Gerard.* 830.

Ivy-leaved Water Crowfoot.

Ditches and flow ftreams. P. May.

Water   CROWFOOT. The leaves which are under the water hair-
Aquatilis   like; thofe above the water have central leaf-ftalks—*Petals have
not a fcale at the bafe but a little hole filled with Honey. Bloffoms
*on fruit-ftalks which arife from the fame fheath with the leaves
do; white, or pale ftraw colour.*
    Ranunculus aquatilis.   *Gerard.* 829. *Ray's Syn.* 249.
    Ranunculus aquaticus, hepaticæ facie.   *Park.* 1216.
    Ranunculus aquaticus folio rotundo et capillaceo. *Bauh pin.*
180.
    Various leaved Water Crowfoot.
    Ditches, ponds, rivers. P. April—May.
    Cows, Horfes, Goats Sheep and Swine eat it.

229 PHEASANT-EYE. 698 Adonis.

EMPAL. *Cup* five leaves; blunt; concave; a little co-
        loured; deciduous.
BLOSS. *Petals* five to fifteen; oblong; blunt; fhining.
CHIVES. *Threads* many; very fhort; awl-fhaped. *Tips*
        oblong; bent inwards.
POINT. *Seedbuds* numerous; forming a knob. *Shafts*
        none. *Summits* fharp; reflected.
S. VESS. None. *Receptacle* oblong; fpiked.
SEEDS. Numerous; irregular; angular; hunched at the
        bafe; crooked at the top; without awns; but hav-
        ing a fmall projection.

Autumnal
Autumnalis   PHEASANT EYE. Flowers with eight petals. Fruit nearly
cylindrical—*Bloffoms fcarlet.*
    Adonis hortenfis flore minore atro-rubente. *Bauh. pin.* 178.
    Flos adonis.   *Park.* 293. *Ray's Syn.* 251.
    Flos adonis. flore rubro. *Gerard.* 387.
    Adonis Flower. Red Maithes. Red Morocco.
    Corn-fields. A. June—July.
    Its beautiful fcarlet bloffoms have caufed it to be admitted in-
to our gardens.

CLASS XIV.

THE effential character of this Clafs confifts in the Flowers being furnifhed with four Chives, two of which are *long*, and two *fhort*. The fhort Chives ftand next together and adjoining to the fhaft of the Pointal. They are covered by the Bloffom, which is irregular in its fhape.——This Clafs comprehends the *Whorled Plants*, the *lipped*, the *mafked*, the *gaping* and the *grinning* flowers of other authors. It admits of the following NATURAL CHARACTER.

EMPAL. *Cup* one leaf; upright; tubular, with five clefts; *Segments* unequal; permanent.

BLOSS. One petal; upright. The bafe tubular, containing honey, and ferving for a honey cup.
    Border generally gaping; the *upper Lip* ftraight; the *lower Lip* expanding, with three *Segments*; the middle one the broadeft.

CHIVES. *Threads* four; ftrap-fhaped; fixed to the tube of the bloffom but leaning towards the back of it. The threads are all parallel; feldom taller than the bloffom. The two middle threads are fhorter than thofe on each fide.
    Tips generally covered by the upper lip of the bloffom, and approaching each other, fo as to ftand in pairs.

POINT. *Seedbud* generally fuperior. *Shaft* fingle; threadfhaped; bent in the fame manner as the chives and generally ftanding in the midft of them, but fomewhat longer, and a little crooked at the top. *Summit* generally cloven.

S. VESS. Either none, (as in the firft order,) but when there is one, (as in the fecond order,) it generally confifts of two cells.

SEEDS. In the *firft Order* four, feated at the bottom of the cup. In the *fecond Order* many; fixed to the receptacle, which is placed in the middle of the feed-veffel.

OBS. *The flowers of this Clafs are for the moft part nearly upright, but leaning a little from the ftem, fo that the bloffom might more effectually cover the tips from the rain, and the duft more eafily fall upon the fummit. The plants in the firft order of this Clafs are odoriferous, cephalic and refolvent. None of them are poifonous.*

CLASS XIV.

TWO CHIVES LONGER.

Order I. Seeds naked.

* Cups generally with five Clefts.

230 MOTHERWORT. *Tips* sprinkled with particles as hard as bone.

231 GILL. - - *Tips* in pairs; each pair forming a crofs.

232 MINT. - - *Threads* diftant; ftraight. *Bloff.* nearly equal.

233 VERVAIN. - *Bloffom* nearly equal. The upper fegment of the cup the fhorteft.

234 GERMANDER. *Bloffom* without any upper lip, but the upper fegment of the petal divided.

235 BUGLE. - - *Bloffom* with the upper lip fhorter than the chives·

236 BETONY. - *Bloffom* with the upper lip flat; afcending. *Tube* cylindrical. *Chives* as long as the mouth of the tube.

237 ARCHANGEL. *Bloffom* with a briftle-fhaped tooth on each fide the lower lip.

238 ALLHEAL. - *Bloffom* with two taper teeth upon the lower lip.

239 CLOWN-

239 Clownheal. *Blossom* with the lateral segments of the lower lip reflected. *Chives* after shedding their dust turned to the sides.

240 Nap. - - *Blossom* with the lower lip scolloped. *Mouth* with the edge reflected.

241 Henbit. - - *Cup* with ten scores. *Blossom* with the upper lip vaulted.

242 Horehound. *Cup* with ten scores. *Blossom* with the upper lip flat and straight.

† *Marjoram.*

** *Cups with two Lips.*

243 Hoodwort. - *Cup* after flowering appears as if closed with a cover, resembling a helmet.

244 Thyme. - - *Cup* closed at the mouth with soft hairs.

245 Selfheal. - *Threads* forked at the end.

246 Marjoram. *Cups* inclosed in a tiled cone.

247 Basil. - - *Cups* inclosed in a fence.

248 Balmleaf. *Cup* wider than the tube of the blossom. The *upper Lip* of the blossom flat and entire. *Tips* crossing each other.

249 Calamint. - *Cup* angular; skinny; upper lip ascending.

Order II. Seeds covered.

* *Caps with two Clefts.*

250 Broomrape. *Capsules* with one cell. *Bloss.* nearly equal; with four clefts. A *Gland* at the base of the seed-bud.

** *Cups with four Clefts.*

251 Toothwort. *Capsules* with one cell. *Bloss.* gaping. A *Gland* at the base of the feed-bud.

252 Paintedcup. *Capsule* with two cells. *Bloss.* gaping. *Cup* coloured.

Z

253 Eye-

253 EYE-BRIGHT.   *Capfule* with two cells. *Bloff.* gaping. Lower *Tips* thorny.

254 RATTLE.   *Capfule* with two cells. *Bloff.* gaping. *Capfule* compreffed.

255 COWWHEAT.   *Capfule* with two cells. *Bloff.* gaping. *Seeds* two; hunched.

*** *Cups with five Clefts.*

256 MUDWEED.   *Capfule* with one cell. *Bloff.* bell-fhaped; regular. *Seeds* many.

257 FIGWORT.   -   *Capfule* with two cells. *Bloff.* facing upwards. The middle fegment hath another within it.

258 COUNTERWORT.   *Capfule* with two cells. *Bloff.* wheel-fhaped. *Chives* placed two and two, approaching.

259 FOXGLOVE.   *Capfule* with two cells. *Bloff.* bell-fhaped, diftended on the under-fide. *Chives* declining.

260 TOADFLAX   -   *Capfules* with two cells. *Bloff.* gaping; with projecting honey cup beneath.

261 LOUSEWORT.   *Capfules* with two cells. *Bloff.* gaping *Seeds* coated.

230 MOTHERWORT. 722 Leonurus.

EMPAL. *Cup* one leaf; tubular; cylindrical, but angular; with five edges, and five teeth; permanent.

BLOSS. One petal; gaping. *Tube* narrow. *Border* opening with a long mouth. *Upper Lip* the longest; semi-cylindrical; concave; hunched; roundish and blunt at the end; entire: covered with soft hairs. *Lower Lip* reflected; with three divisions. *Segments* spear-shaped; nearly equal.

CHIVES. *Threads* four: (two long and two short.) covered by the upper lip. *Tips* oblong; compressed; cloven half way down: fixed sideways to the threads; sprinkled with very small, solid, shining, elevated, globular particles.

POINT. *Seedbuds* four *Shaft* thread-shaped; agreeing in length and situation with the chives. *Summits* cloven; sharp.

S.VESS. None. The *Cup* remaining unchanged, contains the seed within it.

SEEDS. Four; oblong; convex on one side; angular on the other.

MOTHERWORT. The stem leaves spear-shaped, with three Cordial lobes.—*The particles upon the Tips are very conspicuous.* Blossoms Cardiaca purplish.

Cardiaca. *Park.* 41. *Ray's Syn.* 239.
Marrubium cardiaca dictum. *Bauh. pin.* 230.
1. There is a variety with curled leaves.
Dunghills and amongst rubbish. B. July.
The leaves have a strong, but not an agreeable smell, and a bitter taste.
Goats, Sheep and Horses eat it; Cows are not fond of it; Swine refuse it.

231 GILL. 714 Glecoma

EMPAL. *Cup* one leaf; tubular; cylindrical; scored; very small; permanent. Rim with five teeth; taper; unequal.

BLOSS. One petal; gaping, *Tube* slender: compressed. *Upper Lip* upright; blunt; with a shallow cleft. *Lower Lip* expanding; large; blunt; in three segments; the middle segment largest and notched at the end.

CHIVES *Threads* four; (two long and two short;) covered by the upper lip. *Tips* of each pair of chives approaching so as to form a cross.

POINT. *Seedbud* cloven into four. *Shaft* thread-shaped; leaning under the upper lip. *Summit* cloven; sharp.

S. VESS. None. The seeds lye at the bottom of the cup.

SEEDS. Four; egg-shaped.

OBS. *The chives are sometimes imperfect, consisting of threads only half the usual length, and terminated by a reddish blunt point.*

Ground-Ivy
Hederacea

GILL. The leaves kidney-shaped; scolloped—*Little protuberances composed of many cells, are sometimes found upon the leaves and are occasioned by insects.* Stems *four cornered; trailing.* Blossoms blue.

Cultivation hath occasioned the following varieties.
1. Blossoms purple. 2. Blossoms white. 3. Leaves small and elegant.

Hedera terrestris. *Gerard.* 856. vulgaris. *Park.* 676. *Bauh. pin.* 306.

Calamintha humilior, folio rotundiore. *Ray's Syn.* 243

Hedera terrestris montana. *Bauh. jin.* 306. *Park.* 677.

Ground Ivy. Cats-foot. Ale-hoof. Tun-hoof. Robin run in the hedge.

Hedges and shady places. P. May.

The leaves are thrown into the vat with ale, to clarify it and to give it a flavour. Ale thus prepared is often drank as an Antiscorbutic. The expressed juice mixed with a little wine, and applied morning and evening, destroys the white specks upon Horses eyes. The plants that grow near it do not flourish. It is said to be hurtful to Horses if they eat much of it.

Sheep eat it; Horses are not fond of it; Cows, Goats and Swine refuse it.

The furbelow Moth, *Phalæna libatrix* and the *Cynips Glechomæ* live upon it.

SEEDS NAKED. 341

232 MINT. 713 Mentha.

EMPAL. *Cup* one leaf; tubular; upright; with five teeth; equal; permanent.

BLOSS. One petal; upright; tubular; rather longer than the cup. *Border* with four divisions, nearly equal. The *Upper Segment* the broadest and notched at the end.

CHIVES. *Threads* four; awl-shaped; upright; distant; the two next each other the longest. *Tips* roundish.

POINT. *Seedbud* cloven into four. *Shaft* thread-shaped; upright; longer than the blossom. *Summit* cloven; expanding.

S.VESS. None. *Cup* upright; containing the seeds.

SEEDS. Four; small.

OBS. *In the* Water-mint *the chives are all nearly of the same length.*

MINT, with flowers in oblong spikes; leaves oblong; downy; serrated; sitting. Chives longer than the blossom—*Leaves whitish.* Blossoms *pale purple.*   Long-leaved Sylvestris

Mentha sylvestris folio longiore. *Bauh. pin.* 227.

Menthastrum spicatum folio longiore candicante. *Ray's Syn.* 234.

Mentha longifolia. *Hudson.* 221.

1. There is a variety that is hairy with a broader spike. *Ray's Syn.* 234.

Long-leaved Horse-mint.

Marshes. P. August.

Z 3   MINT.

Spear
Viridis

MINT, with flowers in oblong spikes. Leaves spear-shaped; naked; serrated; sitting. Chives longer than the blossom—*Nearly allied to the preceding species but smaller and smoother.* Blossoms *purplish red.*

Mentha angustifolia spicata. *Bauh. pin.* 227.

Mentha angustifolia glabra spicata, folio rugosiore, odore graviore. *Ray's Syn.* 233.

Mentha spicata. *Hudson.* 221.

1. There is one variety with a broad spike. *Ray's Syn.* 233. and another

2. With broader leaves. *Ray's Syn.* 234.

3. Cultivation produces leaves striped with white, or yellow: or curled leaves.

Banks of rivers. P. August.

The flavour of this species being more agreeable than that of the others, it is generally preferred for Culinary and Medicinal purposes. A conserve of the leaves is very grateful, and the distilled waters both simple and spirituous are universally thought pleasant. The leaves are used in spring sallads; and the juice of them boiled up with sugar is formed into tablets. The distilled waters and the essential oil are often given to stop reachings, and frequently with success. From the circumstances noticed under the ninth species, it has been imagined, that Cataplasms and Fomentations of Mint, would dissolve coagulations of milk in the breasts; but *Dr. Lewis* says, that the curd of milk, digested in a strong infusion of Mint, could not be perceived to be any otherwise affected than by common water: however, milk in which Mint leaves were set to macerate, did not coagulate near so soon as an equal quantity of the same milk kept by itself. *Dr. Lewis* says that Dry Mint digested in rectified spirits of wine, gives out a tincture, which appears by day-light of a fine dark green, but by candle-light of a bright red colour. The fact is that a small quantity of this tincture is green either by daylight or by candle-light, but a large quantity of it seems impervious to common day-light; however when held betwixt the eye and a candle, or betwixt the eye and the sun it appears red.

Round-leaved
Rotundifolia

MINT, with flowers in oblong spikes. Leaves circular; wrinkled; scolloped; sitting—*Blossoms pale red; or purple.*

Mentha sylvestris rotundiore folio. *Bauh. pin.* 227.

Menthastrum folio rugoso rotundiore spontaneum, flore spicato, odore gravi. *Ray's Syn.* 234.

Menthastrum. *Gerard.* 683.

Round-leaved Horse-mint.

In marshes. P. August.

MINT.

MINT with flowers in heads., Leaves egg-fhaped; ferrated; Round-headed downy; almoft fitting. Chives longer than the bloffom—*Pointal* Hirfuta *longer than the bloffom, which is nearly equal and deeply divided into four parts.* Cups *fringed.* Bloffoms *purplifh white.*

Sifymbrium hirfutum. *Ray's Syn.* 233.

1. Outfide of the bloffom and the leaves and ftem hairy.

Mentha hirfuta. *Hudfon* 223.

2. There is a variety with leaves more circular. *Ray's Syn.* 233. Tab. 10. fig. 1

In watery places. P. Auguft.

Linnæus fays the chives arc *longer* than the bloffom, and refers to Hudfon's Flora Anglica; but Hudfon fays the chives are *fhorter* than the bloffom.

MINT with flowers in heads. Leaves egg-fhaped; ferrated; Water on leaf-ftalks. Chives longer than the bloffom—*differing but lit-* Aquatica *tle in length*; *not hairy.* Flowers *in terminating whorls, fo crowded together as to refemble a head or a blunt fpike.* Bloffoms *pale red.*

Mentha aquatica feu fifymbrium. *Gerard.* 684. *Ray's Syn.* 233.

Mentha aquatica rubra. *Park.* 1243.

Mentha rotundifolia paluftris, feu aquatica major. *Bauh. pin.* 227.

1. The leaves are fometimes ftriped.

Banks of rivers. P. July—Auguft.

Horfes eat it; Swine refufe it.

MINT with flowers in heads. Leaves egg-fhaped, on leaf- Pepper ftalks. Chives fhorter than the bloffom—*Stem upright.* Bloffoms Piperita *purplifh red. Under each whorl of flowers there are two pointed, fpear-fhaped, hairy floral leaves.*

Mentha fpicis brevioribus et habitioribus, foliis menthæ fufcæ fapore fervido piperis. *Ray's Syn.* 234. Tab. 10. fig. 2.

1. It varies in the leaves being more or lefs pointed.

Brooks and watery places. P. Auguft.

The ftem and leaves are befet with numbers of very minute glands containing the effential oil; which rifes plentifully in diftillation. The Pepper Mint water is well known as a carminative and antifpafmodic. Junipers effence of Pepper Mint is an elegant medicine, and poffeffes the moft active properties of the plant.

MINT with flowers in whorls; leaves egg-fhaped; rather Curled fharp; ferrated. Chives longer than the bloffom—*Stems trailing*; Sativa *leaves often curled at the edges.* Bloffoms *purple.*

Mentha verticillata. *Ray's Syn.* 232. *Hudfon.* 222.

Mentha crifpa. *Park.* 31. verticillata. *Bauh. pin.* 227.

Mentha fativa rubra. *Gerard.* 680.

Marfhes and banks of rivers. P. Auguft.

Reed
Gentilis

MINT, with flowers in whorls. Leaves egg-shaped, sharp;
serrated   Chives shorter than the blossom—*Stems red.* Cups
besprinkled with resinous dots. Leaves *sometimes red.* Blossoms
pale red.

Mentha fusca, seu vulgaris. *Park.* 31. *Ray's Syn.* 232.
Mentha cardiaca. *Gerard.* 680.
Mentha hortensis verticillata, ocymi odore. *Bauh pin.* 227.
1. There is a variety with blunter leaves and a more aromatic
smell. *Ray's Syn.* 239.
In watery places. P. August.

Corn
Arvensis

MINT, with flowers in whorls. Leaves egg-shaped; sharp;
serrated. Chives as long as the blossom—*Stems spreading. The
whole plant hairy.* Cups *white with down.* Blossoms *pale red.*
Mentha arvensis verticillata. *Bauh. pin.* 229.
Mentha seu calamintha aquatica. *Gerard.* 684. *Ray's Syn.* 232.
1. Round-leaved.
Moist corn-fields. P. August—September.
It prevents the coagulation of milk ; and when cows have eat-
en it, as they will do largely at the end of summer, when the
pastures are bare and hunger distresses them, their milk can
hardly be made to yield cheese : a circumstance that puzzles the
dairy maids not a little.
Horses and Goats eat it; Sheep are not fond of it ; Cows and
Swine refuse it.

Smooth
Exigua

MINT, with flowers in whorls. Leaves betwixt spear and
egg-shaped; smooth; sharp; very entire—*Stems trailing.* Blos-
soms *purplish.*
Mentha aquatica exigua. *Ray's Syn.* 232.
Mentha arvensis verticillata, seu aquatica belgarum lobelii.
Park. 36.
Near rivers. P. August.

Pennyroyal
Pulegium

MINT, with flowers in whorls. Leaves egg-shaped; blunt;
slightly scolloped. Stems nearly cylindrical; creeping. Chives
longer than the blossom—*which is of a pale purple : cultivation
will reduce it to a white.*
Pulegium. *Ray's Syn.* 235. vulgare. *Park.* 29.
Pulegium regium. *Gerard.* 671.
Pulegium latifolium. *Bauh. pin* 222.
In places that are flooded. P. August—October.
The expressed juice, with a little sugar is not a bad medicine
in the Hooping Cough. A simple and a spirituous water distilled
from the dried leaves are kept in the shops. They are prescribed in
hysterical Affections, and are not without considerable antispasmodic properties. An infusion of the plant may be used with
the same intentions.

The

SEEDS NAKED.

The Mint Fly, *Musca Pipiens*.

The Green Tortoile Beetle, *Caffida Viridis*, and the Green-filken Moth, *Phalæna Chryfitis*, live upon the different fpecies of mint.

233 VERVAIN. 32 Verbena.

EMPAL. *Cup* of one leaf; angular; tubular; permanent. The tube of equal thicknefs; with five teeth; one of the teeth lopped.

BLOSS. One petal; unequal. *Tube* cylindrical; ftraight; as long as the cup: dilated, and bowed inward towards the top. *Border* expanding; with five fhallow clefts; the *fegments* rounded; nearly equal.

CHIVES. *Threads* four; briftly; very fhort; conc within the tube of the bloffom; two longer thaealed other two. *Tips* crooked: either two or four.n the

POINT. *Seedbud* four cornered. *Shaft* fimple; thread-fhaped; as long as the tube. *Summit* blunt.

S. VESS. Very fine and thin, but generally none; the cup containing the feeds.

SEEDS. Two; or four; oblong.

OBS. *Linnæus attending chiefly to the foreign fpecies, places this in the firft order of the fecond clafs, but the only Britifh fpecies being an exception to that arrangement, we have ventured to put it in its proper fituation.*

VERVAIN, with thread-fhaped panicle fpikes; the leaves with many jagged clefts; the ftem fingle.—*Bloffoms pale blue.* Common Officinalis
Verbena vulgaris. *Ray's Syn.* 236.
Verbena mas f. recta et vulgaris. *Park.* 674.
Verbena communis. *Gerard.* 654. cæruleo flore. *Bauh. pin.* 269.
Simplers Joy.
Ditch-banks and road-fides in gravelly foils. A. July.

It manifefts a flight degree of aftringency. The root worn at the pit of the ftomach, an infufion, and an ointment prepared from the leaves, are faid to produce good effects in Scrophulous cafes; *(Morley's Effay on Scrophula;)* but this wants confirmation from the more rational and lefs enthufiaftic practitioner—Mr. Millar fays it is never found more than a quarter of a mile from a houfe, which is the reafon of the common Englifh name mentioned above - Sheep eat it. Cows, Goats and Horfes refufe it.

234 GERMANDER. 706 Teucrium.

EMPAL. *Cup* one leaf; with five fhallow clefts; nearly equal; hunched on one fide the bafe: fharp; permanent.

BLOSS. One petal; gaping; *Tube* cylindrical; fhort; ending in a crooked mouth. *Upper Lip* upright; fharp; deeply divided; *Segments* ftanding wide. *Lower Lip* with three clefts; expanding. *Lateral Segments* a little upright; the fhape of the upper lip. The middle fegment large; circular.

CHIVES. *Threads* four; awl-fhaped; longer than the upper lip of the bloffom and projecting betwixt its fegments. *Tips* fmall.

POINT. *Seedbud* with four divifions. *Shaft* thread fhaped; agreeing in fize and fituation with the chives. *Summits* two; flender.

S. VESS. None. The *Cup* remaining unchanged contains the feeds within it.

SEEDS. Four: roundifh.

OBS. *The upper lip of the bloffom being deeply divided, and the fegments ftanding wide afunder, make it appear as if the upper lip was wanting.*

Ground Chamæpithys   GERMANDER. The Leaves with three clefts: ftrap-fhaped; very entire. Flowers folitary; lateral; fitting. Stem fpreading—*Bloffoms yellow.*

Chamæpitys vulgaris. *Park.* 283. *Ray's Syn.* 244.
Chamæpitys lutea vulgaris, feu folio trifido. *Bauh. pin.* 249.
Chamæpitys mas. *Gerard.* 525.
Ground Pine.
Corn-fields. A. June—July.

This plant hath a degree of bitternefs and acrimony, but its real ufe is far from being accurately afcertained. It ftands recommended in the Gout, Jaundice and intermitting Fevers.

GERMANDER. The leaves heart-shaped; ferrated; on Sage leaf-stalks. Flowers in lateral bunches, all pointing one way. Scorodonia Stem upright—*Bloffoms greenish ftraw-colour.*

Scorodonia, feu falvia agreftis. *Gerard.* 626. *Ray's Syn,* 245.

Scorodonia, feu fcordium alterum, quibufdam, et falvia agref-tis. *Park.* 111.

Scordium alterum feu falvia agreftis. *Bauh. fin.* 247.

Wood Sage.

Woods, heaths, ditch-banks. P. July.

It poffeffes the bitternefs and a good deal of the aroma of Hops, it is therefore worth while to try if it may not be ufed for the fame purpofe.

GERMANDER. The leaves oblong; fitting; toothed and Water ferrated. Flowers in pairs, on fruit-ftalks at the bafe of the Scordium leaves. Stem fpreading—*Bloffoms red.*

Scordium. *Bauh. pin.* 247. *Ray's Syn.* 246.

Scordium majus et minus. *Gerard.* 661.

Scordium legitimum. *Park.* 111.

Marfhes. P. Auguft.

The frefh leaves are bitter and fomewhat pungent. Powdered they deftroy worms. A decoction of this plant is a good fomentation in gangrenous cafes. If Cows eat it when compelled by hunger, their milk gets a garlic flavour.

Sheep and Goats eat it; Horfes, Cows and Swine refufe it.

GERMANDER. The leaves betwixt wedge and egg-fhaped; Wall jagged; fcolloped; on leaf-ftalks. Flowers growing by threes. Chamædrys Stems trailing; a little hairy—*Bloffoms purplish.*

Chamædryn vulgarem, feu fativam. *Ray's Syn.* 231.

Chamædrys minor repens. *Bauh. pin.* 248.

Chamædrys major latifolia. *Gerard.* 616.

On walls and amongft ruins. P. May—June.

The plant is bitter, with a degree of aroma, and may be ufed with advantage in weak and relaxed conftitutions. It is an ingredient in the celebrated Gout Powders.

235 BUGLE. 705 Ajuga.

EMPAL. *Cup* one leaf; fhort; with five fhallow clefts; nearly equal; permanent.

BLOSS. One petal; gaping. *Tube* cylindrical; crooked. *Upper Lip* very fmall; upright; cloven; blunt. *Lower Lip* large; expanding; with three fegments; blunt. Middle fegment large; inverfely heart-fhaped. Lateral fegments fmall.

CHIVES. *Threads* four; (two fhort and two long), awl-fhaped; upright; taller than the upper lip. *Tips* double.

POINT. *Seedbud* with four divifions. *Shaft* thread-fhaped; agreeing in fize and fituation with the chives. *Summits* two; flender; the lowermoft the fhorteft.

S. VESS. None. The *Cup* clofes and retains the feed.

SEEDS. Four; rather long.

Mountain
Pyramidalis

BUGLE. The flowers in a hairy four-fided pyramid. Root-leaves very large—*The upper leaves are fometimes of a violet colour.* Bloffoms *bluifh*; *red*; or *white*.

Bugula cærulea alpina. *Park.* 525. *Ray's Syn.* 245.
Confolida media cærulea alpina. *Bauh. pin.* 260.
Sicklewort.
Hilly countries. P. June.
Sheep and Goats eat it; Cows are not fond of it; Horfes and Swine refufe it.

Pafture
Reptans

BUGLE. Smooth, with creeping fuckers—*Bloffoms blue*; *red*; *or white*; *in long leafy fpikes*.

Bugula. *Gerard.* 631. *Ray's Syn.* 245.
Bugula vulgaris. *Park.* 525.
Confolida media pratenfis cærulea. *Bauh. pin.* 260.
Wet paftures. P. May.
The roots are aftringent and ftrike a black colour with vitriol of iron.

236 BETONY.

236 BETONY. 718 Betonica.

EMPAL. *Cup* one leaf; tubular; cylindrical; with five teeth; with awns; permanent.

BLOSS. One petal, gaping. *Tube* cylindrical; crooked. *Upper Lip* circular; entire; flat; upright; *Lower Lip* with three segments. The middle segments broader, circular, notched at the end.

CHIVES. *Threads* four, (two long and two short;) as long as the mouth of the blossom, and leaning towards the upper lip. *Tips* roundish.

POINT. *Seedbud* with four divisions. *Shaft* in shape, size and situation resembling the chives. *Summit* cloven.

S. VESS. None. The *Cup* contains the seeds.

SEEDS. Four; egg-shaped.

BETONY. The flowers in an interrupted spike. The middle segment of the lip of the blossom notched—*Root leaves on leaf-stalks. Stem-leaves fitting, heart-shaped, hairy.* Blossoms *purple.* Wood Officinalis

Betonica. *Gerard.* 714. *Ray's Syn.* 238. purpurea. *Bauh. pin.* 235.

Betonica vulgaris flore purpureo. *Park.* 614.

Woods and heaths. P. July—August.

This plant was formerly much used in medicine, but it is discarded from the modern practice; however it is not destitute of virtues, for when fresh it intoxicates, and the dried leaves excite sneezing. It is often smoaked as Tobacco. The roots provoke vomiting.—Sheep eat it; Goats refuse it.

237 ARCHAN-

237 ARCHANGEL. 716 Lamium.

EMPAL. *Cup* one leaf; tubular; wider towards the top; with five teeth, and awns; nearly equal; permanent.

BLOSS. One petal; gaping. *Tube* cylindrical, very short. *Border* open. *Mouth* bladder-shaped; compressed; hunched; with a little tooth turned backwards on each side. *Upper Lip* vaulted; circular; blunt; entire. *Lower Lip* shorter; inversely heart-shaped; notched at the end; reflected.

CHIVES. *Threads* four; awl-shaped; (two long and two short) covered by the upper lip. *Tips* oblong; hairy.

POINT. *Seedbud* with four clefts. *Shaft* thread-shaped; agreeing in length and situation with the chives. *Summit* cloven; sharp.

S. VESS. None. The *Cup* remaining open contains the seeds in its bottom.

SEEDS. Four; short; three-cornered; convex on one side; lopped at each end.

White
Album

ARCHANGEL. The leaves heart-shaped; tapering to a point; serrated; on leaf-stalks. Flowers about twenty in a whorl—Blossoms *white*.

Lamium album. *Gerard.* 782. *Ray's Syn.* 240.
Lamium album non fætens, folio oblongo. *Bauh. pin.* 231.
Lamium vulgare album, seu Archangelicum flore albo. *Park.* 604.
White Dead Nettle.
Ditch-banks. P. May—June.
Goats and Sheep eat it; Cows are not fond of it; Horses and Swine refuse it.

ARCHANGEL. The leaves heart-shaped; blunt; on leaf- Purple
stalks—*Serrated*; *downy*; *but not rough.* *Several flowers in each* Purpureum
whorl. *Lower border of the mouth whitish, with purple streaks;
the other parts pale red.*

Lamium rubrum. *Gerard.* 703. *Ray's Syn.* 240.

Lamium vulgare folio sub-rotundo, flore rubro. *Park.* 604.

Lamium purpureum fætidum, folio subrotundo, seu Galeopsis
Dioscoridis. *Bauh. pin.* 230.

Lamium rubrum. *Hudson.* 225.

1. The leaves are sometimes pretty deeply divided. *Ray's Syn.*
240.

Red Dead Nettle.

Amongst rubbish. A. May—June.

The young leaves both of this and the preceding species may
be eaten with other pot-herbs.—Goats, Sheep and Horses eat it;
Cows refuse it.

ARCHANGEL. The floral leaves sitting; blunt; embrac- Henbit
ing the stem—*Lower leaves on leaf-stalks.* Blossoms *purple.* Amplexicaule

Lamium folio caulem ambiente, majus et minus. *Bauh. pin.*
231 *Ray's Syn.* 240

Alsine hederula altera. *Gerard.* 616.

Alsine hederulæ folio major. *Park.* 762.

Great Henbit.

Corn-fields. A. June.

Sheep, Horses and Goats eat it.

238 ALLHEAL. 717 Galeopſis.

EMPAL. *Cup* one leaf ; tubular ; with five teeth ; ending in ſharp awns as long as the tube : permanent.

BLOSS. One petal ; gaping.- *Tube* ſhort. *Border* open. *Mouth* ſomewhat wider than the tube and as long as the cup. Above the baſe of the lower lip on each ſide lies a little tapering tooth ; hollow on the under ſurface. *Upper Lip* circular ; concave ; ſerrated at the top. *Lower Lip* with three ſegments ; the lateral ones circular ; the *middle* ſegment larger ; ſcolloped ; notched at the end.

CHIVES. *Threads* four, (two long and two ſhort) awl-ſhaped ; covered by the upper lip. *Tips* roundiſh ; cloven.

POINT. *Seedbud* with four clefts. *Shaft* thread-ſhaped ; agreeing in length and ſituation with the chives. *Summit* cloven ; ſharp.

S. VESS. None. The *Cup* ſtiff and ſtraight contains the ſeeds.

SEEDS. Four : three-cornered ; lopped.

OBS. *The firſt ſpecies hath the upper lip of the bloſſom a little reflected, but not very evidently ſcolloped. In the yellow* ALLHEAL *the lower lip of the bloſſom is without teeth ; and divided into three equal ſegments. The upper lip is entire and fringed with a few ſoft hairs.*

Narrowleaved   ALLHEAL, with flowers in remote whorls. Stem equally
Ladanum   thick betwixt the joints. Cups unarmed—*Bloſſoms* red ; ſometimes *white.*

Sideritis arvenſis rubra. *Park.* 587. *Ray's Syn.* 242.
Sideritis VII. *Gerard.* 699.
Sideritis arvenſis anguſtifolia rubra. *Bauh. pin.* 233.
Corn-fields. A. July—Auguſt.
Goats and Cows eat it ; Sheep are not fond of it ; Horſes refuſe it.

SEEDS NAKED. 353

ALLHEAL, with flowers in whorls; the upper whorls Nettle-hemp almoft cont guous. Stem thickeft juft beneath the joints; Cups Tetrahit fomewhat prickly—*and much fmaller than the* Bloffoms, *which are yellow, purple, or white.*

Lamium Cannabino folio vulgare. *Ray's Syn.* 240.
Cannabis fpuria. *Gerard.* 709. *Park.* 599.
Urtica aculeata, foliis ferratis. *Bauh. pin.* 232.

1. It varies in having white, purple or yellow flowers. *Ray's Syn.* 241.
Hemp-leaved Dead Nettle.
Ditch-banks and corn-fields. A. Auguft.
Sheep and Goats eat it; Horfes, Cows and Swine refufe it.

ALLHEAL, with fix flowers in a whorl. Fence of four Yellow leaves—*Bloffoms yellow* ; *fometimes as many as ten in a whorl. The* Galeobdolon *fence fometimes confifts of fix leaves. There is a difficulty in af-certaining the genus of this plant, as the lip of the bloffom is equal and without teeth. The leaves are fometimes fpotted with white.*

Lamium luteum. *Gerard.* 671. *Park.* 606. *Ray's Syn.* 240.
Lamium folio oblongo, luteum. *Bauh. pin.* 231.

1. Leaves fpotted.
Yellow Nettle-hemp. Yellow Archangel.
Woods and fhady moift hedges. P. May.
The green filken Moth, *Phalæna Chryfitis,* and the Goofe-berry Moth, *Phalæna Wauvaria,* live upon the different fpecies.

A 2   239 CLOWN-

239 CLOWNHEAL. 719 Stachys.

EMPAL. *Cup* one leaf; tubular; angular; with five shallow clefts; taper; permanent. Teeth awl-shaped; taper; nearly equal.

BLOSS. One petal; gaping. *Tube* very short. *Mouth* oblong; hunched downwards towards the base. *Upper Lip* upright; somewhat egg-shaped; vaulted; generally notched at the end. *Lower Lip* large; with three segments; the two outer segments reflected; the *middle* segment, which is the largest, notched at the end, and folded back.

CHIVES. *Threads* four, (two long and two short) awl-shaped; bent to the sides of the mouth. *Tips* simple.

POINT. *Seedbud* with four divisions. *Shaft* thread shaped; agreeing in length and situation with the chives. *Summit* cloven: sharp.

S. VESS. None. The *Cup* but little changed contains the seeds.

SEEDS. Four; egg-shaped; angular.

OBS. *In some species the upper lip is vaulted; in others it stands upright.*

Nettle
Sylvatica

CLOWNHEAL. The flowers six or eight in a whorl. Leaves heart-shaped; on leaf-stalks—*Spikes leafy.* Blossoms *deep purple, with white spots.*
Galeopsis legitima Dioscoridis. *Park.* 908. *Ray's Syn.* 237.
Galeopsis vera. *Gerard.* 709.
Lamium maximum sylvaticum fœtidum. *Bauh. pin.* 231.
1. There is a variety that is smaller and the leaves angular.
Hedge Nettle.
Woods and hedges. A. July—August.
It will dye yellow. The whole plant has a fœtid smell, and Toads are thought to be fond of living under its shade.—Sheep and Goats eat it; Horses, Cows and Swine refuse it.

Betony
Recta

CLOWNHEAL. The flowers in spiked whorls; leaves betwixt oval and heart-shaped; rough, scolloped; stems ascending—*Stiff; beset with a few hairs. Cups thorny.* Flowers *yellow.*
Stachys vulgaris hirsuta erecta. *Bauh. pin.* 233.
Sideritis arvensis latifolia hirsuta lutea. *Ray's Syn.* 242.
Betonica hirta. *Hudson.* 220.
Yellow Betony.
Corn-fields. P. August.

CLOWN-

CLOWNHEAL. The flowers about six in a whorl. Leaves Water
betwixt strap and spear-shaped, half embracing the stem—*Blof-* Palustris
soms purple; in a leafy spike.

Stachys palustris fœtida. *Bauh. pin.* 236.
Sideritis anglica strumosa radice. *Park.* 587. *Ray's Syn.* 242.
Panax coloni. *Gerard.* 1005.
Clowns Allheal.
Banks of rivers. P. August.

The roots when dried and powdered will make bread. Swine eat
them—Sheep eat it ; Cows, Horses, Goats and Swine refuse it:

CLOWNHEAL. The flowers several in a whorl. Leaves Horehound
serrated ; the teeth lapping over each other. Stem woolly— Germanica
The whole plant is white. Tips *yellowish white. Floral leaves
spear-shaped; reflected.* Blossoms *white ; with a purplish tinge
within.*

Stachys Fuchsii. *Gerard.* 795. *Ray's Syn.* 239.
Stachys major Germanica. *Bauh. pin.* 236.
Base Horehound.
Hedges. P. July.

CLOWNHEAL. With six flowers in a whorl. Stem feeble. Gill
Leaves blunt; almost naked. Blossoms as long as the cup.— Arvensis
Stem four-cornered; blunt: branched. Leaves *heart-shaped, bluntly
serrated.* Cup *sitting ; with five shallow clefts, equal ; sharp pointed.* Blossoms *white. Middle segment of the lower lip with a
purplish spot.*

Sideritis humilis, lato obtuso folio. *Gerard.* 699. *Ray's Syn.*
242.
Sideritis hederulæ folio. *Park.* 587.
Sideritis alsines trixaginis folio. *Bauh. pin.* 233.
Glechoma arvensis. *Hudson.* 224.
Upright Ground Ivy.
Corn-fields. A. August.

240 NAP. 710 Nepeta.

EMPAL. *Cup* one leaf; tubular; cylindrical. *Mouth* with five teeth; sharp; upright; the *upper* teeth the longest; *lower* teeth most expanded.

BLOSS. One petal; gaping. *Tube* cylindrical; crooked. *Border* open. *Mouth* expanding; heart-shaped; terminated by two very short, reflected, blunt segments. *Upper Lip* upright; circular; notched at the end. *Lower Lip* circular; concave; large and entire; a little scolloped at the edge.

CHIVES. *Threads* four; (two long and two short) awl-shaped; approaching; covered by the upper lip. *Tips* fixed side-ways to the threads.

POINT. *Seedbud* with four clefts. *Shaft* thread-shaped; agreeing in length and situation with the chives. *Summit* cloven; sharp.

S. VESS. None. The *Cup* standing upright contains the seeds.

SEEDS. Four; somewhat egg shaped.

OBS. *If you reckon the segments of the mouth as part of the lower lip, that lip will appear to have three segments.*

Cats
Cataria

NAP. The flowers growing in spikes; forming whorls round the spikes and supported upon little fruit-stalks. Leaves heart-shaped, toothed and serrated; on leaf-stalks—*Stem and leaves white with down.* Plossoms *pale purple*; or blue.

Nepeta major vulgaris. *Park.* 38. *Ray's Syn.* 237.

Mentha cataria vulgaris et major. *Bauh. pin.* 228.

Mentha selina, seu Cataria. *Gerard.* 682.

1. There is a variety in which all the parts of the plant are smaller.

Nap. Catmint.

Hedges. P. July.

An infusion of it is deemed a specific in Chlorotic cases. Two ounces of the expressed juice may be given for a dose. Cats are so delighted with this plant that they can hardly be kept out of the garden wherein it grows. Mr. Millar says that Cats will not meddle with it if it is raised from seeds, and in support of this opinion quotes an old saying, " If you set it, the Cats will eat : if you sow it the Cats will not know it." The fact seems to be, that they will attack a single plant, but not a number together —Sheep eat it ; Cows, Horses, Goats and Swine refuse it.

241 HENBIT. 720 Ballota.

EMPAL. *Cup* one leaf; tubular: falver-fhaped; with five
edges; oblong: with ten fcores; upright; permanent; equal. *Rim* fharp; open; plaited; with five
teeth. *General Fence* to the whorls formed of ftrap-
fhaped leaves.

BLOSS. One petal; gaping. *Tube* cylindrical; as long
as the cup. *Upper Lip* upright; Egg-fhaped; entire;
fcolloped; concave. *Lower Lip* with three fegments; blunt. The middle fegment the largeft;
notched at the end.

CHIVES. *Threads* four, (two long and two fhort) awl-
fhaped; leaning towards and fhorter than the upper
lip. *Tips* oblong; lateral.

POINT. *Seedbud* with four clefts. *Shaft* thread-fhaped; in
fhape and fituation fimilar to the chives. *Summit*
flender; cloven.

S. VESS. None. The *Cup* unchanged contains the feeds.

SEEDS. Four; egg-fhaped.

OBS. *This genus poffeffeth the fence of the* BASIL, *the cup of the*
HOREHOUND; *the bloffom of the* CLOWNHEAL.

HENBIT. The leaves heart-fhaped; undivided; ferrated. Black
Cups with fharp points—*Bloffoms purple; in whorls.* Nigra
 Ballota. *Ray's Syn.* 244.
 Marrub um n grum. *Gerard.* 701.
 Marrubium nigrum fætidum, Ballotte dictum. *Park.* 1230.
 Marrubium nigrum fætidum, Ballotte Diofcoridis. *Bauh. pin.*
230.
1. There is a variety in which the cups appear lopped, and the
flowers white.
 Stinking Horehound. Hedge Nettle.
 Amongft rubbifh. A. July.
 It ftands recommended in Hyfterical Cafes. The Swedes
reckon it almoft a univerfal remedy in the difeafes of their cattle.
Horfes, Cows, Sheep and Goats refufe it.

242 HOREHOUND. 721 Marrubium.

EMPAL. *Cup* one leaf; funnel-fhaped; tubular; with ten fcores. *Rim* equal; open; generally with ten teeth; alternately fmaller.

BLOSS. One petal; gaping. *Tube* cylindrical. *Border* open; *Mouth* long; tubular. *Upper Lip* upright; narrow; fharp: cloven a little way down. *Lower Lip* broader; reflected; with three fhallow fegments; the *middle* fegment broad; notched at the end; the *lateral* fegments fharp.

CHIVES. *Threads* four, (two long and two fhort) fhorter than the bloffom; covered by the upper lip. *Tips* fimple.

POINT. *Seedbud* with four clefts. *Shaft* thread-fhaped; agreeing in length and fituation with the chives. *Summit* cloven.

S. VESS. None. The *Cup* clofed at the neck, but open at the rim, contains the feeds.

SEEDS. Four; rather oblong.

White
Vulgare

HOREHOUND. The teeth of the cup like briftles, but hooked—*Leaves white and woolly. Bloffoms white.*
Marrubium album. *Gerard.* 693. *Ray's Syn.* 239.
Marrubium album vulgare. *Bauh. pin.* 230. *Park.* 44.
Road-fides and among rubbifh. P. July.

It is very bitter to the tafte and not altogether unpleafant to the fmell. It was a favourite medicine with the ancients in Obftructions of the Vifcera. In large dofes it loofens the belly. It is a principal ingredient in the Negro Cæfar's remedy for vegetable poifons.—A young man who had occafion to take Mercurial Medicines, was thrown into a falivation which continued for more than a year. Every method that was tried to remove it, rather increafed the complaint. At length Linnæus prefcribed an infufion of this plant, and the patient got well in a fhort time Horfes, Cows, Sheep and Goats refufe it.

243 HOOD.

SEEDS NAKED. 359

243 HOODWORT. 734 Scutellaria.

EMPAL. *Cup* one leaf; very short; tubular. *Rim* almost
entire ; covered by a scale, lying over it like a lid ;
which is formed by an expansion of the upper part
of the cap.

BLOSS. One petal; gaping. *Tube* very short ; bent backwards. *Mouth* long; compressed. *Upper Lip* concave; divided into three segments. *Midd'e segment*
concave ; notched at the end. *Lateral segments* flat,
rather sharp ; placed under the middle segment.
Lower Lip broad; notched at the end, keel-shaped
beneath.

CHIVES. *Threads* four (two long and two short ;) concealed under the upper lip. *Tips* small.

POINT. *Seedbud* with four divisions. *Shaft* thread-shaped;
agreeing in length and situation with the chives.
Summit simple; crooked; taper.

S. VESS. None. The *cup* three cornered ; covered with
a lid resembling a Helmet; which answers the purpose of a capsule. It opens at the lower margin.

SEEDS. Four; roundish.

HOODWORT. The leaves betwixt heart and spear-shaped ; Willow
scolloped; oppositc. Flowers at the base of the leaves—*On fruit-* Galericulata
stalks; *pendant.* Blossoms *blue*; *hairy on the outside. Chives and
pointal white.* Tips *purple. Branches numerous*; *opposite.*
Cassida palustris vulgatior, flore cæruleo. *Ray's Syn.* 244.
Lysimachia cærulea galericulata, seu gratiola cærulea. *Bauh.
pin.* 246.
Lysimachia galericulata. *Gerard.* 477.
Lysimachia cærulea, seu latifolia major. *Park.* 221.
Hooded Willowherb.
Banks of Rivers. P. August.
When the blossom falls off, the cup closes upon the seeds;
which when ripe, being still smaller than the cup, could not
possibly open its mouth or overcome its elastic force as the down
of the seeds do in the compound flowers, and must consequently
remain useless without a possibility of escaping. But nature,
ever full of resources, finds a method to discharge them. The
cup grows dry, and then divides into two distinct parts ; so that
the seeds already detached from the receptacle, fall to the ground,
Cows, Goats and Sheep eat it, Horses, and Swine refuse it.

A a 4　　　　SCULL-

Lesser Minor HOODWORT. The leaves betwixt heart and egg-shaped; nearly entire. Flowers at the base of the leaves—*Blossoms purple.*

Cassida palustris minima, flore purpurascente. *Ray's Syn.* 244.
Gratiola latifolia. *Gerard.* 581, sive nostras minor. *Park.* 221.
Lesser hooded Willowherb.
Marshes. P. July—August.

244 THYME. 727 Thymus.

EMPAL. *Cup* one leaf; tubular; cloven halfway down into two lips; permanent. The *mouth* closed by soft hairs. *Upper lip* broad; flat; upright; with three teeth. *Lower lip* like two bristles, of equal length.

BLOSS. One petal; gaping. *Tube* as long as the cup. *Mouth* small. *Upper lip* short; flat; upright; notched at the end; blunt. *Lower lip* long; expanding; broad; in three segments; blunt. *Midale segment* broadest.

CHIVES. *Threads* four, (two long and two short,) crooked. *Tips* small.

POINT. *Seedbud* with four divisions. *Shaft* thread-shaped. *Summit* cloven; sharp.

S.VESS. None. The *cup* narrow at the neck incloses the seeds.

SEEDS. Four; small; roundish.

OBS. *In the* common Thyme *the stems are woody. In the* basil Thyme *the middle segment of the lower lip of the blossom is notched.*

Common Serpyllum THYME. The flowers in heads. Stems creeping. Leaves flat; blunt; fringed at the base—*Blossoms pale red, or white. Teeth of the cup fringed.* Chives *as long as the blossom; and nearly all of the same length.*

Serpyllum vulgare. *Gerard.* 570. *Ray's Syn.* 230.
Serpyllum vulgare minus. *Bauh. pin.* 220. *Park.* 8.
Mother of Thyme.

1. Blossoms white; leaves variegated.
2. Flowers large. *Ray's Syn.* 230.
3. *Hoary Thyme,* Leaves narrow and hairy. *Ray's Syn.* 231.
4. *Lemon Thyme,* Smells like citron. *Gerard.* 575.
5. Leaves narrow and smooth. *Bauh. pin.* 220.
6. More shrubby hairy Mother of Thyme, with pale red Blossoms.

On hills. 6, on Snowden in Wales. P. July—August.

The

The whole plant is fragrant, and yields an essential oil that is very heating. An infusion of the leaves removes the head-ach occasioned by the debauch of the preceding evening. The fourth variety is frequently used in sauces.

A general opinion prevails that the flesh of sheep that feed upon aromatic plants, particularly upon Thyme, is much superior in flavour to common Mutton: but the ingenious author of the account of the Sheep-walks in Spain, (Gent. Mag. 1764.) considers this as a vulgar error. He says Sheep are not fond of aromatic plants; that they will carefully push aside the thyme to get at the grafs growing beneath it; and that they never touch it unless when walking apace, and then they will catch at any thing—the attachment of Bees to this and other aromatic plants is well known.—In the experiments made at Upsal, Sheep and Goats were observed to eat it, and Swine to refuse it. —The green broom Moth, *Phalæna papilionaria*, lives upon it.

THYME. The flowers in whorls; one upon each fruit-stalk. Stems upright; somewhat branched. Leaves sharp, serrated—*The middle segment of the lower lip of the Blossom notched at the end.* Blossoms *purple*.

Acinos multis. *Ray's Syn.* 238.
Ocymum sylvestre. *Gerard.* 675.
Clinopodium arvense ocymi facie. *Bauh. pin.* 225.
Clinopodium minus seu vulgare. *Park.* 21.
Wild Basil.
On Chalk or Gravelly Hills. A. July—August.
Horses eat it; Cows are not fond of it. Sheep and Goats refuse it.

245 SELFHEAL. 735 Prunella.

EMPAL. *Cup* one leaf; with two lips; mouth fhort; permanent. *Upper lip* flat, broad, lopped; with three very fmall teeth. *Lower lip* upright, narrow; fharp; with a fhallow cleft.

BLOSS. One petal; gaping. *Tube* fhort; cylindrical. *Mouth* oblong. *Upper lip* concave; entire; nodding. *Lower lip* reflected; blunt; with three fegments. The *middle fegment* broadeft; notched at the end; ferrated.

CHIVES. *Threads* four (two a little longer than the other two) awl-fhaped, forked at the end. *Tips* fimple, fixed to the threads beneath the top, and only to one of the divifions of the fork.

POINT. *Seedbud* with four divifions. *Shaft* thread-fhaped, leaning along with the chives towards the upper lip. *Summit* notched at the end.

S. VESS None. The *cup* clofes and contains the feeds.

SEEDS. Four, fomewhat egg-fhaped.

OBS. *The effential character of this genus confifts in the threads being forked.*

Common Vulgaris.

SELFHEAL. With all the leaves oblong egg-fhaped; on leaf-ftalks—*In open' funny fituations it grows trailing and not above a fingers length, but in woods it is upright and near a foot high.* Floral leaves *heart fhaped.* Bloffoms *blue, purplifh,* or *white.*

Prunella. *Gerard.* 631. *Ray's Syn.* 238. vulgaris. *Park.* 1680.

Prunella major folio non diffecto. *Bauh. pin.* 260.

Paftures. P. Auguft.

Cows, Goats and Sheep eat it; Horfes are not fond of it.

246 MARJORAM. 726 Origanum.

EMPAL. *Fence* fpiked, tiled with *Floral leaves*, egg-fhaped, coloured. *Cup* unequal, various.

BLOSS. One petal, gaping. *Tube* cylindrical, compreff-ed. *Upper lip* upright, flat, blunt. imperfect at the margin. *Lower lip* with three fegments, nearly equal.

CHIVES. *Threads* four, (two long and two fhort ;) thread-fhaped ; as long as the bloffom. *Tips* fimple.

POINT. *Seedbud* with four clefts. *Shaft* thread-fhaped, leaning towards the upper lip of the bloffom. *Summit* flightly cloven.

S. VESS. None. The *cup* clofing a little contains the feeds.

SEEDS. Four ; egg fhaped.

OBS. *The fence to the cups conflitutes the effential character of this genus.*
The *cup* in fome fpecies is nearly equal ; with five teeth : In others it confifts of two lips ; the upper lip large and entire, the lower lip hardly perceptible. In others again the cup is formed of two leaves.

MARJORAM. The flowers in clofe, roundifh, panicled Wild fpikes. Floral leaves egg-fhaped ; longer than the cups—Vulgare *Leaves* oval; *pointed*; *on fhort leaf-flalks*; *dark green*. Bloffoms *purple*.

Origanum vulgare fpontaneum. *Ray's Syn.* 236.
Origanum anglicum. *Gerard.* 666.
Origanum fylveftre. *Bauh. pin.* 223.
Marjorana fylveftris. *Park.* 12.
Among Brambles and Hedges. ˙ P. July.

The whole plant is a warm aromatic. The dried leaves ufed inftead of tea are exceedingly grateful—the effential oil of this plant is fo acrid that it may be confidered as a cauftic, and is much ufed with that intention by Farriers.˙ A little cotton wool moiftened with it, and put into the hollow of an aching tooth, frequently relieves the pain—the country people ufe the tops to dye purple.—Goats and Sheep eat it ; Horfes are not fond of it ; Cows refufe it.

Pot
Onites

MARJORAM. The flowers in oblong, hairy; incorporated ed fpikes. Leaves heart-fhaped; downy—*Three flowers upon eacb fruit-flalk; woolly. The one in the middle, fitting.* Bloffoms *white; fometimes tinged with red.*

 Origanum onites. *Bauh. pin.* 223. *Ray's Syn.* 236.
 Majorana major anglica. *Gerard.* 664.
 Majorana latifolia, feu major anglica. *Park.* 12.
 Dry Fields. P. Auguft.

247 BASIL. 725 Clinopodium.

EMPAL. *Fence* of many briftle-fhaped leaves, as long as the cup; placed under the whorls. *Cup* one leaf; cylindrical; a little curved. *Mouth* with two lips. *Upper lip* broad; with three fegments; fharp; reflected. *Lower lip* deeply divided; flender; bent inwards.

BLOSS. One petal, gaping. *Tube* fhort; gradually widening into a mouth. *Upper lip* upright; concave; blunt; notched. *Lower lip* with three fegments: blunt. *Middle fegment* broad; notched.

CHIVES. *Threads* four; (two long and two fhort;) covered by the upper lip. *Tips* roundifh.

POINT. *Seedbud* with four divifions. *Shaft* thread-fhaped; agreeing in length and fituation with the chives. *Summit* fimple; fharp; compreffed.

S. VESS. None. The *cup*, clofing at the neck, and fwelling out in the body, contains the feeds.

SEEDS. Four; egg-fhaped.

Common
Vulgare

BASIL. The flowers in roundifh hairy heads. Floral leaves like briftles—*Bloffoms purple. Cultivation will turn them white.*
 Clinopodium origano fimile. *Bauh. pin.* 224. *Ray's Syn.* 239.
 Acinos. *Gerard.* 675. feu clinopodium majus. *Park.* 22.
 Great wild Bafil.
 Hedges and dry Paftures. P. July.
 Goats and Sheep eat it; Horfes refufe it.

248 BALMLEAF. 731 Melittis.

EMPAL. *Cup* one leaf; bell-shaped; cylindrical; straight.
Mouth with two lips. *Upper lip* tall; notched; sharp.
Lower lip shorter; cloven; sharp. *Segments* standing
wide.

BLOSS. One petal; gaping. *Tube* much more slender
than the cup. *Mouth* but little thicker than the
tube. *Upper lip* upright; roundish; flat. *Lower lip*
expanding; with three segments; blunt. *Middle
segment* large, scolloped.

CHIVES. *Threads* four; the middle ones shorter than the
outer ones; awl-shaped; standing under the upper
lip. *Tips* blunt; cloven; each pair forming a
cross.

POINT. *Seedbud* blunt; with four clefts; covered with
soft hairs. *Shaft* thread-shaped; agreeing in length
and situation with the chives. *Summit* cloven;
sharp.

S. VESS. None. The *cup* unchanged contains the
seeds.

SEEDS. Four.

OBS. *The lower lip of the cup is sometimes scolloped.*

BALMLEAF. As there is only one species known Linnæus
gives no description of it—*Blossoms on single fruit-stalks, six in
each whorl; red, with purple spots, or white with red spots.*
 Melissa fuchsii. *Gerard.* 690. *Ray's Syn.* 242.
 Melissophyllon fuchsii. *Park.* 41.
 Lamium montanum melissæ folio. *Bauh. pin.* 231.
 Bastard Balm.
 Woods and Hedges. P. June.

Bastard
Melissophyl-
lum.

249 CALAMINT. 728 Meliſſa.

EMPAL. *Cup* one leaf; ſomewhat bell-ſhaped; dry and
    ſkinny; a little expanding; angular; ſcored; per-
    manent. *Mouth* with two lips. *Upper lip* with three
    teeth; reflected; expanding; flat. *Lower lip* ſhort;
    a little ſharp; divided.

BLOSS. One petal; gaping. *Tube* cylindrical. *Mouth*
    open. *Upper lip* ſhort; upright; vaulted; roundiſh;
    notched at the end. *Lower lip* with three ſegments;
    the *middle ſegment* largeſt; and inverſely heart-ſhaped.

CHIVES. *Threads* four: awl-ſhaped; two as long as the
    bloſſom; the other two but half as long. *Tips*
    ſmall; ſtanding together in pairs.

POINT. *Seedbud* with four clefts. *Shaft* thread-ſhaped; as
    long as the bloſſom; leaning along with the chives
    under the upper lip of the bloſſom. *Summit* ſlender;
    cloven; reflected.

S. VESS. None.   The *cup* unchanged, but enlarging
    contains the ſeeds.

SEEDS. Four; egg-ſhaped.

Common
Calamintha   CALAMINT. The fruit-ſtalks forked; as long as the
leaves; growing at the baſe of the leaves. *Bloſſoms bluiſh white.*
Calamintha vulgaris. *Park.* 36. *Ray's Syn.* 243. *Gerard.* 687.
Calamintha vulgaris, vel officinarum germaniæ. *Bauh. pin.* 228.
Roads and corn-fields. P. Auguſt.

Field
Nepeta   CALAMINT. The fruit-ſtalks forked; longer than the
leaves. Growing at the baſe of the leaves. Stem aſcending;
hairy. Stems hard. -- *The whole plant hairy. Leaves ſomewhat*
heart-ſhaped; ſmooth on the upper ſurface. Floral leaves awl-ſhaped.
Bloſſoms *bluiſh. Palate white with a blue border.*
Calamintha odore pulegii. *Gerard.* 687. *Ray's Syn.* 243.
Calamintha pulegii odore, ſeu nepeta. *Bauh. pin.* 228.
Calamintha altera odore pulegii, foliis maculoſis *Park.* 36.
1. There is a variety with indented leaves.
Road-ſides and round corn-fields. P. Auguſt.
The Green Tortoiſe Beetle, *Caſſida viridis*, feeds upon it.

Order

Order- II. Seeds Covered.

250 BROOMRAPE. 779 Orobanche.

EMPAL. *Cup* one leaf, with two or five clefts; upright; coloured; permanent.

BLOSS. One petal; gaping. *Tube* leaning; large; diftended. *Border* expanding. *Upper lip* concave; open; notched at the end. *Lower lip* reflected; with three fegments, unequal at the edge. *Segments* nearly equal.

CHIVES. *Threads* four, (two long and two fhort;) awl-fhaped; concealed under the upper lip. *Tips* upright; approaching; fhorter than the border.
Honeycup a gland at the bafe of the feedbud.

POINT. *Seedbud* oblong. *Shaft* fimple; agreeing in length and fituation with the chives. *Summit* with a fhallow cleft; blunt; thick; nodding.

S. VESS. *Capfule* oblong egg-fhaped; taper; with one cell and two valves.

SEEDS. Numerous; very fmall. *Receptacles* four; ftrap-fhaped; lateral; connected.

OBS. *In one fpecies the cup hath five clefts, and each fegment of the fummit is notched at the end.*

BROOMRAPE. The ftem undivided; downy. Chives al- Great moft appearing above the bloffom—*Bloffoms in fpikes; purplifh* Major *ftraw coloured.*
Orobanche major garyophyllum olens. *Bauh. pin.* 87. *Ray's Syn.* 288.
Rapum geniftæ. *Gerard.* 1311. *Park.* 229.
1. There is a variety with a fmaller flower.
Dry Paftures, amongft Broom. P. May—June.

BROOMRAPE. The ftem branched. Bloffoms with five Branched clefts—*pale red.* Ramofa
Orobanche ramofa. *Bauh. pin.* 88. *Park.* 1363. *Ray's Syn.* 28
In corn-fields. June.

251 TOOTH-

251 TOOTHWORT.   743 Lathræa.

EMPAL. *Cup* one leaf ; bell-ſhaped; ſtraight. *Mouth* with
    four deep clefts.

BLOSS. One petal ; gaping. *Tube* longer than the cup.
    Border gaping, diſtended. *Upper lip* concave ; hel-
    met-ſhaped ; broad; with a narrow hooked top.
    Lower lip. ſmaller; reflected ; blunt; with three ſeg-
    ments.
        Honey-cup a gland notched at the end ; depreſſed
    on each ſide; very ſhort ; ſituated upon the receptacle
    of the flower at one corner of the ſeedbud.

CHIVES. *Threads* four ; awl-ſhaped ; as long as the bloſ-
    ſom ; concealed under the upper lip. *Tips* blunt ;
    depreſſed ; approaching.

POINT. *Seedbud* globular; compreſſed. *Shaft* thread-ſhap-
    ed ; agreeing in length and ſituation with the chives.
    Summit lopped ; nodding.

S. VESS. *Capſule* roundiſh ; blunt ; but furniſhed with a
    ſmall point ; with one cell ; and two elaſtic valves ;
    ſurrounded by the cup which is large and expand-
    ing.

SEEDS. Few ; globular.

Scaled
Squamaria

TOOTHWORT. The ſtem undivided ; bloſſoms pendant.
Lower lip with three ſegments—*Root beaded.*   Root *leaves none.*
Stem leaves *membranaceous, coloured.*   Branches *none.   Lower lip
of the* bloſſom *white, the other parts purple.*
    Anblatum cordi, ſeu aphyllon. *Ray's Syn.* 288.
    Orobanche radice dentata major. *Bauh. ſin.* 88.
    Orobanche radice dentata, ſeu dentaria major matthiolo.
Gerard. 1322.
    Shady places and valleys. P. April—May.
    Goats, Sheep and Swine eat it ; Cows and Horſes refuſe it.

252 PAINT-

252 PAINTEDCUP. 739 Bartſia.

EMPAL. *Cup* one leaf ; tubular ; permanent. *Mouth* blunt, cloven. *Segments* notched at the end; points coloured.

BLOSS. One petal ; gaping. *Upper lip* upright ; ſlender ; entire ; long. *Lower lip* reflected ; with three ſegments ; blunt ; very ſmall.

CHIVES. *Threads* four, (two a little ſhorter than the other two ;) briſtle ſhaped ; as long as the upper lip. *Tips* oblong ; approaching; ſtanding under the top of the upper lip.

POINT. *Seedbud* egg-ſhaped. *Shaft* thread-ſhaped ; longer than the chives. *Summit* blunt ; nodding.

S. VESS. *Capſule* egg-ſhaped ; compreſſed ; taper ; with two cells and two valves ; partition oppoſite to the valves.

SEEDS. Numerous ; angular ; ſmall.

OBS. *This genus is a ſort of connecting link betwixt the* RATTLE *the* EYEBRIGHT *and the* LOUSEWORT, *but it is diſtinguiſhed by the* coloured *cup.*

PAINTEDCUP. The upper leaves alternate ; ſerrated. Marſhy
Flowers diſtant ; lateral—*Bloſſoms ſolitary, yellow.* Viſcoſa
Euphraſia major lutea latifolio paluſtris. *Ray's Syn.* 285.
Marſh Eyebright Cow-wheat.
Boggy wet places in Cornwall. A. Auguſt.

PAINTEDCUP. The leaves oppoſite ; heart-ſhaped ; Mountain
bluntly ſerrated—*Bloſſoms in leaſy ſpikes ; purple.* Alpina
Clinopodium alpinum hirſutum. *Bauh. pin.* 225.
Euphraſia rubra Weſtmorlandica, foliis brevibus obtuſis.
Ray's Syn. 285.
Mountain Eyebright Cow-wheat.
Near rivulets in hilly countries. P. Auguſt.
Sheep and Goats eat it.

253 EYEBRIGHT. 741 Euphrasia.

EMPAL. *Cup* one leaf; cylindrical; with four clefts; unequal; permanent.

BLOSS. One petal; gaping. *Tube* as long as the cup.
Upper Lip concave; notched at the end. *Lower Li*p expanding; divided into three segments which are
equal, blunt.

CHIVES. *Threads* four; thread-shaped; leaning under
the upper lip. *Tips* two lobed. The lower lobe of
the lower tips pointed with a thorn.

POINT. *Seedbud* egg shaped. *Shaft* threadshaped; agreeing in shape and situation with the chives. *Summit*
blunt; entire.

S. VESS. *Capsule* egg-shaped, but oblong; compressed.
Cells two.

SEEDS. Numerous; very small; roundish.

Common
Officinalis

EYEBRIGHT. The leaves egg-shaped; streaked; sharply
and elegantly toothed—*Blossoms bluish white, with purple streaks.*
Euphrasia. *Gerard.* 563. *Ray's Syn.* 284.
Euphrasia vulgaris. *Park.* 1329.
Euphrasia officinarum. *Bauh. pin.* 233.
Meadows and pastures. A. August—September.
It is a weak astringent, and was formerly in repute as a reme-
dy for impaired vision. It will not grow but when surrounded
by plants taller than itself.—Cows, Horses, Goats and Sheep
eat it; Swine refuse it.

Red
Odontites

EYEBRIGHT. The leaves spear shaped; all serrated—
Sometimes reddish; blossoms dusky red; or purple.
Euphrasia pratensis rubra. *Bauh. ʃin.* 234. *Ray's Syn.* 284.
Euphrasia pratensis rubra major. *Park.* 1329.
Cratæogonon Euphrosyne. *Gerard.* 91.
Corn-fields and pastures. A. August—September.
Cows, Horses, Sheep and Goats eat it.

254 RATTLE. 740 Rhinanthus.

EMPAL. *Cup* one leaf; roundifh; bladder-fhaped; com-preffed; with four teeth; permanent.

BLOSS. One petal; gaping. *Tube* fomewhat cylindrical, as long as the cup. *Border* open; compreffed at the bafe. *Upper Lip* fhaped like a helmet; compreff-ed; notched; narrow. *Lower Lip* open; flat; with three flight fegments; blunt; the *middle fegment* the broadeft.

CHIVES, *Threads* four; (two long and two fhort;) *nearly* as long as the upper lip which conceals them. *Tips* fixed fideways to the threads; cloven at one end; hairy.

POINT. *Seedbud* egg-fhaped; compreffed. *Shaft* thread-fhaped; agreeing in fituation with the chives, but longer. *Summit* blunt; bent inwards.

S. VESS. *Capfule* blunt; upright; compreffed; cells two: Valves two: Partition oppofite to the valves: open-ing at the edges.

SEEDS. Many; compreffed.

OBS. *In the firft fpecies the edges of the capfule are bordered with a membranaceous margin: the feeds are encompaffed by a loofe membrane; the cup is equal;) with four clefts.*

RATTLE. The upper lip of the bloffom compreffed and Yellow fhort—*When the feeds are ripe they rattle in the capfule.* Bloffoms Crifta galli yellow.

Pedicularis, feu crifta galli lutea. *Bauh. pin.* 163.
Crifta galli. *Gerard.* 1071,

1. There is a variety which is much branched, with very nar-row leaves, yellow bloffoms and purple lips. *Ray's Syn.* 284. Cockfcomb.

Meadows and paftures. A. June.—July.
Horfes, Goats and Sheep eat it; Cows refufe it.

255 COW-WHEAT.   742 Melampyrum.

EMPAL. *Cup* one leaf, tubular; with four shallow clefts. *Segments* slender; permanent.

BLOSS. One petal; gaping. *Tube* oblong, bent back. *Border* compressed. *Upper Lip* helmet-shaped; compressed; notched at the end; lateral margins bent back. *Lower Lip* flat; upright; as long as the upper lip; with three shallow segments; blunt; marked with two projections in the middle.

CHIVES. *Threads* four, (two long and two short;) awl-shaped; crooked; concealed under the upper lip. *Tips* oblong.

POINT. *Seedbud* tapering. *Shaft* simple; agreeing in length and situation with the chives. *Summit* blunt.

S. VESS. *Capsule* oblong; oblique; taper; compressed; the upper edge convex, the lower edge straight. Cells two. Valves two. *Partition* placed in a contrary direction to the valves. Opening at the upper seam.

SEEDS. Two; egg-shaped; hunched; enlarged at the base.

Crested
Cristatum

COW-WHEAT. The flowers in quadrangular spikes. Floral leaves heart-shaped; firm; tiled; with little teeth—*Blossoms generally yellow but sometimes white or purplish.*
Melampyrum cristatum flore albo et purpureo. *Ray's Syn.* 286.
Melampyrum luteum angustifolium. *Bauh. pin.* 234.
In woods. A. July.
Cows, Goats and Sheep eat it.

Purple
Arvense

COW-WHEAT. The flowers in flexible conical spikes. Floral leaves with bristle-shaped teeth; coloured—*Blossoms yellow and dusky purple.*
Melampyrum purpurascente coma. *Bauh. pin.* 234. *Ray's Syn.* 286.
Corn-fields. A. July.
The seeds when ground with corn, give a bitterness and greyish cast to the bread, but do not make it unwholesome.
Cows and Goats eat it; Sheep refuse it.

COW WHEAT. The flowers in spikes pointing all one way Meadow
Leaves in distant pairs. Blossoms closed—*White*; *with yellow* Pratense
spots on the lower lip.
Melampyrum latifolium flore albo; labio inferiore duabus ma-
culis luteis distincto. *Ray's Syn.* 286.
Meadows and woods. A July—August.
Where this plant abounds, the butter is yellow, and uncom-
monly good—Swine are very fond of the seeds.
Sheep and Goats eat it; Cows are very fond of it; Horses and
and Swine refuse it.

COW-WHEAT. The flowers in spikes pointing all one way. Yellow
Leaves in distant pairs. Blossoms gaping—*The blossoms in this are* Sylvaticum
*only half as long as those in the preceding species and are entirely yel-
low, tube and all.*
Melampyrum sylvaticum flore luteo, sive satureja lutea sylves-
tris. *Ray's Syn.* 286.
Melampyrum luteum latifolium. *Bauh. pin.* 234.
Cratæogonon album. *Gerard.* 91.
Cratæogonon vulgare. *Park.* 1326.
In woods A. June—August,
Cows, Sheep and Goats eat it; and with a plentiful allow-
ance of it soon grow fat.

256 MUDWEED. 776 Limosella.

EMPAL. *Cup* one leaf; with five shallow clefts; sharp;
upright; permanent.
BLOSS. One petal; bell-shaped; upright; equal; with
five shallow clefts; sharp; small. *Segments* expanding.
CHIVES. *Threads* four; upright; two leaning to the same
side. shorter than the blossom. *Tips* simple.
POINT. *Seedbud* oblong; blunt. *Shaft* simple; as long as
the chives; declining. *Summit* globular.
S.VESS. *Capsule* egg-shaped; half inclosed in the cup;
with one cell and two valves.
SEEDS. Many; oval. *Receptacle* egg-shaped; very large.

MUDWEED. As there is only one species known, Linnæus Plantain
gives no description of it—*Stems trailing.* Leaves *spear shaped*; on Aquatica
very long leaf-stalks. Blossoms *solitary*; *white on the outside*; *reddish within.*
Plantaginella palustris. *Bauh. pin.* 190. *Ray's Syn.* 278.
Bastard Plantain.
In places that are liable to be flooded. P. August—September.

257 FIGWORT.   756 Scrophularia.

EMPAL. *Cup* one leaf; with five clefts; permanent. *Segments* rounded, ſhorter than the bloſſom.

BLOSS. One petal; unequal. *Tube* globular; large; bladder-ſhaped. *Border* very ſmall; with five diviſions. The two *Upper Segments* larger than the others; upright: two *Lateral Segments* open. *Lower Segment* reflected.

CHIVES. *Threads* four; ſtrap-ſhaped; declining; as long as the bloſſoms; two of them ripening later than the other two. *Tips* double.

POINT. *Seedbud* egg-ſhaped. *Shaft* ſimple; agreeing in length and ſituation with the chives. *Summit* ſimple.

S.VESS. *Capſule* roundiſh; tapering; cells two; valves two; partition formed by the edges of the valves turning in: opening at the top.

SEEDS. Many; ſmall. *Receptacle* ſingle; roundiſh; extending itſelf into each cell.

OBS. *In the mouth of the bloſſom, beneath the upper ſegment; lies another little ſegment reſembling a lip; but this is not common to every ſpecies.*

Knobby-rooted Nodoſa FIGWORT. The leaves three fibred; heart-ſhaped. Corners of the ſtem not membranaceous—*Root knotty.* Bloſſoms *greeniſh purple; terminating.*
Scrophularia major, *Gerard.* 715. *Ray's Syn.* 283.
Scrophularia vulgaris. *Park.* 610.
Scrophularia nodoſa fœtida. *Bauh. pin.* 235.
Kernelwort.
Woods and moiſt hedges. P. July.
This plant is hardly known in modern practice; but the rank ſmell, and bitter taſte of the leaves, ſeem to indicate ſome active properties.—Swine that have the ſcab are cured by waſhing them with a decoction of the leaves.—Waſps reſort greatly to the flowers.
Goats eat it; Cows, Horſes, Sheep and Swine refuſe it.

Water Aquatica FIGWORT. The leaves heart-ſhaped; blunt; on leaf-ſtalks, which run along the ſtem. Flowers in bunches; terminal. Stem angular, membranaceous at the corners—*Bloſſoms purple.*
Scrophularia aquatica major. *Bauh. pin.* 235. *Ray's Syn.* 283.
Betonica aquatica. *Gerard.* 715. major. *Park.* 613.
Water-betony.
In watery places. P. July.

FIGWORT.

FIGWORT. The leaves heart-fhaped; doubly ferrated. Balm-leaved Flowers in compound bunches, with leaves intorfperfed.—*Stem* Scorodonia *very hairy.* Bloffoms *dufky purple.*
Scrophularia Scorodoniæ folio. *Ray's Syn.* 283.
Hedges and rivulets. P. Auguft.

FIGWORT. The leaves heart-fhaped. Fruit-ftalks folitary; Yellow forked; growing within the bafe of the leaves—*Bloffoms yellow.* Vernalis
Scrophularia flore luteo. *Gerard.* 717. *Bauh. pin.* 236. *Park.* 610.
Hedges. B. April—May.
The different fpecies of Figwort, afford nourifhment to the Water Betony Moth, *Phalæna Verbafci,* the Figwort Weevil, *Curculio Scrophulariæ,* and the *Tenthredo Scrophulariæ.*

258 COUNTERWORT. 775 Sibthorpia.

EMPAL. *Cup* one leaf; turban-fhaped; with five divifions; expanding. *Little leaves* egg-fhaped; permanent.
BLOSS. One petal; with five divifions; expanding; equal; as long as the cup. *Segments* rounded.
CHIVES. *Threads* four; hair-like; two of them approaching. *Tips* heart-fhaped, but oblong.
POINT. *Seedbud* roundifh; compreffed. *Shaft* cylindrical; thicker than the threads; as long as the bloffom. *Summit* a fimple knob, depreffed.
S. VESS. *Capfule* compreffed; round and flat; fwelling out in two places; edges fharp; valves two; cells two; partition tranfverfe.
SEEDS. Several; oblong: fomewhat roundifh; convex on one fide; flat on the other.

COUNTERWORT. The leaves fomewhat kidney-fhaped; Baftard fcolloped; with leaf-ftalks nearly central—*Stem thread-fhaped;* Europæa *creeping;* but little branched. Leaves *alternate; very remote; with a few hairs at the edges.* Flowers *on fruit-ftalks; folitary; arifing from the bafe of the leaves.* Fruit-ftalk *generally fhorter than the leaf-ftalks, with a fomewhat egg-fhaped floral leaf near the end.* Bloffoms *pale red.*
Did not this plant originate from the feedbud of the alternate leaved Saxifrage *fertilized by the duft of the* Marfh Penny-wort?
Alfine fpuria pufilla repens foliis faxifragæ aureæ. *Ray's Syn.* 352.
Baftard Money-wort.
Rivulets and fprings. P. Auguft.

259 FOXGLOVE.   758 Digitalis.

EMPAL. *Cup* with five divisions. *Segments* roundish; sharp; permanent. The upper segment narrower than the rest.

BLOSS. One petal: bell-shaped. *Tube* large; expanding; distended on the underside; cylindrical and slender at the base. *Border* small; with four clefts. *Upper Segment* the most expanded; notched at the end. *Lower Segment* largest.

CHIVES. *Threads* four. (two long and two short;) awl-shaped; fixed to the base of the blossom; declining. *Tips* cloven; tapering to a point at one end.

POINT. *Seedbud* taper. *Shaft* simple; standing along with the chives. *Summit* sharp

S. VESS. *Capsule* egg-shaped; as long as the cup; taper; cells two; valves two; opening in two directions.

SEEDS. Many; small.

OBS. *In the only British species the blossom is rather* tubular *than bell-shaped, and the upper segment is notched at the end.*

Common
Purpurea

FOXGLOVE. The leaves of the cup egg-shaped; sharp. Blossoms blunt; upper lip entire—*Flowers in long terminating spikes, all pointing one way.* Blossom *purple; elegantly mottled on the inside.*

Digitalis purpurea. *Gerard.* 790. *Ray's Syn.* 283.
Digitalis purpurea, folio aspero. *Bauh. pin.* 243.
Digitalis purpurea vulgaris. *Park.* 1653.
1. There is a variety with a white blossom. *Bauh. pin.* 243.
Purple Fox-glove.
Gravelly soil. B. July.
A dram of it taken inwardly excites violent vomiting. It is certainly a very active medicine, and merits more attention than modern practice bestows upon it.

260 TOADFLAX. 750 Antirrhinum.

EMPAL. *Cup* with five divisions; permanent. *Segments* oblong; permanent. *Two lower Segments* more expanding.

BLOSS. One petal; gaping. *Tube* oblong; hunched. *Border* with two lips. *Upper Lip* cloven; reflected side-ways. *Lower Lip* with three segments, blunt. *Palate* convex; mouth closed by a projection of the lower lip, which is channelled on the under side.

Honey-cup extending backwards from the base of the blossom.

CHIVES. *Threads* four; (two short and two long;) nearly as long as the blossom and inclosed by the upper lip. *Tips* approaching.

POINT. *Seedbud* roundish. *Shaft* simple; agreeing in length and situation with the chives. *Summit* blunt.

S. VESS. *Capsule* roundish; blunt; cells two. Figure and manner of opening different in different species.

SEEDS. Many; *Receptacles* kidney-shaped; solitary; fixed to the partition.

OBS. *The* Honey-cup *and the* Seed-vessel *differ greatly in the different species. In some the former is long and awl-shaped, and the latter opens equally. In others the honey-cup is blunt, scarcely protuberating; the capsule is unequal at the base; opening at the top obliquely; and in others again still different.*

* Leaves angular.

TOADFLAX. The leaves alternate; heart-shaped; with five lobes. Stems trailing.—*Is not this plant produced by the dust of the* Ivy-leaved Bell-flower *fertilizing the seedbud of the* Toad-flax ? *Spur of the honey-cup slender; crooked; purple.* Blossoms *purple and yellow.*   Ivy-leaved Cymbalaria Fluellin

Linaria hederaceo folio glabro, seu Cymbalaria. *Ray's Syn.* 282.

Cymbalaria. *Bauh. pin.* 305.

1. There is a variety with woolly leaves.
On old walls. P. June—October.

Fluellin   TOADFLAX. The leaves halberd-fhaped; alternate. Stems
Elatine   trailing—*Bloffoms yellow and purple; or blue.*
    Linaria Elatine dicta, folio acuminato. *Ray's Syn.* 282.
    Elatine folio acuminato, in bafi auriculato, flore luteo. *Bauh.*
    pin. 253.
    Elatine folio acuminato. *Park.* 553.
    Elatine altera. *Gerard.* 623.
    Sharp-pointed Fluellin.
    Corn-fields. A. Auguft—October.
    This is confiderably more bitter than the other fpecies, and
    is faid to have been ufed fuccefsfully in cafes of foul Ulcers and
    in cutaneous eruptions.

Round-leaved   TOADFLAX. The leaves egg-fhaped; alternate. Stems
Spurium   trailing—*It is highly probable that this plant originated from the*
    duft of the Fluellin Toad-flax *fertilizing the feedbud of the* Ivy-
    leaved Toad-flax. *Bloffoms yellow and purple.*
    Linaria Elatine dicta folio fubrotundo. *Ray's Syn.* 282.
    Elatine folio fubrotundo. *Bauh. pin.* 252. *Park.* 553.
    Veronica fæmina Fuchfii, feu Elatine. *Gerard.* 625.
    Round-leaved Fluellin.
    Corn-fields. A. Auguft.

* * *Leaves oppofite.*

Creeping   TOADFLAX. The leaves ftrap-fhaped; crowded together.
Repens   Lower leaves growing by fours. Cups as long as the
    Capfules—*Not odoriferous.* Bloffoms *purple and yellow.* Spur
    of the honey-cup, long and pointed.
    In hedges in Cornwall. P. Auguft.

Sweetfmelling   TOADFLAX. The leaves ftrap-fhaped; crowded together.
Monfpeffula-   Stem and fruit ftalks fhining. Flowers in fpiked panicles—*The*
num   *fpur of the bloffom ftraight and very fhort: odoriferous.* Bloffoms
    blue.
    Linaria odorata monfpeffulana. *Ray's Syn.* 282.
    Linaria capillaceo folio odora. *Bauh. pin.* 213.
    Corn-fields. P. July—Auguft.

TOAD-

TOADFLAX. The leaves nearly ftrap-fhaped : lower leaves Corn growing by fours. Cups hairy and clammy. Flowers in fpikes. Arvenfe Stem upright—*Bloſſoms blue.*

The varieties are : 1. Dwarf fleſhy-leaved yellow. 2. Violet. 3. Four-leaved yellow.

Linaria arvenſis cærulea. *Bauh. pin.* 213.

Linaria cærulea foliis brevioribus et anguſtioribus. *Ray's Syn.* 282.

Corn blue Toad-flax.

Dry Corn-fields. A. July—Auguſt.

* * * *Leaves alternate.*

TOADFLAX. The leaves for the moſt part alternate; fpear- Leaſt fhaped; blunt. Stem much branched; fpreading—*Bloſſoms pur-* Minus *ple and yellowiſh white; with a very ſhort ſpur.*

Antirrhinum arvenfe minus. *Bauh. pin.* 212.

Antirrhinum minimum repens. *Gerard.* 549.

Antirrhinum fylveftre minimum. *Park.* 1334.

Linaria Antirrhinum dicta. *Ray's Syn.* 283.

Gravelly corn-fields. A. June—October.

Cows and Sheep eat it; Swine are not fond of it; Horfes and Goats refufe it.

TOADFLAX. The leaves betwixt fpear and ftrap-fhaped; Common crowded together. Stem upright. Flowers tiled, in fitting ter- Linaria minating fpikes—*Spur long and crooked.* Blotfom *yellow and orange; fometimes whitiſh.*

Linaria lutea vulgaris. *Gerard.* 550. *Ray's Syn.* 281.

Linaria vulgaris noftras. *Park.* 458.

Linaria vulgaris lutea, flore majore. *Bauh. pin.* 212.

Common yellow Toad-flax.

In barren ground and road-fides. P. July.

Cows, Horfes and Swine refufe it; Sheep and goats are not fond it.

1. PELORIA. This extraordinary variety feems to have originated from an impregnation of the feed-bud of the *Common*
   TOADFLAX, by the duft from the tips of fome other plant:
   perhaps from one of the GENTIANS. If it had produced feeds
   capable of vegetation, it would have conftituted a new genus,
   as is evident from the following defcription.

P E L O R I A.

Cup one leaf; with five divisions; equal; very short; permanent.

Bloss. one petal; funnel-shaped: with five honey-cups at the base.

Tube long; straight; cylindrical, but distended from the middle downwards.

Border with five divisions; blunt; equal; expanding.

Honey-cups five, awl-shaped; flat; resembling a horn; growing in a circle round the base of the tube.

Chives. Threads five; hair-like; equal; half as long as the tube and fixed to the receptacle. *Tips* roundish; fixed side-ways to the threads.

Point. Seedbud egg-shaped; *Shaft* thread-shaped; as long as the chives. *Summit* thick and blunt.

S. Vess. Capsule egg-shaped; with two cells and two valves.

Receptacles convex; growing to the partition.

Seeds. Several; angular.

OBS. Stem *simple; upright; cylindrical: sometimes, though rarely, furnished with one or two branches.* Leaves *numerous; scattered; strap-shaped; smooth; upright but expanding.* Flowers *in a spike, terminating; nine, twelve or eighteen together.* Blossoms *yellow; paler towards the base; the inner side filled with tawney hairs.*

In sandy fields near Clapham in the county of Surry. P. July.

It hath the same taste and smell that the *Common* TOADFLAX possesses, among which it is found to grow. The seeds are barren, but it may be propagated by roots. *Amœnitates Academicœ* vol. 1. p. 282. tab. 3.

An infusion of the leaves of the Common Toad-flax is diuretic and purgative. An ointment prepared from them gives relief in the Piles. The expressed juice mixed with milk is a poison to Flies, as is likewise the smell of the flowers.

* * * * *Blossom*

* * * * *Bloſſoms gaping; or without a ſpur.*

TOADFLAX. The bloſſoms without a ſpur, forming ſpikes. Snap-dragon
Cups roundiſh.—*Leaves on leaf-ſtalks.* Bloſſoms *purple.*   Majus
    Cultivation or accident hath produced the following varieties.
1. Long-leaved purple and white.
2. Reddiſh and white.
3. Red and white.
4. Yellow and white.
5. Yellow
    Antirrhinum majus alterum, folio longiore. *Bauh. ſin.* 211.
    Antirrhinum purpureum. *Gerard.* 549.
    Greater Snap-dragon.
    On old walls and chalky cliffs. P. June.

TOADFLAX. The bloſſoms without ſpurs, forming a ſort Snout
of ſpike. Cups longer than the bloſſoms—*The capſule when* Orontium
open reſembles the ſkull of a monkey. Bloſſom *purple; with a little*
yellow.
    Antirrhinum anguſtifolium ſylveſtre. *Ray's Syn.* 283.
    Antirrhinum arvenſe majus. *Bauh. ſin.* 212.
    Antirrhinum ſylveſtre medium. *Park.* 1334.
    Antirrhinum minus. *Gerard.* 549.
    Laſt Snap-dragon. Calfs-ſnout.
    Corn-fields. A. July—Auguſt.
    This plant is poiſonous.

261 LOUSEWORT.   746 Pedicularis

EMPAL. *Cup* one leaf; roundiſh; diſtended. *Mouth* with five clefts.; equal; permanent.

BLOSS. One petal; gaping. *Tube* oblong; hunched. *Upper Lip* helmet-ſhaped: upright; compreſſed; narrow; notched. *Lower Lip* expanding; flat; with three ſhallow ſegments; blunt. *Middle Segment* the narroweſt.

CHIVES. *Threads* four; (two long and two ſhort;) nearly as long as the upper lip under which they lie concealed. *Tips* fixed ſide-ways to the threads: roundiſh; compreſſed.

POINT. *Seedbud* roundiſh. *Shaft* thread-ſhaped; agreeing in ſituation with the chives, but longer. *Summit* blunt; bent inwards.

S. VESS. *Capſule* roundiſh: tapering; oblique. Cells two. Partition oppoſite to the valves; opening at the top.

SEEDS. Many; roundiſh; compreſſed; covered with a coat. *Receptacles* oblong, ſupported by foot-ſtalks.

OBS. *The Capſule is for the moſt part oblique. In ſome ſpecies the Cup is cloven at the rim into two parts.*

Marſh   LOUSEWORT. The ſtem branched; cups beſet with callous
Paluſtris   points. Lips of the bloſſom oblique—*Floſſoms purple.*
            Pedicularis paluſtris rubra elatior. *Ray's Syn.* 284.
            Wet meadows. A. June.
            This is an unwelcome gueſt in meadows, being very diſagreeable to cattle.
            Goats eat it; Horſes, Sheep and Cows refuſe it; Swine are not fond of it.

LOUSE-

LOUSEWORT. The ftem branched; cups oblong; angular; Common fmooth. Lip of the bloffom heart-fhaped—*Bloffoms purple.* Sylvatica
Pedicularis. *Gerard.* 1071. pratenfis purpurea. *Bauh. pin.* 163.
Pedicularis pratenfis rubra vulgaris. *Park.* 713. *Ray's Syn.* 284.
Wet heaths and paftures. P. May—June.
The expreffed juice, or a decoction of this plant hath been ufed with advantage as an injection for finuous Ulcers—If the healthieft flock of fheep are fed with it, they become fcabby and fcurfy in a fhort time; the wool gets loofe, and they will be over-run with vermin.
Cows and Swine refufe it.

The End of the Firft Volume.

Printed in the United States
By Bookmasters